Nanofluids for Efficient Energy Conservation and Process Intensification

Nanofluids for Efficient Energy Conservation and Process Intensification presents mathematical, numerical, and experimental methodologies of applying nanofluids in renewable and non-renewable energy conservation and process intensification. It explores the mechanisms for developing high-performance heat transfer and mass transfer hybrid nanofluids.

Beginning with thermophysical and optical properties enhancement of nanofluids, the book covers nanoparticle synthesis, stabilization techniques, and life-cycle assessment (LCA). It delves into the use of ionanofluids in environmental remediation, hybrid nanofluids for carbon capture, and the role of nanofluids in solar thermal systems. The book also discusses the use of Response Surface Methodology (RSM) and Artificial Neural Networks (ANN) to facilitate optimization and predictive modeling of nanofluid performance.

This book will interest researchers, graduate students, and industry practitioners studying the advances in applications of nanofluids within mechanical and chemical engineering.

Dr. Shriram S. Sonawane is Professor of Chemical Engineering at Visvesvaraya National Institute of Technology (VNIT), Nagpur, India. With over two decades of academic and research experience, he has made pioneering contributions in the fields of polymer nanocomposites, nanofluids, nano-separation technology, and process modeling and simulation. He has published more than 250 research papers in various national and international journals of repute, and has over 12 patents to his credit. He is featured in the 2023, 2024 and 2025 editions of Stanford University's "World's Top 2% Scientists" list. He is also the author of four books, editor of five books, and has contributed more than 50 book chapters. Dr. Sonawane is a Fellow of the Maharashtra Academy of Sciences and the Indian Institute of Chemical Engineers (IIChE). He has been nominated twice for the prestigious Shanti Swarup Bhatnagar Award.

Dr. Parag P. Thakur is Assistant Professor at SVNIT, Surat, India. He previously served as an ad-hoc faculty at NIT, Warangal. Following the completion of his Ph.D. at VNIT, Nagpur, India, he has been invited to conduct his postdoc research at Chonnam National University, South Korea. His research areas include nanofluids, microfluids, nanocomposites, and modeling and simulation. Dr. Thakur has published

36 book chapters, two books, three patents, and 15 articles in reputed journals, receiving more than 800 citations for his research. He has attended over 50 international conferences, where he received several best presentation awards. He also serves as a reviewer for the *Journal of Mass and Heat Transfer* and *Separation Science and Technology.*

Dr. Manjakuppam Malika is Assistant Professor in the Department of Chemical Engineering, BV Raju Institute of Technology, Narsapur, Telangana, India. Her current research focuses on nanofluid applications in chemical engineering and technology. Dr. Malika has published 25 research articles in reputed journals as well as 22 book chapters, 3 authored books and 3 granted patents on Nanofluid applications. She has received more than 800 citations for her work. She is featured in the 2024 and 2025 editions of Stanford University's "World's Top 2% Scientists" list.

Nanofluids for Efficient Energy Conservation and Process Intensification

Shriram S. Sonawane, Parag P. Thakur,
and Manjakuppam Malika

CRC Press
Taylor & Francis Group
Boca Raton London New York

CRC Press is an imprint of the
Taylor & Francis Group, an **informa** business

Designed cover image: Shutterstock

First edition published 2026
by CRC Press
2385 NW Executive Center Drive, Suite 320, Boca Raton FL 33431

and by CRC Press
4 Park Square, Milton Park, Abingdon, Oxon, OX14 4RN

CRC Press is an imprint of Taylor & Francis Group, LLC

ISBN: 978-1-041-04368-3 (hbk)
ISBN: 978-1-041-04369-0 (pbk)
ISBN: 978-1-003-62807-1 (ebk)

DOI: 10.1201/9781003628071

Typeset in Times
by Apex CoVantage, LLC

Contents

Foreword

It gives me immense pleasure to write the foreword for the book *Nanofluids for Efficient Energy Conservation and Process Intensification*, authored by Professor Shriram S. Sonawane, Dr. Parag P. Thakur, and Dr. Manjakuppam Malika. This timely and comprehensive volume addresses one of the most critical global challenges of our era—sustainable energy management through the deployment of advanced materials and intelligent process technologies.

Nanofluids, as an emerging class of *engineered smart fluids*, have demonstrated exceptional potential in enhancing heat and mass transfer characteristics, thereby enabling significant advancements in energy conservation, renewable energy utilization, and process intensification.

The present book provides a meticulous exploration of this transformative field, integrating fundamental principles, experimental investigations, modeling frameworks, and real-world applications into a unified narrative.

The chapters span a broad scientific and technological spectrum—beginning with the fundamentals of nanofluid synthesis and the mechanisms of thermal and rheological enhancement, and extending to hybrid nanofluids, ionanofluids, and multifunctional nanocomposites for environmental remediation, carbon capture, and clean energy systems. Equally impressive are the discussions on mathematical modeling, molecular dynamics simulations, and AI-driven predictive approaches (RSM, ANN), which strengthen the theoretical foundations and bridge them with practical applications. The coverage of solar thermal systems, emulsion-based nanofluid membranes, and thermal management in industrial operations showcases the breadth and interdisciplinarity of the authors' approach.

What truly distinguishes this book is its holistic perspective—it seamlessly integrates scientific innovation, engineering design, sustainability, and techno-economic feasibility. The thoughtful emphasis on scale-up challenges, environmental impacts, and life cycle assessment ensures that the content is not merely academic but also industrial and policy-relevant, serving as a catalyst for sustainable technological advancement.

For students, researchers, and professionals across chemical engineering, materials science, nanotechnology, and sustainable energy systems, this book serves as both a comprehensive reference and a practical guide. It bridges the gap between academic research and industrial implementation, offering a visionary roadmap for integrating nanofluid technologies into next-generation energy and process platforms.

I wholeheartedly congratulate the authors for this outstanding scholarly contribution, which reflects deep scientific insight and remarkable clarity of thought. I am confident that this book will inspire future innovations, enrich global research discourse, and serve as an indispensable resource for all those dedicated to advancing energy efficiency and sustainability in the 21st century.

Professor Aniruddha B. Pandit
Professor of Chemical Engineering & Vice Chancellor,
Institute of Chemical Technology (ICT), Matunga, Mumbai, India

1 Introduction to Nanofluids for Efficient Energy Conservation and Process Intensification

1.1 INTRODUCTION

Nanofluids are a class of engineered fluids that consist of nanometer-sized particles (typically of 1–100 nm) dispersed in conventional base fluids such as water, ethylene glycol, oils, or other heat transfer fluids. These nanoparticles are commonly made of metals (e.g., Cu, Al), metal oxides (e.g., Al_2O_3, TiO_2), carbon-based materials (e.g., graphene, carbon nanotubes), or advanced 2D materials (e.g., MXenes). When suspended stably in a base fluid, these nanoparticles significantly enhance the thermal, rheological, and sometimes even chemical properties of the respective base fluid. The resulting fluid, termed a *nanofluid,* exhibits superior heat transfer characteristics, improved thermal conductivity, better convective heat transfer coefficients, and, in some cases, increased boiling heat transfer rates compared to the base fluid alone. Application of nanofluids is not only limited to heat transfer applications but can be used for other applications like mass transfer enhancements, enhanced oil recovery, enhanced hydrogen production, and biomedical applications [1–3].

Conventional heat transfer fluids have already reached their limits. Traditional fluids such as water, oil, and ethylene glycol possess inherently low thermal conductivities, limiting their effectiveness in high-performance thermal systems. The addition of thermally conductive nanoparticles alters this property, allowing for more rapid heat dissipation and improved overall system efficiency [4–5].

Nanofluids also often demonstrate non-Newtonian behavior, altered viscosity, and, in certain formulations, can be tailored to exhibit phase-change behavior or photothermal responses, opening up new avenues for their use in renewable energy systems, electronics cooling, automotive engines, biomedical devices, and chemical process intensification [6].

1.1.1 Historical Background and Development

The evolution of nanofluids as a distinct research field is relatively recent, gaining momentum only in the past two decades. The concept was first introduced in the mid-1990s by Choi and Eastman at Argonne National Laboratory (United States), who highlighted the substantial potential of suspending nanoparticles in conventional liquids to enhance thermal conductivity. Their seminal work in 1995 demonstrated

DOI: 10.1201/9781003628071-1

that even a small fraction of nanoparticles (less than 1% by volume) could result in appreciable improvements in thermal conductivity. This finding sparked an exponential surge of interest from researchers across disciplines like mechanical, chemical, materials science, and nanotechnology [7–8].

In the early stages, the focus was primarily on exploring different types of nanoparticles and base fluids to measure enhancements in thermal conductivity and convective heat transfer coefficients. Commonly studied nanoparticles included Al_2O_3, CuO, and TiO_2, often dispersed in water or ethylene glycol. The key challenges were stabilization (preventing agglomeration), understanding the influence of particle size and shape, and establishing reliable preparation methods such as two-step and one-step synthesis routes [9–10].

The 2000s witnessed the development of theoretical models attempting to explain the mechanisms of enhanced heat transfer in nanofluids. Although classical models based on effective medium theory (such as the Maxwell model) were initially used, they often failed to predict the observed enhancements accurately. This led to the exploration of various mechanisms such as Brownian motion, thermophoresis, particle–fluid interaction, micro-convection, and clustering effects [11].

From 2010 onward, the application domain expanded significantly. Researchers began exploring the utility of nanofluids in solar thermal collectors, nuclear cooling systems, data center thermal management, automotive heat exchangers, and biomedical applications like targeted drug delivery and hyperthermia. Recent advancements have also seen the integration of smart nanofluids that can respond to external stimuli (magnetic, electric, or thermal), making them attractive for process intensification and smart heat management systems [12–13].

1.1.2 Relevance in the Context of Energy Crisis and Sustainability

The 21st century has brought with it an urgent need to address challenges associated with energy efficiency, global warming, and sustainable industrial development. With rising population levels, rapid industrialization, and increasing energy demand, efficient utilization of thermal systems and process energy has become more critical than ever. In this context, nanofluids have emerged as a transformative solution with direct implications for energy conservation and environmental sustainability [14].

One of the major areas where energy losses occur is during heat exchange processes in power plants, automobiles, refrigeration systems, and electronic cooling units. Even minor improvements in the efficiency of heat transfer fluids can result in significant energy savings across sectors. Nanofluids offer a pathway to achieve this without requiring major modifications to existing systems. For instance, using nanofluids in solar collectors can enhance the absorption of solar radiation due to their tailored optical properties, thereby improving the conversion efficiency of solar thermal energy systems [15]. Moreover, nanofluids are being explored for use in heat recovery systems where waste heat from industrial processes can be recycled more effectively. In systems where water is used as a coolant, the replacement with nanofluids can lead to higher heat flux handling, allowing for downsizing of equipment and reduction in operational costs [16].

From a sustainability perspective, nanofluids align with several Sustainable Development Goals (SDGs) proposed by the United Nations, particularly SDG 7 (Affordable and Clean Energy), SDG 9 (Industry, Innovation, and Infrastructure), and SDG 13 (Climate Action). By improving the efficiency of energy systems and reducing carbon footprints, nanofluids can play a significant role in achieving decarbonization goals. Environmental considerations are also being taken into account with the development of biodegradable or bio-based nanofluids. For instance, the use of vegetable oil-based nanofluids containing naturally derived nanoparticles like cellulose or chitosan offers an eco-friendly alternative to petroleum-based coolants. Additionally, researchers are examining life-cycle analysis (LCA) and techno-economic analysis (TEA) to ensure that the benefits of nanofluids are not offset by their environmental or economic costs [17].

1.1.3 Scope and Objectives of the Book

This book aims to provide a comprehensive introduction to the field of nanofluids with a particular focus on their role in energy conservation and process intensification. The primary objective is to furnish readers—students, researchers, engineers, and policymakers—with foundational knowledge that bridges the gap between laboratory-scale research and real-world industrial applications. This chapter is important to introduce the concept of nanofluids to readers [18].

The following key objectives define the scope of this chapter:

1. **Conceptual understanding**: To offer a detailed overview of the physicochemical basis behind nanofluids, including nanoparticle types, dispersion methods, thermophysical properties, and behavior under different flow regimes.
2. **Historical insight**: To present a chronological development of nanofluid research and its transition from academic curiosity to industrial relevance.
3. **Mechanistic perspectives**: To discuss the underlying mechanisms that govern enhanced thermal and mass transfer in nanofluids, such as Brownian motion, nanoparticle clustering, and interfacial thermal resistance.
4. **Energy relevance**: To highlight the role of nanofluids in improving the efficiency of energy systems, including solar collectors, heat exchangers, nuclear reactors, and electronics cooling systems.
5. **Process intensification**: To illustrate how nanofluids can aid in reducing equipment size, increasing process throughput, and enhancing mass/heat transfer rates in chemical and process industries.
6. **Sustainability considerations**: To address the environmental and economic viability of nanofluid usage, including life-cycle impacts, recyclability, and regulatory concerns.
7. **Research trends and gaps**: To identify current research trends, unresolved challenges, and future directions in the domain of nanofluid development and application.

The chapter sets the stage for the subsequent chapter that will dig deeper into synthesis strategies, characterization techniques, application case studies, and experimental validation of nanofluid-based systems. Special attention is paid to interdisciplinary integration, where nanofluids intersect with renewable energy technologies, green chemistry, computational modeling, and data-driven optimization tools [19].

By framing the chapter around both the scientific fundamentals and the application-driven motivations for nanofluids, this introduction intends to serve as a gateway for understanding their transformative potential in the modern energy landscape [20–21]. Figure 1.1 represents the various applications of nanomaterials in various sectors.

Nanofluids represent a frontier area in fluid engineering with far-reaching implications for energy conservation and industrial process improvement. Their ability to enhance heat transfer performance, combined with the flexibility in formulation and adaptability to various thermal systems, makes them promising candidates for next-generation heat transfer applications. As the global community moves toward a sustainable, low-carbon future, technologies like nanofluids that offer both technical and environmental benefits are poised to play a central role. Nanofluids can also be used for better mass transfer applications, biomedical applications, and hydrogen production applications [22].

1.2 FUNDAMENTALS OF NANOFLUIDS

As discussed in the previous section, nanofluids represent an innovative class of heat transfer fluids engineered by dispersing nanometer-sized solid particles into conventional base fluids [23]. The inclusion of these nanoscale particles significantly modifies the thermal, rheological, and optical characteristics of the fluid, making nanofluids highly suitable for energy-intensive applications. Understanding the fundamental aspects of nanofluids is essential for optimizing their design, formulation, and application in real-world systems.

1.2.1 Composition: Base Fluids and Nanoparticles

The formulation of nanofluids involves two key components: a *base fluid* and *nanoparticles.*

1.2.1.1 Base Fluids

Base fluids are the continuous phase in which nanoparticles are dispersed. They serve as the medium for heat or mass transport. Commonly used base fluids include [24]:

- **Water**: Universally available, high specific heat, and widely used in cooling systems.
- **Ethylene glycol (EG)** and **propylene glycol (PG)**: Used in automotive and industrial systems due to their lower freezing point and higher boiling point than water. This allows the higher boiling point and lower freezing point for the coolants. Thus, these chemicals are also known as anti-knocking agents.

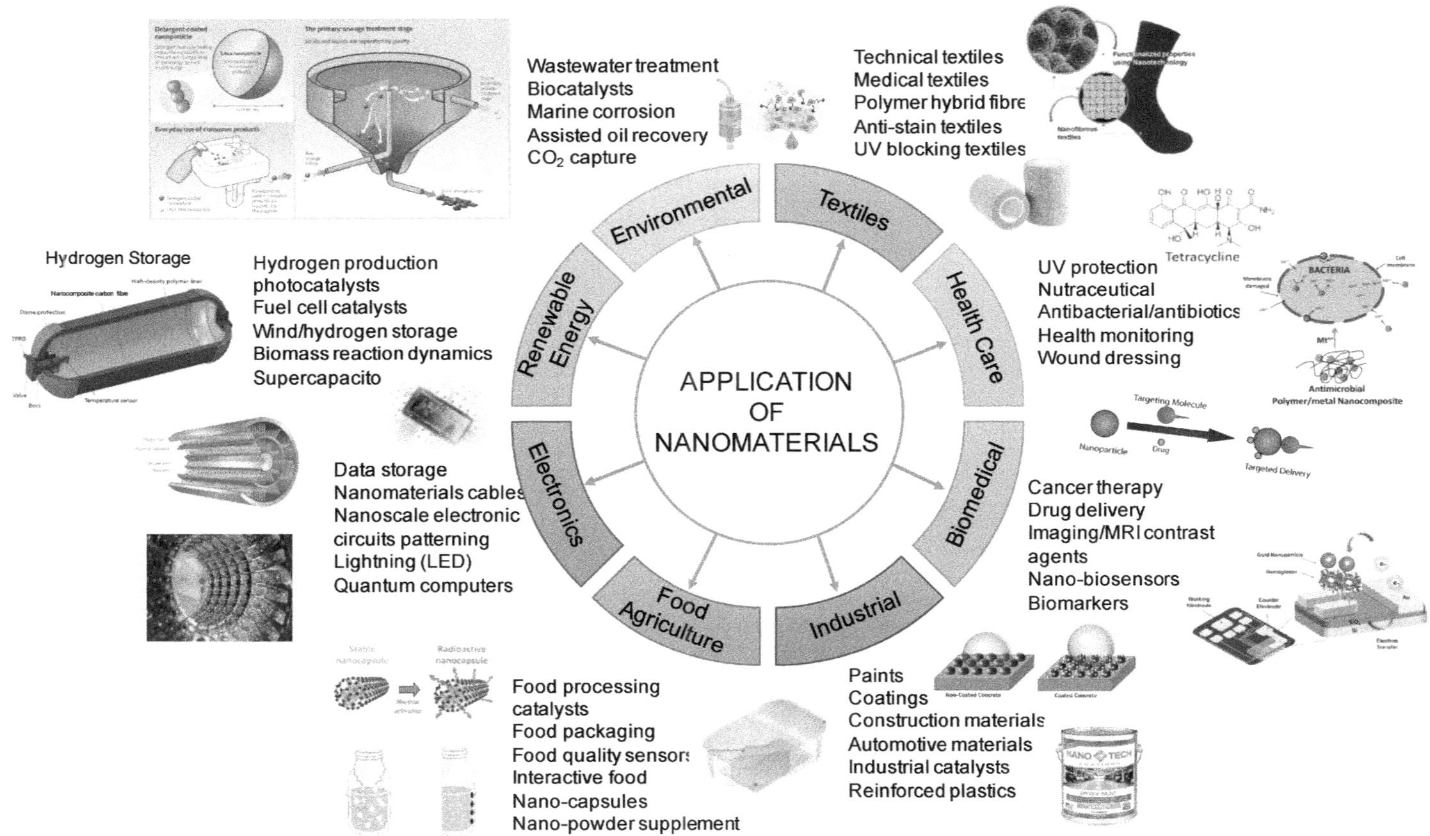

FIGURE 1.1 Application of nanomaterials in various sectors [20].

- **Oils**: Engine oils, mineral oils, and synthetic oils are employed in high-temperature applications.
- **Refrigerants**: Used in Heating, Ventilation, and Air Conditioning (HVAC) and refrigeration systems.
- **Ionic liquids and molten salts**: Recently explored for their superior thermal stability in high-temperature applications.

The choice of base fluid depends on the specific application, operating temperature, chemical compatibility, and environmental constraints.

1.2.1.2 Nanoparticles

Nanoparticles are the dispersed phase and typically range from 1 to 100 nm in size. They are selected based on their high thermal conductivity, chemical stability, and surface characteristics. Common nanoparticle materials include:

- **Metals**: Such as copper (Cu), silver (Ag), and gold (Au)—known for high thermal conductivity.
- **Metal oxides**: Including aluminum oxide (Al_2O_3), titanium dioxide (TiO_2), zinc oxide (ZnO), and silicon dioxide (SiO_2)—stable and cost-effective.
- **Carbon-based nanomaterials**: Such as carbon nanotubes (CNTs), graphene, and fullerene—exhibit extraordinary thermal and electrical properties.
- **Two-dimensional materials**: For example, boron nitride (BN) and MXenes emerging as high-performance thermal additives [25].

The morphology (spherical, rod-like, platelets), crystal structure, and surface chemistry of nanoparticles also influence nanofluid behavior.

Surface functionalization (e.g., with surfactants or polymer coatings) is often employed to enhance compatibility with the base fluid and improve dispersion stability.

1.2.2 Types of Nanofluids

Nanofluids can be categorized based on the nature and composition of the dispersed nanoparticles. These classifications impact their thermophysical behavior and applicability in various domains.

1.2.2.1 Metal Nanofluids

Metal nanofluids contain pure metallic nanoparticles such as copper (Cu), silver (Ag), or gold (Au). These materials offer exceptionally high thermal conductivity, with Cu nanoparticles being especially popular due to their cost-effectiveness and excellent heat transfer properties.

However, metallic nanofluids can suffer from issues such as oxidation, sedimentation, and chemical reactivity with the base fluid. Stabilization strategies are essential to prevent particle agglomeration.

1.2.2.2 Metal Oxide Nanofluids

These are the most extensively studied class of nanofluids. Common oxides include Al_2O_3, TiO_2, Fe_3O_4, and ZnO. Metal oxides are favored due to the following properties:

- Chemical inertness
- Relatively low cost
- Stability in aqueous media
- Biocompatibility (for biomedical use)

Despite having lower thermal conductivities compared to metals, they offer better long-term stability and are suitable for industrial heat transfer systems.

1.2.2.3 Carbon-Based Nanofluids

Carbon-based materials such as CNTs, graphene, and nanodiamonds are known for superior thermal conductivities, often an order of magnitude higher than traditional oxides. Their high aspect ratio (especially in CNTs) facilitates efficient thermal percolation networks.

Key advantages:

- Lightweight
- Chemically modifiable
- Electrically conductive (useful in electronic cooling)

However, dispersing carbon nanomaterials uniformly in polar solvents remains a technical challenge due to hydrophobicity.

1.2.2.4 Hybrid Nanofluids

Hybrid nanofluids combine two or more types of nanoparticles, in the form of either mixed particles or core-shell structures. The goal is to synergistically enhance heat transfer and tailor specific properties such as viscosity, stability, or electrical conductivity [26].

Examples:

- Cu-Al_2O_3 in water or oil
- Graphene-TiO_2 nanofluids
- Fe_3O_4-SiO_2 nanocomposites

Hybrid nanofluids provide flexibility in tuning thermophysical properties but increase formulation complexity.

1.2.3 Key Thermophysical Properties

The performance of nanofluids in any application is governed by their thermophysical properties, which are often different from the base fluid alone. These include the following.

1.2.3.1 Thermal Conductivity

Arguably the most critical property, thermal conductivity determines a fluid's ability to transfer heat via conduction. Nanofluids generally exhibit improved thermal conductivity due to the high intrinsic conductivity of nanoparticles and mechanisms like Brownian motion [27].

Factors affecting thermal conductivity:

- Nanoparticle material, size, and morphology
- Nanoparticle loading
- Temperature
- Particle–fluid interaction
- Aggregation and clustering

Empirical correlations and models, such as the Maxwell model, are used but often fail to fully capture the complex behavior of nanofluids.

1.2.3.2 Viscosity

Viscosity impacts the fluid's flow behavior and pumping power requirements. While nanoparticle addition enhances heat transfer, it can also increase viscosity, leading to higher energy costs for circulation.

Key factors affecting viscosity:

- Particle loading
- Temperature
- Base fluid nature
- Presence of surfactants

The goal is to optimize nanoparticle concentration to maximize thermal enhancement with minimal increase in viscosity.

1.2.3.3 Specific Heat Capacity

Specific heat determines the amount of heat required to raise the temperature of a fluid. Nanoparticle addition generally decreases the specific heat of the fluid due to the lower specific heat of solids compared to liquids.

This property is critical in systems where temperature variation or thermal buffering is required.

1.2.3.4 Density

Nanofluids typically have slightly higher density than their base fluids, especially at higher nanoparticle loadings [28]. Accurate density measurements are essential for thermal design calculations and simulation models.

1.2.4 Mechanisms of Enhanced Heat Transfer

The observed enhancement in nanofluid heat transfer properties arises from a combination of *macroscopic* and *microscopic* mechanisms, many of which are still under investigation [29].

1.2.4.1 Brownian Motion

Brownian motion refers to the random movement of nanoparticles suspended in a fluid. This motion can induce micro-convection and enhance energy transport at the molecular level.

- More significant in smaller particles and lower viscosity fluids
- Increases with temperature

1.2.4.2 Liquid Layering

A thin ordered layer of liquid molecules forms around nanoparticles, modifying the thermal conductivity of the interfacial region. This "nanolayer" acts as a bridge for energy transfer between the solid and liquid phases.

1.2.4.3 Agglomeration and Clustering

In controlled environments, nanoparticles may form stable clusters or chains that act as conductive bridges, increasing the thermal transport pathway. But excessive aggregation is undesirable as it can lead to sedimentation and clogging.

1.2.4.4 Thermophoresis

Nanoparticles tend to migrate from high-temperature regions to low-temperature regions, influencing mass and heat transport. This phenomenon plays a role in non-uniform temperature fields [30].

1.2.4.5 Micro-Convection

The relative movement between particles and fluid due to shear or temperature gradients can generate micro-level convection currents that supplement heat transfer.

1.2.4.6 Ballistic Phonon Transport

At the nanoscale, heat can be transferred by phonons (quantized lattice vibrations), particularly in high-conductivity particles like carbon nanotubes or graphene. This ballistic transport contributes to ultra-fast energy transfer rates [31].

1.2.5 Stability of Nanofluids and Dispersion Challenges

The stability of nanofluids is a major practical concern, particularly for long-term or industrial applications. Poor dispersion or sedimentation leads to reduced performance, inconsistent results, and possible system damage [32].

1.2.5.1 Key Challenges

- **Agglomeration**: Nanoparticles tend to attract each other due to van der Waals forces.
- **Sedimentation**: Denser particles may settle under gravity, especially in low-viscosity fluids.
- **pH Sensitivity**: Stability can change dramatically with pH variations.
- **Thermal Cycling**: Repeated heating and cooling can destabilize dispersions.

1.2.5.2 Stabilization Strategies

To enhance the dispersion and stability of nanofluids, several methods are employed:

- **Surfactants/Dispersants**: Chemical additives that reduce surface tension and improve wettability. Examples include SDS, CTAB, and PVP.
- **pH Adjustment**: Tuning the pH to ensure maximum zeta potential and electrostatic repulsion.
- **Ultrasonication**: High-frequency energy is used to break agglomerates and achieve homogeneous dispersion.
- **Functionalization**: Chemically modifying the nanoparticle surface to improve compatibility with the base fluid (e.g., grafting hydrophilic groups).

1.2.5.3 Evaluation Techniques

Nanofluid stability can be assessed using the following:

- Zeta potential analysis
- UV-Vis spectroscopy
- Turbidity measurements
- Sedimentation rate studies

The development of nanofluids, which can self-stabilize under varying conditions (smart nanofluids), is a growing research area aimed at overcoming long-standing dispersion issues.

An understanding of nanofluid fundamentals spanning composition, types, thermophysical properties, enhancement mechanisms, and stability is essential to fully exploit their potential in energy conservation and process intensification. Each property or design choice influences the fluid's behavior under thermal stress, flow dynamics, and system integration.

In subsequent chapters, these fundamental principles will serve as the foundation for exploring the practical implementation of nanofluids in various domains, including solar energy harvesting, electronic cooling, and chemical process intensification.

1.3 PREPARATION AND CHARACTERIZATION OF NANOFLUIDS

For successful application of nanofluids in various thermal and chemical processes, the preparation and characterization of nanofluids are crucial. Proper synthesis ensures the desired thermophysical properties, while accurate characterization enables a better understanding of performance parameters such as stability, dispersion, and flow behavior.

1.3.1 Synthesis Techniques

The synthesis of nanofluids typically follows one of two methodologies: the one-step method or the two-step method.

1.3.1.1 One-Step Method

The one-step method involves the simultaneous synthesis and dispersion of nanoparticles directly into the base fluid. This technique minimizes the risk of particle agglomeration since nanoparticles are never dried and are immediately suspended in the fluid medium. One-step methods can be classified into various sub-categories, including the following:

- Chemical Vapor Condensation (CVC)
- Direct Evaporation Condensation (DEC)
- Microwave-assisted synthesis
- Laser ablation

For instance, in the CVC technique, nanoparticles are generated in a reactor by condensing metal vapor in a fluid stream. The laser ablation method uses high-powered lasers to vaporize a target material submerged in the base fluid, leading to in situ formation of nanoparticles.

Advantages of One-Step Method

- Reduced agglomeration
- Enhanced stability
- Better control over particle size and dispersion

Disadvantages of One-Step Method

- Limited scalability
- Higher equipment and operational costs
- Complex to operate

1.3.1.2 Two-Step Method

The two-step method is the most widely used technique for nanofluid preparation due to its simplicity and scalability. It involves:

1. **Synthesis or purchase of dry nanoparticles** using conventional techniques like sol-gel, hydrothermal synthesis, or chemical precipitation.
2. **Dispersing nanoparticles into the base fluid** using mechanical stirring, ultrasonication, and surfactant addition.

This method is particularly suited for industrial applications as it allows bulk production and easy manipulation of fluid properties [33].

Advantages of Two-Step Method

- Cost-effective
- Scalable
- Wide selection of nanoparticle types and base fluids

Disadvantages of Two-Step Method

- Tendency for particle agglomeration
- Requires post-processing stabilization techniques

1.3.2 Surfactants and Dispersion Techniques

Achieving a stable and homogeneous nanofluid requires effective dispersion of nanoparticles in the base fluid. This is typically facilitated using surfactants and physical dispersion techniques.

1.3.2.1 Use of Surfactants

Surfactants are surface-active agents that enhance the wettability of nanoparticles, thereby improving dispersion and preventing re-agglomeration. Commonly used surfactants include the following:

- **Anionic:** Sodium dodecyl sulphate (SDS)
- **Cationic:** Cetyltrimethylammonium bromide (CTAB)
- **Non-ionic:** Polyvinylpyrrolidone (PVP), Tween-80

Surfactants reduce the surface tension between the nanoparticles and the base fluid, promoting better stability [34].

Selection Criteria

- Compatibility with the base fluid and nanoparticles
- Thermal and chemical stability
- Minimal influence on the thermophysical properties of the nanofluid

1.3.2.2 Ultrasonication

Ultrasonication is a widely used method to break down agglomerates into smaller clusters or even individual nanoparticles. The process uses high-frequency sound waves (typically 20 kHz to several MHz) to produce cavitation bubbles that collapse violently, imparting high shear forces.

Parameters Affecting Dispersion

- Sonication time
- Power and frequency of the ultrasonic probe
- Volume of the nanofluid

Extended ultrasonication may lead to degradation of nanoparticles or breakdown of surfactant molecules. Hence, optimization is critical.

1.3.2.3 pH Control

pH adjustment can influence the surface charge on nanoparticles, which in turn affects the electrostatic repulsion among them. At a pH where the zeta potential is maximized (either positive or negative), particles tend to remain better dispersed.

Applications

- Stabilizing oxide nanoparticles (e.g., Al_2O_3, TiO_2) in aqueous media
- Enhancing colloidal stability without excessive surfactant use

1.3.3 Characterization Techniques

Characterizing nanofluids involves evaluating their physical, chemical, and flow-related properties to understand their behavior under application-specific conditions [35]. Figure 1.2 represents SEM images of melanin and TiO_2.

1.3.3.1 Particle Size Analysis

Particle size significantly influences the thermal conductivity, stability, and viscosity of nanofluids.

Dynamic Light Scattering (DLS), also known as photon correlation spectroscopy (PCS), is a non-invasive technique used to determine the hydrodynamic diameter of particles in suspension. It measures the time-dependent fluctuations in scattered light intensity caused by Brownian motion.

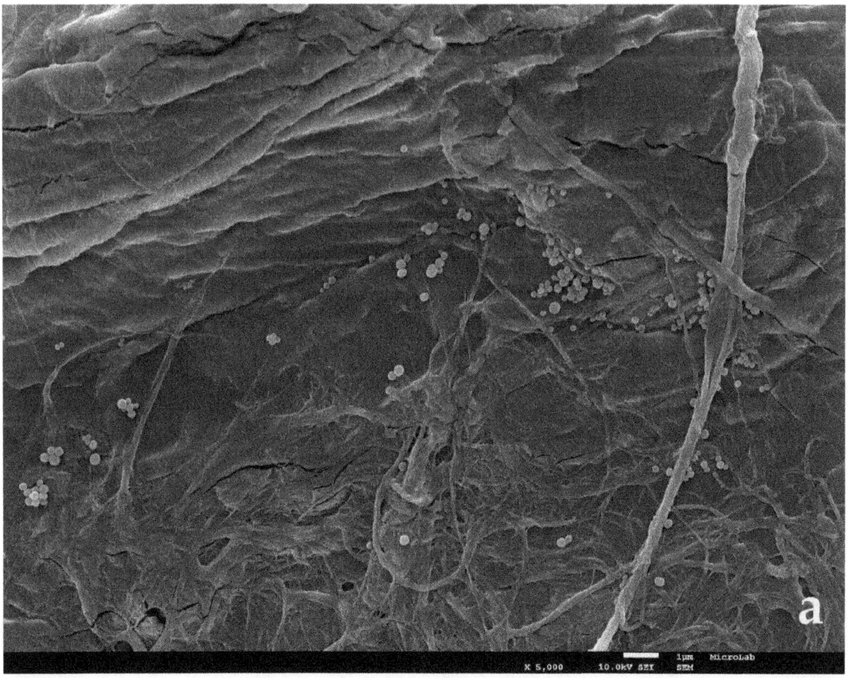

FIGURE 1.2 SEM images of melanin and TiO_2 nanoparticles used. (a–d), for melanin; (e) for TiO_2 [35].

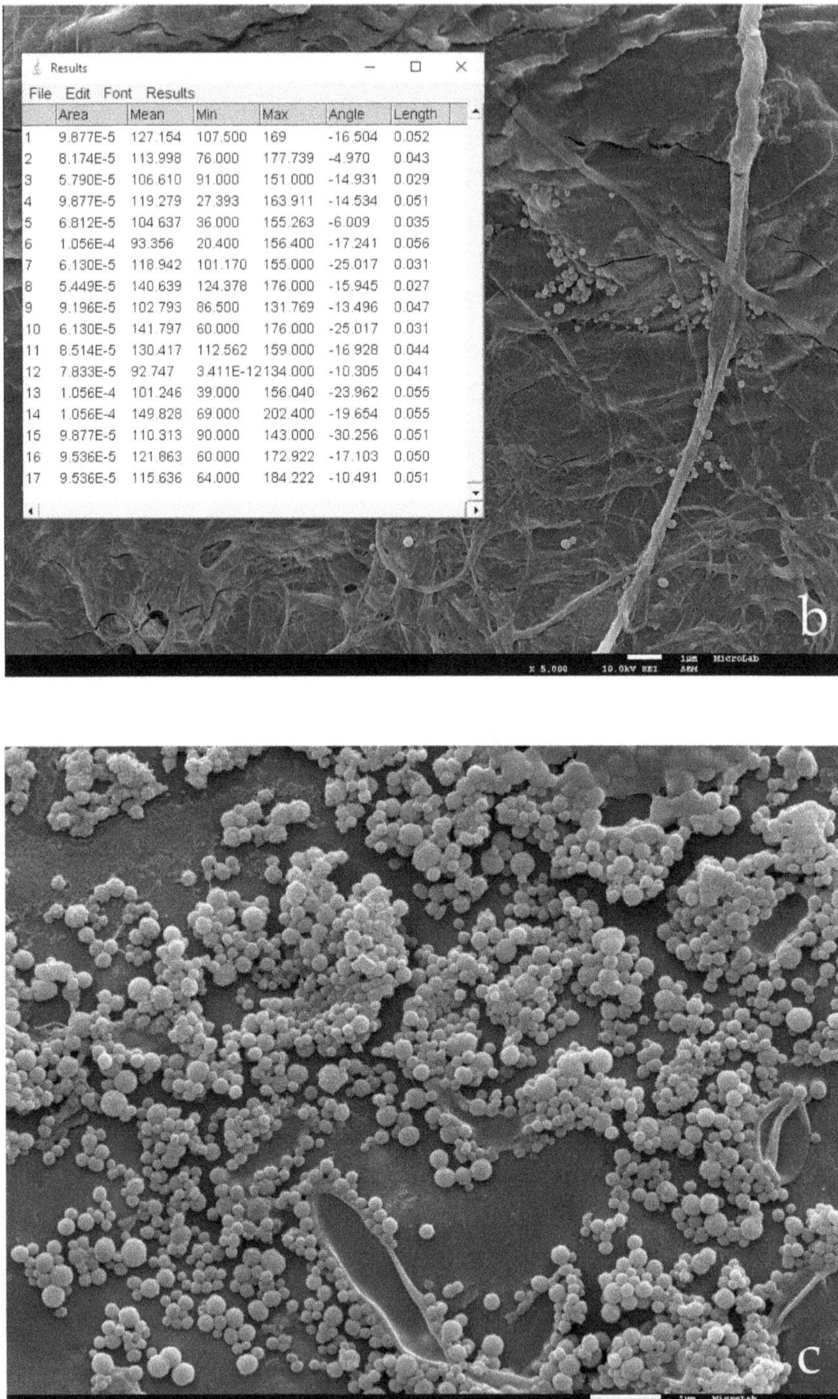
Results

File Edit Font Results

	Area	Mean	Min	Max	Angle	Length
1	9.877E-5	127.154	107.500	169	-16.504	0.052
2	8.174E-5	113.998	76.000	177.739	-4.970	0.043
3	5.790E-5	106.610	91.000	151.000	-14.931	0.029
4	9.877E-5	119.279	27.393	163.911	-14.534	0.051
5	6.812E-5	104.637	36.000	155.263	-6.009	0.035
6	1.056E-4	93.356	20.400	156.400	-17.241	0.056
7	6.130E-5	118.942	101.170	155.000	-25.017	0.031
8	5.449E-5	140.639	124.378	176.000	-15.945	0.027
9	9.196E-5	102.793	86.500	131.769	-13.496	0.047
10	6.130E-5	141.797	60.000	176.000	-25.017	0.031
11	8.514E-5	130.417	112.562	159.000	-16.928	0.044
12	7.833E-5	92.747	3.411E-12	134.000	-10.305	0.041
13	1.056E-4	101.246	39.000	156.040	-23.962	0.055
14	1.056E-4	149.828	69.000	202.400	-19.654	0.055
15	9.877E-5	110.313	90.000	143.000	-30.256	0.051
16	9.536E-5	121.863	60.000	172.922	-17.103	0.050
17	9.536E-5	115.636	64.000	184.222	-10.491	0.051

FIGURE 1.2 (Continued)

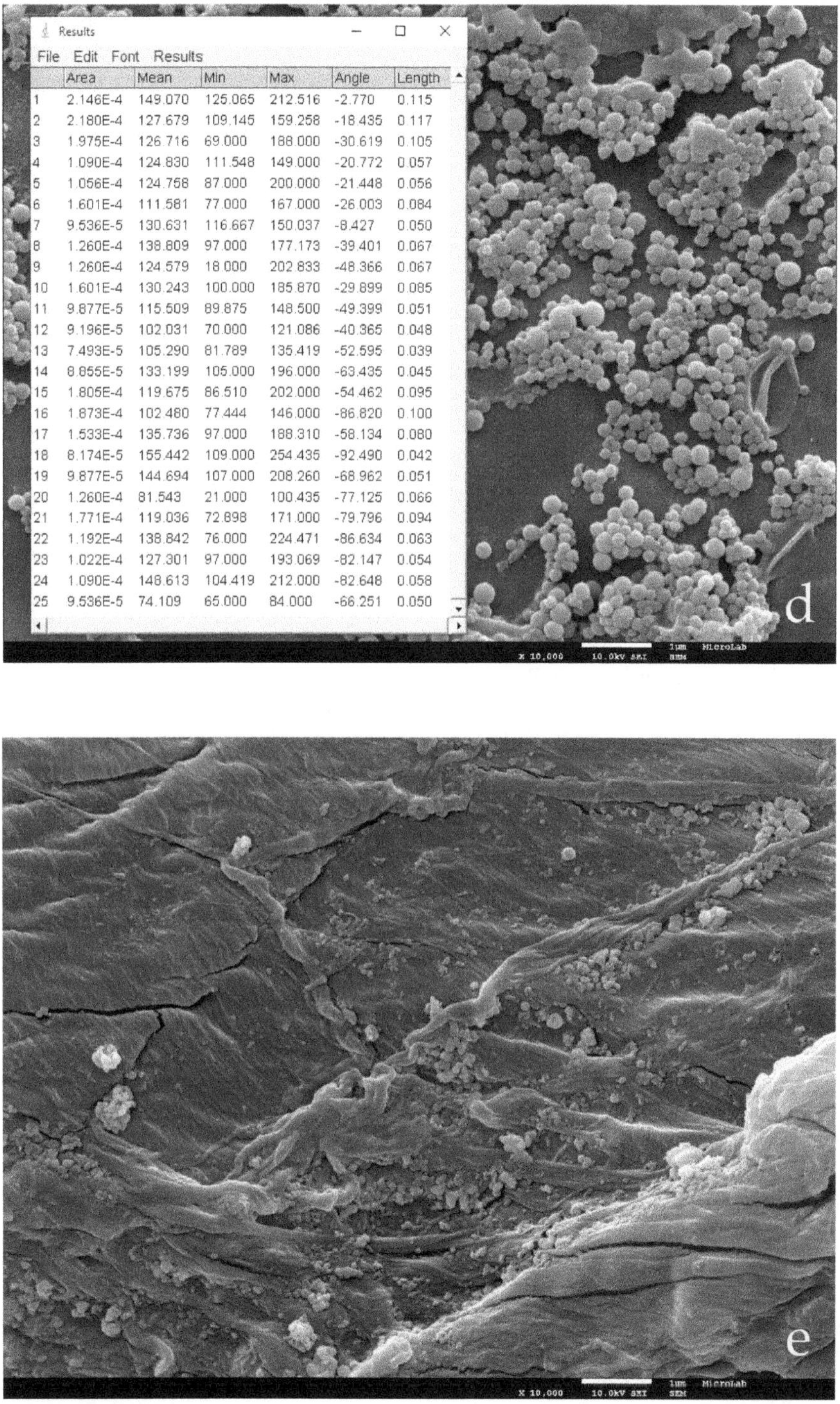

	Area	Mean	Min	Max	Angle	Length
1	2.146E-4	149.070	125.065	212.516	-2.770	0.115
2	2.180E-4	127.679	109.145	159.258	-18.435	0.117
3	1.975E-4	126.716	69.000	188.000	-30.619	0.105
4	1.090E-4	124.830	111.548	149.000	-20.772	0.057
5	1.056E-4	124.758	87.000	200.000	-21.448	0.056
6	1.601E-4	111.581	77.000	167.000	-26.003	0.084
7	9.536E-5	130.631	116.667	150.037	-8.427	0.050
8	1.260E-4	138.809	97.000	177.173	-39.401	0.067
9	1.260E-4	124.579	18.000	202.833	-48.366	0.067
10	1.601E-4	130.243	100.000	185.870	-29.899	0.085
11	9.877E-5	115.509	89.875	148.500	-49.399	0.051
12	9.196E-5	102.031	70.000	121.086	-40.365	0.048
13	7.493E-5	105.290	81.789	135.419	-52.595	0.039
14	8.855E-5	133.199	105.000	196.000	-63.435	0.045
15	1.805E-4	119.675	86.510	202.000	-54.462	0.095
16	1.873E-4	102.480	77.444	146.000	-86.820	0.100
17	1.533E-4	135.736	97.000	188.310	-58.134	0.080
18	8.174E-5	155.442	109.000	254.435	-92.490	0.042
19	9.877E-5	144.694	107.000	208.260	-68.962	0.051
20	1.260E-4	81.543	21.000	100.435	-77.125	0.066
21	1.771E-4	119.036	72.898	171.000	-79.796	0.094
22	1.192E-4	138.842	76.000	224.471	-86.634	0.063
23	1.022E-4	127.301	97.000	193.069	-82.147	0.054
24	1.090E-4	148.613	104.419	212.000	-82.648	0.058
25	9.536E-5	74.109	65.000	84.000	-66.251	0.050

FIGURE 1.2 (Continued)

Advantages of DLS Technique

- Rapid analysis
- Suitable for particles in the range of 1 nm to several microns
- Provides particle size distribution

Limitations of DLS Technique

- Sensitive to aggregates
- Requires clear, dilute samples

Transmission Electron Microscopy (TEM) provides high-resolution images of nanoparticles, offering direct visualization of size, shape, and dispersion status.

Features of TEM

- Resolution up to 0.1 nm
- Accurate assessment of core particle size
- Can identify crystalline phases via electron diffraction

Drawbacks of TEM

- Labor-intensive sample preparation
- Not representative of bulk behavior

1.3.3.2 Zeta Potential Measurement

Zeta potential measures the electrical potential at the slipping plane of nanoparticles in a fluid. It is an indicator of electrostatic stability:

- High zeta potential ($>|30|$ mV) suggests strong repulsive forces, promoting stability.
- Low zeta potential ($<|15|$ mV) indicates a risk of aggregation.

Measurement Techniques

- Electrophoretic light scattering (ELS)
- Laser Doppler velocimetry

Zeta potential values can vary with pH, ionic strength, and presence of surfactants.

Applications

- Selecting optimum pH for stability
- Evaluating the effect of surfactants or functionalization

1.3.3.3 Rheological Properties (Viscometry)

Understanding the rheological behavior of nanofluids is essential for designing flow systems. Rheology studies how a fluid deforms under shear stress.

Viscosity is typically measured using the following:

- Rotational viscometers (Brookfield-type)
- Capillary viscometers
- Oscillatory rheometers

Factors influencing viscosity are as follows:

- Particle concentration
- Temperature
- Particle shape and aspect ratio
- Base fluid characteristics

Behavior types:

- **Newtonian:** Viscosity remains constant with shear rate (common at low particle loadings).
- **Non-Newtonian:** Viscosity varies with shear rate (e.g., shear-thinning or shear-thickening behavior).

1.3.3.4 Yield Stress and Thixotropy

Some nanofluids exhibit a yield stress, the minimum stress required for flow initiation. Thixotropic behavior (time-dependent viscosity changes) is also observed, particularly in high-viscosity or gel-forming nanofluids.

Understanding these behaviors is vital for:

- Pump and piping design.
- Process modeling and simulation.
- Ensuring uniform heat and mass transfer.

The preparation and characterization of nanofluids are pivotal in determining their performance and stability. While the one-step method offers superior dispersion at the cost of complexity, the two-step method is more practical and scalable. Effective use of surfactants, ultrasonication, and pH control enhances dispersion stability.

Characterization techniques such as DLS, TEM, zeta potential analysis, and rheological measurements provide essential insights into particle behavior, stability, and flow properties. A rigorous understanding of these parameters is necessary for tailoring nanofluids to meet the demands of advanced energy systems and process intensification applications.

In the next chapter, we will explore the application domains of nanofluids in greater depth, highlighting their role in enhancing the efficiency of thermal systems, solar energy devices, and industrial heat exchangers.

1.4 ENERGY CONSERVATION APPLICATIONS

Nanofluids, owing to their remarkable thermal properties and adaptability, are emerging as transformative agents in various energy conservation applications. By enhancing heat transfer capabilities in traditional and advanced thermal systems, nanofluids contribute to significant energy savings, reduced operational costs, and improved system efficiency. This chapter explores how nanofluids are applied in diverse sectors such as HVAC systems, renewable energy, automotive industries, electronics cooling, and process heat recovery, emphasizing their role in addressing global energy demands sustainably [36].

1.4.1 Nanofluids in Heat Exchangers

Heat exchangers are critical components in energy systems, employed to transfer heat between two or more fluids without direct contact. Conventional fluids often limit the heat transfer rate due to their relatively low thermal conductivities. Nanofluids address this limitation by providing improved thermal conductivity and heat transfer coefficients.

Applications

- *Shell-and-tube heat exchangers* using Al_2O_3 or CuO nanofluids demonstrate 15–40% better thermal performance [37].
- *Compact plate heat exchangers* benefit from enhanced convective heat transfer, enabling miniaturization.

Energy Benefits

- Increased energy efficiency
- Reduced exchanger size and material cost
- Lower pumping power (with optimized viscosity) [38]

1.4.2 Solar Thermal Collectors

Nanofluids enhance solar thermal energy systems by improving optical absorption and thermal conductivity.

Types

- *Flat plate collectors* with TiO_2 or carbon nanotube (CNT) nanofluids
- *Parabolic trough collectors* employing hybrid nanofluids (e.g., Al_2O_3-Cu)

Mechanisms

- Nanoparticles increase solar energy absorption through scattering and surface plasmon resonance.
- Higher heat transfer efficiency boosts thermal storage and reduces losses.

Case Studies

- Efficiency increase of 10–20% reported in solar collectors using graphene-water nanofluids [39]
- Reduced payback period and operating costs in solar-assisted desalination systems [40]

1.4.3 Refrigeration and Heating, Ventilation, and Air Conditioning (HVAC Systems)

In HVAC systems, nanofluids are employed in both refrigerants and heat transfer fluids.

Enhancements

- Improved heat exchange in condensers and evaporators
- Higher coefficient of performance (COP)

Examples

- CuO and ZnO nano-refrigerants used in R134a and R22 improve system COP by up to 15%
- TiO_2-based nanofluids reduce compressor work and enhance cooling rates [41–43]

Benefits

- Energy savings in commercial buildings and industrial cooling systems
- Reduced refrigerant charge and greenhouse gas emissions

1.4.4 Automotive Thermal Management

Nanofluids are being integrated into vehicle thermal systems to manage engine heat, improve fuel efficiency, and enhance passenger comfort.

Applications

- *Radiator coolants* with Al_2O_3, SiO_2, or carbon-based nanoparticles [44]
- *Battery cooling systems* in electric vehicles using nanofluid-based phase change materials

Performance Gains

- Up to 25% increase in radiator efficiency
- Improved engine life and reduced emission levels
- Faster heat dissipation in EV battery packs

1.4.5 Electronics and Microelectronics Cooling

The miniaturization of electronics has led to increased heat densities, necessitating advanced cooling methods. Nanofluids offer superior thermal conductivity and convective performance for electronics cooling.

Technologies

- *Microchannel heat sinks* using SiC or diamond nanofluids
- *Spray cooling systems* with nanofluid sprays

Benefits

- Enhanced thermal management in CPUs, GPUs, and data centers
- Prevention of thermal throttling and component failure [45–47]

1.4.6 Industrial Process Heat Recovery

Nanofluids facilitate more efficient heat recovery from industrial processes, where large quantities of waste heat are typically lost.

Systems

- Heat exchangers in chemical and petroleum industries
- Steam generation systems using nanofluid-based heat transfer fluids

Advantages

- Better energy capture from low-grade heat
- Reduced fuel consumption and carbon footprint [48]

1.4.7 Nuclear Reactor Cooling

In nuclear systems, nanofluids enhance safety and performance by providing superior heat transfer under extreme conditions.

Roles

- Coolant fluids in pressurized water reactors (PWRs)
- Emergency core cooling systems using nanofluids

Nanoparticles Used

- Al_2O_3, SiO_2, and TiO_2 for their radiation stability and thermal conductivity

Impact

- Improved heat removal during high-power operations
- Enhanced safety margins in accident scenarios [49]

1.4.8 Renewable Energy Integration

Nanofluids support the efficiency and reliability of renewable energy systems, especially in solar, geothermal, and wind applications [50].

Examples

- *Geothermal systems* using nanofluids for enhanced subsurface heat extraction
- *Wind turbine gear cooling* with nano-lubricants

Benefits

- Improved energy conversion efficiency
- Reduced maintenance and longer operational life

1.4.9 Building-Integrated Thermal Systems

Nanofluids contribute to energy savings in smart buildings via advanced thermal regulation.

Applications

- Radiant floor heating using nanofluid-based water circuits
- Building-integrated photovoltaic/thermal (BIPV/T) systems [51–53]

Benefits

- Increased energy autonomy
- Reduction in peak load demand

1.4.10 Desalination and Water Heating

Solar-assisted desalination and water heating systems are more efficient with nanofluids.

Contributions

- Higher evaporation rates in solar stills
- Faster heat transfer in solar water heaters [54]

The adoption of nanofluids across multiple energy systems is transforming the landscape of thermal management and energy efficiency. Their role in reducing energy consumption, enabling compact system designs, and supporting the deployment of renewable technologies aligns strongly with global sustainability targets. However, challenges such as long-term stability, cost-effectiveness, and environmental impact must be addressed to unlock their full potential in mainstream energy conservation applications. The next chapter will explore the limitations, environmental concerns, and future directions for nanofluid technology.

1.5 PROCESS INTENSIFICATION USING NANOFLUIDS

Process intensification (PI) is a transformative strategy in chemical engineering aimed at making chemical manufacturing significantly more efficient, cleaner, and safer. It involves the development of novel apparatuses and techniques that provide substantial improvements in process efficiency and sustainability. Among the emerging tools for achieving PI, nanofluids have gained considerable attention due to their enhanced thermophysical properties and potential to boost heat and mass transfer in various unit operations. This chapter presents an in-depth exploration of the role of nanofluids in process intensification, focusing on their applications in heat and mass transfer enhancement, chemical reactors, multiphase systems, absorption/desorption cycles, and membrane separation processes [55].

1.5.1 Role in Enhancing Heat and Mass Transfer

The fundamental appeal of nanofluids in process intensification lies in their ability to significantly augment heat and mass transfer rates.

1.5.1.1 Enhanced Heat Transfer

Nanofluids exhibit higher thermal conductivities and convective heat transfer coefficients compared to conventional fluids. This improvement arises from:

- Increased surface area of nanoparticles.
- Brownian motion enhancing micro-convection.
- Nanoparticle-induced disruption of thermal boundary layers [56–58].

1.5.1.2 Impact on Mass Transfer

While most studies focus on heat transfer, nanofluids also influence mass transfer through:

- Improved diffusivity due to microscale mixing.
- Increased interfacial area in multiphase systems.
- Reduction in mass transfer resistance.

These enhancements are particularly beneficial in processes such as absorption, distillation, and reactive mass transfer operations [59].

1.5.2 Applications in Chemical Reactors

Nanofluids are being increasingly used in chemical reactors to improve performance, reduce footprint, and lower energy demands.

1.5.2.1 Tubular and Batch Reactors

In both continuous and batch reactors, nanofluids improve internal heat transfer, leading to better temperature control and reaction rates. Examples include:

- Water-based Al_2O_3 nanofluids used in exothermic reactions for uniform heat removal.
- CNT-based nanofluids enhancing catalyst surface interactions in polymerization reactors [60].

1.5.2.2 Microreactors

Microreactors are a prime candidate for nanofluid integration due to their small dimensions and high surface area-to-volume ratios. Nanofluids help in:

- Rapid heat dissipation, preventing hot spots.
- Enhanced mixing and reaction kinetics.

Studies show up to 50% increase in yield and reaction speed with nanofluid-cooled microreactors [61–63].

1.5.2.3 Photocatalytic Reactors

In photocatalytic processes, nanofluids containing TiO_2 or ZnO serve not only as heat transfer fluids but also as active photocatalysts. This dual functionality supports advanced oxidation processes (AOPs) for pollutant degradation [64].

1.5.3 Use in Multiphase Systems (Boiling, Condensation, Evaporation)

Multiphase processes are inherently limited by interfacial resistances and phase change inefficiencies. Nanofluids have been shown to mitigate these issues.

1.5.3.1 Boiling Heat Transfer

Nanofluids improve nucleate boiling performance due to:

- Increased number of active nucleation sites.
- Improved surface wettability.
- Modification of bubble dynamics.

Reported benefits include:

- Enhanced boiling heat transfer coefficients by up to 40% [65].
- Delayed onset of critical heat flux (CHF), improving safety margins.

1.5.3.2 Condensation Heat Transfer

In condensation processes, nanofluids modify surface tension and condensate film behavior:

- Promotion of dropwise rather than film-wise condensation
- Improved condensate drainage and surface renewal

Enhanced condensation coefficients result in energy savings in HVAC and power plant condensers [66].

1.5.3.3 Evaporation and Desalination

Nanofluids improve solar and industrial evaporation processes:

- Increased solar absorption and thermal gradient
- Higher evaporation flux in membrane distillation and solar stills

Graphene and CNT nanofluids demonstrate superior performance in solar-enhanced evaporation systems [67].

1.5.4 Nanofluids in Absorption and Desorption Systems

Absorption refrigeration and gas separation systems rely heavily on the thermophysical properties of working fluids. Nanofluids provide significant improvements in these processes.

1.5.4.1 Absorption Refrigeration

In lithium bromide-water (LiBr-H_2O) systems, the addition of nanoparticles:

- Increases heat and mass transfer rates.
- Reduces cycle time and enhances COP.

Nanoparticles such as CuO and Al_2O_3 lower the solution viscosity and improve heat absorption from the environment.

1.5.4.2 Desorption Enhancement

Desorption processes benefit from:

- Higher thermal conductivity enabling faster regeneration.
- Improved mass diffusion at interfaces.

Nanofluids are also investigated for ammonia-water and ionic liquid-based absorption cycles [68].

1.5.4.3 Gas Absorption Columns

Packed and spray columns benefit from the enhanced interfacial area and wettability provided by nanofluids, improving CO_2 and SO_2 capture efficiencies [69].

1.5.5 Role in Membrane and Separation Processes

Nanofluids are increasingly integrated into membrane processes, either as feed additives or as part of composite membranes.

1.5.5.1 Membrane Distillation

Nanoparticle-enhanced feed solutions improve:

- Permeate flux.
- Thermal efficiency.

Additionally, nanoparticles reduce membrane fouling and improve lifespan [70, 71].

1.5.5.2 Ultrafiltration and Nanofiltration

Pre-treatment with nanofluids enhances particle removal and reduces membrane clogging. Functional nanoparticles such as silver or ZnO also impart antimicrobial properties.

1.5.5.3 Pervaporation

Nanofluids in pervaporation systems facilitate selective separation by:

- Modifying vapor-liquid equilibrium.
- Enhancing driving force for mass transfer [72].

Nanofluids have emerged as vital enablers of process intensification in chemical engineering. Their ability to enhance both heat and mass transfer, combined with their adaptability to various unit operations, makes them ideal candidates for upgrading existing processes and designing new ones. From chemical reactors to multiphase and separation systems, the application of nanofluids offers reduced energy consumption, increased efficiency, and more compact and sustainable process designs. Future research should focus on optimizing nanofluid formulations, understanding long-term behavior under process conditions, and scaling up successful lab-scale demonstrations into industrial practice.

1.6 RECENT ADVANCES AND CASE STUDIES

The field of nanofluids has evolved rapidly over the past two decades, with significant research and development in both fundamental understanding and real-world applications. Recent trends emphasize hybrid nanofluids, magnetic nanofluids, and smart thermal management systems, enabled through synergistic formulations and advanced modeling. Furthermore, computational and experimental studies have provided insights into their behavior under diverse conditions, while pilot-scale and industrial demonstrations signal a transition from lab-scale promise to practical deployment [73]. This chapter elaborates on these advancements and presents compelling case studies that highlight the effectiveness of nanofluids in energy conservation and process intensification.

1.6.1 Hybrid Nanofluids: Synergistic Enhancements

1.6.1.1 Concept and Composition

Hybrid nanofluids are composed of two or more types of nanoparticles dispersed in a base fluid. These nanoparticles can be metal-metal, metal-metal oxide, or metal-carbon based, providing synergistic benefits by combining the advantageous properties of each component.

1.6.1.2 Advantages over Conventional Nanofluids

- **Improved thermal conductivity**: Synergistic interactions between nanoparticles enhance energy transport mechanisms.
- **Better stability**: Proper formulation can prevent agglomeration and sedimentation.
- **Optimized viscosity**: Dual-component systems allow tuning of rheological behavior.
- **Enhanced photothermal properties**: Useful in solar thermal applications.

1.6.1.3 Notable Combinations

- Al_2O_3-Cu/water for balanced thermal and rheological properties
- TiO_2-SiO_2/ethylene glycol for solar collector applications
- Graphene-CNT hybrid for high thermal conductivity and electrical performance

1.6.1.4 Application Highlights

- Hybrid nanofluids in parabolic solar collectors show efficiency gains of 25–35%.
- In automotive radiators, hybrid nanofluids reduce coolant volume and increase engine thermal efficiency [74].

1.6.2 Magnetic Nanofluids for Smart Heat Transfer

1.6.2.1 Fundamentals of Magnetic Nanofluids

Magnetic nanofluids (also called ferrofluids) are colloidal suspensions of magnetic nanoparticles such as Fe_3O_4 or $CoFe_2O_4$ in a base fluid. They exhibit unique behavior under external magnetic fields, enabling tunable thermal properties.

1.6.2.2 Smart and Controlled Heat Transfer

- **Field-responsive thermal conductivity**: Varies with the direction and intensity of the applied magnetic field.
- **Directional heat flow control**: Enables heat steering in advanced thermal systems.
- **Magneto-convection**: Enhanced convection due to magnetic forces.

1.6.2.3 Applications

- Targeted cooling in microelectronics
- Localized heating in biomedical hyperthermia treatments
- Adaptive thermal barriers in space applications

1.6.2.4 Challenges and Solutions

- **Stability under field exposure**: Use of surfactants and core-shell nanoparticles
- **Compatibility with system materials**: Development of non-reactive, biocompatible formulations [75]

1.6.2.5 Recent Studies

- Magnetite (Fe_3O_4)/water nanofluids demonstrated a 40% enhancement in thermal conductivity under 0.5 T magnetic field.
- In microchannels, magnetic nanofluids increased Nusselt number by 20–30% with controlled field application [76].

1.6.3 Computational and Experimental Studies

1.6.3.1 Modeling Approaches

Computational studies play a key role in predicting nanofluid behavior, optimizing design, and reducing trial-and-error in experiments.

1.6.3.1.1 CFD Simulations

- Numerical models using ANSYS Fluent or COMSOL Multiphysics
- Single-phase and two-phase models to simulate heat transfer
- Incorporation of thermophysical property models based on nanoparticle shape, size, and concentration

1.6.3.1.2 Lattice Boltzmann Method (LBM)

- Ideal for simulating micro and nanoscale flow behavior
- Captures complex boundary conditions and interfacial interactions

1.6.3.1.3 Molecular Dynamics (MD) Simulations

- Useful in studying fundamental particle–fluid interactions
- Provides insight into thermal conductivity enhancement mechanisms

1.6.3.2 Experimental Techniques and Innovations

- **High-speed infrared thermography** for transient heat transfer measurement
- **MEMS sensors** for microscale temperature mapping
- **Micro-PIV** (Particle Image Velocimetry) for flow visualization

1.6.3.3 Validation and Correlation

Combining experimental and computational studies improves accuracy and predictive capability. Benchmark data from lab studies validate CFD models, leading to reliable system designs [77].

1.6.4 Industrial Demonstrations and Pilot-Scale Applications

1.6.4.1 Solar Thermal Energy Systems

Pilot-scale installations using nanofluids in solar collectors show improved energy capture and thermal efficiency.

- **Case Study: India**—Solar-assisted air conditioning system using TiO_2-water nanofluid reported 20% efficiency improvement and 15% cost savings.
- **Case Study: UAE**—Solar desalination using CNT-based nanofluids yielded 30% higher distillate output.

1.6.4.2 HVAC and Refrigeration

- **Case Study: South Korea**—Commercial HVAC unit retrofitted with Al_2O_3-R134a nano-refrigerant showed 17% COP improvement and 10% energy savings.
- **Case Study: United States**—Supermarket refrigeration using ZnO nanofluids achieved more uniform cooling with lower compressor workload.

1.6.4.3 Automotive and Aerospace Cooling

- **Case Study: Germany**—Electric vehicle battery thermal management system with hybrid graphene/Al_2O_3 nanofluid extended battery life and reduced hotspot formation.

- **Case Study: Japan**—Jet engine testing with nanofluid-based cooling showed up to 22% better heat dissipation.

1.6.4.4 Process Industry Integration

- *Chemical reactors* equipped with nanofluid jackets for exothermic heat removal
- *Evaporative cooling towers* using nanofluid sprays for higher latent heat capture [78]

1.6.4.5 Barriers to Adoption

- Cost of nanoparticles and surfactants
- Long-term stability and clogging risks
- Need for regulatory and environmental safety evaluation

Recent advances in nanofluids underscore their transformative potential across industries. Hybrid and magnetic nanofluids enable smart, tunable, and highly efficient thermal management. Computational tools coupled with experimental validation have accelerated their design and application. Industrial case studies confirm real-world benefits, including energy savings, process enhancement, and environmental advantages. While challenges remain, continued innovation and supportive policy frameworks are likely to drive widespread commercialization of nanofluid technologies in the coming decade [79, 80].

In the following chapter, we will summarize the contributions of nanofluids in energy conservation and process intensification and explore future directions and research opportunities.

1.7 CHALLENGES AND LIMITATIONS OF APPLICATIONS OF HYBRID NANOFLUIDS

Hybrid nanofluids, as advanced thermal fluids composed of two or more types of nanoparticles suspended in a base fluid, offer remarkable potential in thermal management and energy applications. Despite the demonstrated improvements in heat and mass transfer performance, numerous technical, economic, and environmental challenges still hinder their widespread industrial adoption. This chapter provides an in-depth discussion on the key challenges and limitations associated with hybrid nanofluid applications, focusing on stability and agglomeration, production scalability and cost, material compatibility and corrosion, environmental and health impacts, and the current lack of standardized data and methodologies.

1.7.1 Stability and Agglomeration Issues

1.7.1.1 Nature of Agglomeration

The major technical hurdle in hybrid nanofluid application is the tendency of nanoparticles to agglomerate due to high surface energy and Van der Waals attraction forces. Agglomeration leads to:

- Sedimentation of particles.
- Loss of enhanced thermal properties.
- Uneven distribution within the fluid.
- Clogging of microchannels or heat exchanger passages.

In hybrid systems, the interaction between dissimilar nanoparticles (e.g., metal and metal oxide) may exacerbate these effects due to mismatched zeta potential and particle sizes.

1.7.1.2 Influence on Performance

Instability affects:

- Thermal conductivity.
- Viscosity and flow behavior.
- Pressure drop across heat exchangers.

Experimental studies show that even minor sedimentation over time can lead to a 10–20% drop in thermal enhancement, especially in static or low-flow systems.

1.7.1.3 Mitigation Strategies

- **Surfactant addition**: Using compatible surfactants such as SDS, CTAB, or PEG to improve dispersion.
- **Ultrasonication**: High-energy ultrasonic treatment to break down agglomerates.
- **pH optimization**: Adjusting pH to maximize zeta potential and electrostatic repulsion.
- **Functionalization**: Surface modification of nanoparticles to increase hydrophilicity or hydrophobicity.

Despite these efforts, long-term stability under high-temperature or cyclic conditions remains an unresolved issue in hybrid nanofluids.

1.7.2 Cost and Scalability of Production

1.7.2.1 High Cost of Nanoparticles

Nanoparticle synthesis, especially with high-purity metals or carbon-based materials (e.g., graphene, CNTs), is cost-intensive due to:

- Energy-intensive synthesis methods (e.g., chemical vapor deposition).
- Post-processing steps such as drying, grinding, and surface modification.
- Quality control and batch consistency requirements.

Hybrid nanofluids require two or more types of nanoparticles, further increasing raw material and processing costs.

1.7.2.2 Dispersion and Handling Costs

- Use of high-energy ultrasonication equipment
- Need for specialized surfactants or stabilizers
- Handling and storage under inert atmospheres for reactive materials

1.7.2.3 Scalability Issues

Most hybrid nanofluid studies are confined to lab-scale volumes (≤1 L). Scaling up introduces:

- Increased sedimentation challenges.
- Equipment wear due to abrasive particles.
- Higher viscosity, impacting pump and piping systems.

1.7.2.4 Economic Feasibility

Cost–benefit analysis often does not favor hybrid nanofluids unless:

- Energy savings are substantial and consistent.
- Lifespan is long, without the need for frequent replacement.
- Maintenance costs are not significantly increased.

Bulk commercialization requires cost-effective synthesis methods such as green synthesis or scalable sol-gel techniques.

1.7.3 Corrosion and Compatibility with Materials

1.7.3.1 Corrosive Effects of Nanoparticles

Nanoparticles may accelerate corrosion of system components through:

- Galvanic coupling (especially with metallic nanoparticles).
- Increased ion exchange reactions.
- Catalysis of oxidation/reduction reactions on surfaces.

Studies report increased corrosion rates in aluminum, copper, and steel surfaces exposed to nanofluids over extended periods.

1.7.3.2 Hybrid Systems and Complexity

In hybrid nanofluids, the presence of dissimilar materials complicates corrosion chemistry. For example:

- Carbon-based particles may form passive films.
- Metal oxide particles may catalyze electrolyte degradation.

1.7.3.3 Material Compatibility

The fluid's impact on seals, gaskets, pumps, and heat exchanger materials must be considered. High particle loading can:

- Erode surfaces.
- Increase mechanical stress.
- Lead to micro-abrasion in narrow flow channels.

1.7.3.4 Mitigation Approaches

- Coating or passivation of metal surfaces
- Use of corrosion inhibitors compatible with nanofluids
- Selecting corrosion-resistant alloys (e.g., Inconel, Hastelloy)

1.7.4 Environmental and Health Safety Concerns

1.7.4.1 Nanoparticle Toxicity

Nanoparticles possess high reactivity and bioavailability, raising concerns over:

- **Inhalation risks**: Particularly during synthesis, handling, or accidental spillage.
- **Skin and eye contact**: Leading to irritation or allergic responses.
- **Ingestion and accumulation**: Causing organ damage or systemic toxicity.

Carbon-based nanoparticles (CNTs, graphene) and metal oxides (ZnO, TiO_2) are known to cause oxidative stress in living cells. Figure 1.3 represents nanomaterials toxicity flow.

1.7.4.2 Environmental Release

- Accidental leakage or disposal into water bodies can disrupt aquatic ecosystems.
- Nanoparticles can persist in the environment and accumulate in food chains.
- Current water treatment plants are not fully equipped to filter nanoparticles.

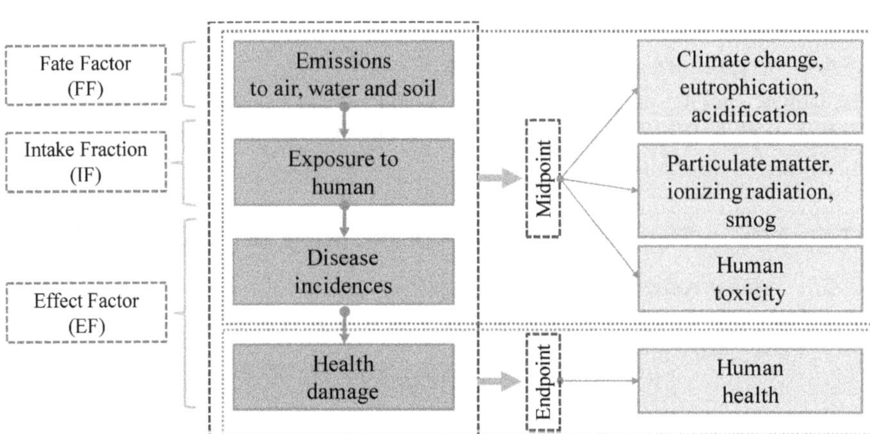

FIGURE 1.3 Nanomaterial's toxicity flow [20].

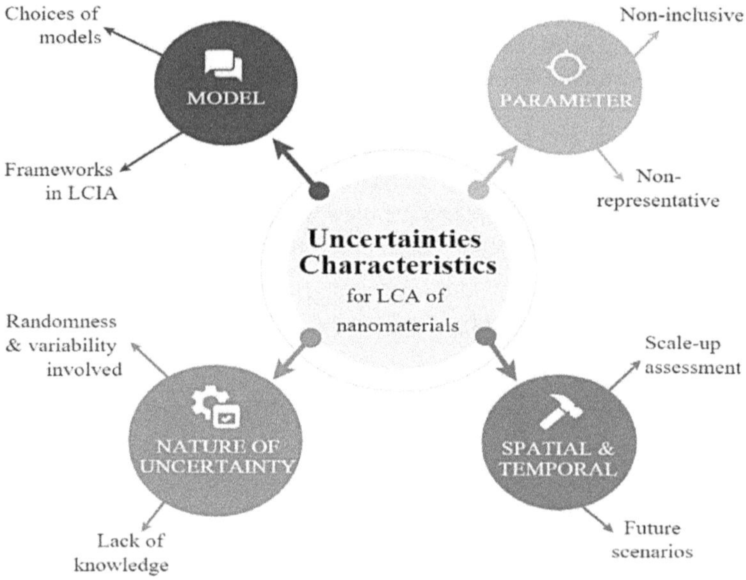

FIGURE 1.4 Uncertainties of life cycle assessment studies [20].

1.7.4.3 Life cycle Analysis

There is a need for comprehensive assessment of:

- Raw material extraction.
- Synthesis and operational phase emissions.
- End-of-life treatment and recyclability.

But, there are a few uncertainties also in these studies. These uncertainties deviate the actual results from the calculated parameters. Figure 1.4 shows these uncertainties.

1.7.4.4 Regulatory and Ethical Issues

The lack of specific regulatory frameworks for hybrid nanofluids complicates safe implementation. Ethical concerns about exposure in poorly ventilated or low-safety environments also arise.

1.7.5 Lack of Standardization in Data and Methods

1.7.5.1 Measurement Variability

Thermophysical properties like thermal conductivity, viscosity, and specific heat vary widely across studies due to:

- Different nanoparticle sizes and concentrations
- Inconsistent stabilization protocols
- Diverse measurement instruments and conditions

1.7.5.2 Reproducibility Challenges

Results often lack reproducibility due to proprietary nanoparticle synthesis methods and differing experimental setups.

1.7.5.3 Absence of Benchmarking Standards

There are no universally accepted:

- Preparation procedures for hybrid nanofluids
- Stability testing protocols
- Thermal performance evaluation metrics

1.7.5.4 Need for Standardization Bodies

- Organizations like ASTM, ISO, and ASME must develop relevant standards.
- Creation of a centralized database for nanofluid performance data is critical.

1.7.5.5 Impact on Industrial Uptake

Lack of standardization creates skepticism among engineers and decision-makers, delaying technology transfer from research to application.

While hybrid nanofluids hold considerable promise in enhancing energy efficiency and process performance, numerous challenges must be addressed for their widespread adoption. Stability and agglomeration remain major technical concerns, particularly in long-term or high-temperature operations. Economic feasibility is limited by the high cost of nanoparticles and dispersion equipment, while scalability remains a bottleneck. Corrosion, compatibility, and environmental safety raise operational and ethical questions. The lack of standardized procedures for preparation, testing, and benchmarking continues to undermine confidence in their applicability.

Overcoming these barriers will require interdisciplinary collaboration between researchers, industry stakeholders, and regulatory agencies. Prioritizing green synthesis methods, establishing international standards, and conducting comprehensive life cycle and safety assessments will pave the way for the responsible and impactful integration of hybrid nanofluids into industrial and energy systems. The concluding chapter will summarize the transformative role of nanofluids and outline the strategic roadmap for future research and implementation.

1.8 FUTURE PROSPECTS AND RESEARCH DIRECTIONS

Nanofluids have demonstrated their potential as transformative agents for enhancing heat and mass transfer across diverse sectors, from energy conservation to advanced manufacturing. Despite these achievements, various areas of nanofluid research are still not yet explored. The following are a few areas where the research needs to be done in detail.

1.8.1 Green and Bio-Based Nanofluids

Conventional nanofluids often use chemically synthesized nanoparticles and base fluids that may pose environmental or health hazards. There is a growing emphasis on sustainable materials that align with green chemistry principles like biodegradability, low toxicity, and renewable sourcing.

Research is increasingly focusing on nanoparticles derived from plant extracts (e.g., green synthesis of Ag, ZnO, TiO_2 using neem, aloe vera, or tea extracts), biopolymers (chitosan, cellulose, and alginate), and agricultural waste (as precursors for carbon-based nanoparticles). These offer lower toxicity and enhanced compatibility with biological and environmental systems. These nanofluids can also be synthesized by replacing petroleum-based fluids with water–glycerol mixtures, vegetable oils, and ionic liquids with low environmental impact.

1.8.2 AI/ML-Assisted Design and Optimization

Formulating nanofluids involves optimizing multiple variables like nanoparticle type, size, shape, concentration, surfactants, dispersion methods, and base fluid compatibility.

Conventional experimental approaches are time-consuming and costly. Artificial intelligence (AI) and machine learning (ML) offer powerful tools to model complex interactions and identify optimal formulations. Artificial Neural Networks (ANN) are useful for predicting thermal conductivity and viscosity. Support Vector Machines (SVM) are useful for the classification of stable and unstable formulations of nanofluids. Random Forest and Decision Trees are models that can be used for sensitivity analysis and feature importance. Similarly, optimization algorithms can be used for multi-objective optimization, and Bayesian Optimization can be used for probabilistic predictions.

1.8.3 Advanced Modeling and Simulation (CFD, ML-CFD)

Computational fluid dynamics (CFD) remains a cornerstone of nanofluid research, allowing prediction of flow behavior, heat transfer, and pressure drops in complex geometries. In conventional CFD analysis, Newtonian behavior is assumed, and it is not able to capture nanoparticle dynamics in detail. The computational cost is very high for multiphase and transient simulations. Machine learning models are now integrated for CFD analysis. These surrogate ML models reduce computational burden. Real-time optimization during simulations is also possible in these models. In these models, adaptive meshing based on predicted hotspots is also possible. Molecular dynamics (MD) can be coupled with CFD for resolving nanoparticle behavior at different spatial and temporal scales. Real-time system modeling and diagnostics using digital twins is also possible. Integration with IoT devices for condition monitoring is also possible. Smart cooling systems in electronics and data centers are also possible. Next-gen heat exchangers with dynamic performance tuning are also possible.

1.8.4 Commercialization Pathways and Policy Support

Scale-up of synthesis methods is important for long-term performance testing. The research on techno-economic and life cycle assessments and certification and quality control is also important for the commercialization pathways. Engagement with industries is essential for the development of the nanofluid-based infrastructure. Partnership with energy, automotive, and HVAC sectors is very important in this regard. Piloting in renewable energy platforms and manufacturing lines is also very crucial.

1.8.4.1 Standardization and Regulation

- Development of guidelines for nanofluid synthesis, handling, and testing
- Regulatory frameworks addressing safety, environmental impact, and workplace exposure
- ISO and ASTM involvement in drafting protocols

1.8.4.2 Policy Incentives

- Government support through grants, subsidies, and green innovation funds
- Carbon credit schemes for energy-efficient systems using nanofluids
- Public–private partnerships to promote deployment in critical infrastructure

1.8.4.3 Roadmap for Adoption

- **Short-term**: Pilot projects, data repositories, training modules
- **Medium-term**: Industrial integration, cost reduction strategies, public engagement
- **Long-term**: Circular economy integration, biodegradable nanofluids, global standardization

The future of nanofluids lies in sustainable development, intelligent design, and system-level integration. Green and bio-based nanofluids will address the ecological footprint of current formulations. AI/ML will revolutionize how nanofluids are formulated, tested, and optimized, while advanced simulations will deepen understanding and guide scalable designs. Successful commercialization will depend on policy support, industry collaboration, and standardization. These directions will ensure nanofluids fulfill their promise as high-impact agents for energy conservation, process intensification, and sustainable industrial innovation.

1.9 CONCLUSION

Nanofluids have emerged as a revolutionary class of heat transfer and process enhancement fluids, drawing immense attention from researchers, engineers, and industry stakeholders. As global challenges related to energy efficiency, sustainability, and process optimization intensify, nanofluids offer novel solutions that integrate science, technology, and application-oriented innovation. This chapter presents a comprehensive synthesis of key points, emphasizing the strategic roles of nanofluids in energy and process industries, and explores pathways to bridge the existing gap between cutting-edge research and industrial-scale adoption.

As this book chapter illustrates, a coordinated effort involving scientific innovation, technological scaling, policy support, and societal engagement is essential to realizing the full potential of nanofluids. When these elements come together, nanofluids will not only revolutionize thermal systems but also significantly contribute to a more energy-efficient, sustainable, and technologically advanced future.

REFERENCES

1. Das, S. K., Choi, S. U., & Patel, H. E. (2006). Heat transfer in nanofluids—a review. *Heat Transfer Engineering*, *27*(10), 3–19.
2. Godiwal, R., & Mandal, M. M. (2024). Modeling of enhanced oil recovery using nanofluid. *Chemical Engineering & Technology*, *47*(3), 568–577.
3. Peng, B., Zhang, L., Luo, J., Wang, P., Ding, B., Zeng, M., & Cheng, Z. (2017). A review of nanomaterials for nanofluid enhanced oil recovery. *RSC Advances*, *7*(51), 32246–32254.
4. Li, S., Sng, A., Daniel, D., Lau, H. C., Torsæter, O., & Stubbs, L. P. (2021). Visualizing and quantifying wettability alteration by silica nanofluids. *ACS Applied Materials & Interfaces*, *13*(34), 41182–41189.
5. Durgam, S., & Kadam, G. (2021). Investigation on thermal conductivity and viscosity of nanofluids using analytical and machine learning models. *arXiv preprint arXiv:2102.12120.*
6. Tuok, L. P., Elkady, M., Zkria, A., Yoshitake, T., & Nour Eldemerdash, U. (2023). Evaluation of stability and functionality of zinc oxide nanofluids for enhanced oil recovery. *Micro and Nano Systems Letters*, *11*(1), 12.
7. Choi, S. U., & Eastman, J. A. (1995). *Enhancing thermal conductivity of fluids with nanoparticles* (No. ANL/MSD/CP-84938; CONF-951135-29). Argonne National Lab. (ANL), Argonne, IL (United States).
8. Jang, S. P., & Choi, S. U. (2004). Role of Brownian motion in the enhanced thermal conductivity of nanofluids. *Applied Physics Letters*, *84*(21), 4316–4318.
9. Prasher, R. (2005, January). Brownian-motion-based convective-conductive model for the thermal conductivity of nanofluids. In *Heat Transfer Summer Conference* (Vol. 47314, pp. 343–353).
10. Koo, J., & Kleinstreuer, C. (2005). Laminar nanofluid flow in microheat-sinks. *International Journal of Heat and Mass Transfer*, *48*(13), 2652–2661.
11. Kumar, P. M., Kumar, J., Tamilarasan, R., Sendhilnathan, S., & Suresh, S. (2015). Review on nanofluids theoretical thermal conductivity models. *Engineering Journal*, *19*(1), 67–83.
12. Razzaq, I., Xinhua, W., Rasool, G., Sun, T., Shflot, A. S., Malik, M. Y., . . . & Ali, A. (2025). Nanofluids for advanced applications: A comprehensive review on preparation methods, properties, and environmental impact. *ACS Omega*, *10*(6), 5251–5282.
13. Bhatti, M. M., Vafai, K., & Abdelsalam, S. I. (2023). The role of nanofluids in renewable energy engineering. *Nanomaterials*, *13*(19), 2671.
14. Tao, Q., Zhong, F., Deng, Y., Wang, Y., & Su, C. (2023). A review of nanofluids as coolants for thermal management systems in fuel cell vehicles. *Nanomaterials*, *13*(21), 2861.
15. Wanatasanappan, V. V., Rezman, M., & Abdullah, M. Z. (2022). Thermophysical properties of vegetable oil-based hybrid nanofluids containing Al2O3-TiO2 nanoparticles as insulation oil for power transformers. *Nanomaterials*, *12*(20), 3621.
16. Haeri, S. Z., Khiadani, M., Ramezanzadeh, B., Mohammed, H. A., & Zargar, M. (2024). Review on stability, thermophysical properties, and solar harvesting applications of titanium nitride-based nanofluids: Current status and outlook. *Energy & Fuels*, *38*(4), 2548–2572.

17. Malika, M., & Sonawane, S. S. (2024). Ecological optimization and LCA of TiO2-SiC/water hybrid nanofluid in a shell and tube heat exchanger by ANN. *Proceedings of the Institution of Mechanical Engineers, Part E: Journal of Process Mechanical Engineering*, *238*(1), 45–55.
18. Rahman, M. A., Hasnain, S. M., Pandey, S., Tapalova, A., Akylbekov, N., & Zairov, R. (2024). Review on nanofluids: Preparation, properties, stability, and thermal performance augmentation in heat transfer applications. *ACS Omega*, *9*(30), 32328–32349.
19. Malika, M., Jhadav, P. G., Parate, V. R., & Sonawane, S. S. (2023). Synthesis of magnetite nanoparticle from potato peel extract: Its nanofluid applications and life cycle analysis. *Chemical Papers*, *77*(2), 1081–1094.
20. Nizam, N. U. M., Hanafiah, M. M., & Woon, K. S. (2021). A content review of life cycle assessment of nanomaterials: Current practices, challenges, and future prospects. *Nanomaterials*, *11*(12), 3324.
21. Chronis, I., Tang, C., & Psomopoulos, C. S. (2024). An improved approach for the evaluation of the environmental impact of nano-modified insulating oils in electrical transformers. *IET Nanodielectrics*, *7*(4), 216–225.
22. Barberio, G., Scalbi, S., Buttol, P., Masoni, P., & Righi, S. (2014). Combining life cycle assessment and qualitative risk assessment: The case study of alumina nanofluid production. *Science of the Total Environment*, *496*, 122–131.
23. Kalsi, S., Kumar, S., Kumar, A., Alam, T., & Dobrotă, D. (2023). Thermophysical properties of nanofluids and their potential applications in heat transfer enhancement: A review. *Arabian Journal of Chemistry*, *16*(11), 105272.
24. Kakaç, S., & Pramuanjaroenkij, A. (2009). Review of convective heat transfer enhancement with nanofluids. *International Journal of Heat and Mass Transfer*, *52*(13–14), 3187–3196.
25. Kalsi, S., Kumar, S., Kumar, A., Alam, T., Sharma, A., & Yadav, A. S. (2025). A review on hybrid nanofluids for heat transfer: Advancements, synthesis, challenges and applications. *Discover Applied Sciences*, *7*(7), 698.
26. Kumar, K. R., & Shaik, A. H. (2023). Synthesis, thermophysical characterization and thermal performance analysis of novel Cu-MXene hybrid nanofluids for efficient coolant applications. *RSC Advances*, *13*(42), 29536–29560.
27. Rahman, M. A., Hasnain, S. M., Pandey, S., Tapalova, A., Akylbekov, N., & Zairov, R. (2024). Review on nanofluids: Preparation, properties, stability, and thermal performance augmentation in heat transfer applications. *ACS Omega*, *9*(30), 32328–32349.
28. Rudyak, V. Y., & Minakov, A. V. (2018). Thermophysical properties of nanofluids. *The European Physical Journal E*, *41*(1), 15.
29. Choi, T. J., Park, M. S., Kim, S. H., & Jang, S. P. (2021). Experimental study on the effect of nanoparticle migration on the convective heat transfer coefficient of EG/water-based Al2O3 nanofluids. *International Journal of Heat and Mass Transfer*, *169*, 120903.
30. Yu, Y., Du, J., Hou, J., Jin, X., & Wang, R. (2024). Investigation into the underlying mechanisms of the improvement of thermal conductivity of the hybrid nanofluids. *International Journal of Heat and Mass Transfer*, *226*, 125468.
31. Yang, N., Xu, X., Zhang, G., & Li, B. (2012). Thermal transport in nanostructures. *Aip Advances*, *2*(4).
32. Ali, A. R. I., & Salam, B. (2020). A review on nanofluid: Preparation, stability, thermophysical properties, heat transfer characteristics and application. *SN Applied Sciences*, *2*(10), 1636.
33. Sidik, N. A. C., Mohammed, H. A., Alawi, O. A., & Samion, S. (2014). A review on preparation methods and challenges of nanofluids. *International Communications in Heat and Mass Transfer*, *54*, 115–125.

34. Yu, F., Chen, Y., Liang, X., Xu, J., Lee, C., Liang, Q., . . . & Deng, T. (2017). Dispersion stability of thermal nanofluids. *Progress in Natural Science: Materials International, 27*(5), 531–542.
35. Lourenço, M. J., Alexandre, J., Huisman, C., Paredes, X., & Nieto de Castro, C. (2021). The balance between energy, environmental security, and technical performance: The regulatory challenge of nanofluids. *Nanomaterials, 11*(8), 1871.
36. Bocanegra, J. A., Marchitto, A., & Misale, M. (2025). Nanofluids in solar collectors: A comprehensive review focused on its sedimentation. *Clean Technologies and Environmental Policy, 27*(4), 1753–1784.
37. Al-Obaidi, M. A., Rashid, F. L., Rasheed, M. K., Aljibori, H. S. S., Mohammed, H. I., Mahdi, A. J., . . . & Mujtaba, I. M. (2024). Recent achievements in heat transfer enhancement with hybrid nanofluid in heat exchangers: A comprehensive review. *International Journal of Thermophysics, 45*(9), 133.
38. Baraı, R., Kumar, D., & Wankhade, A. (2023). Heat transfer performance of nanofluids in heat exchanger: A review. *Journal of Thermal Engineering, 9*(1), 86–106.
39. Gupta, S. K., & Gupta, S. (2021). The role of nanofluids in solar thermal energy: A review of recent advances. *Materials Today: Proceedings, 44*, 401–412.
40. Mahian, O., Kianifar, A., Kalogirou, S. A., Pop, I., & Wongwises, S. (2013). A review of the applications of nanofluids in solar energy. *International Journal of Heat and Mass Transfer, 57*(2), 582–594.
41. Thakur, P., Sonawane, S. S., Sonawane, S. H., & Bhanvase, B. A. (2020). Nanofluids-based delivery system, encapsulation of nanoparticles for stability to make stable nanofluids. In *Encapsulation of active molecules and their delivery system* (pp. 141–152). Elsevier.
42. Bethi, B., Sonawane, S. H., Bhanvase, B. A., & Sonawane, S. S. (2021). Textile industry wastewater treatment by cavitation combined with Fenton and ceramic nanofiltration membrane. *Chemical Engineering and Processing-Process Intensification, 168*, 108540.
43. Sidik, N. A. C., Kean, T. H., Chow, H. K., Rajaandra, A., Rahman, S., & Kaur, J. (2018). Performance enhancement of cold thermal energy storage system using nanofluid phase change materials: A review. *International Communications in Heat and Mass Transfer, 94*, 85–95.
44. Ukueje, W. E., Abam, F. I., & Obi, A. (2022). A perspective review on thermal conductivity of hybrid nanofluids and their application in automobile radiator cooling. *Journal of Nanotechnology, 2022*(1), 2187932.
45. Malika, M., & Sonawane, S. S. (2020). Effect of nanoparticle mixed ratio on stability and thermo-physical properties of CuO-ZnO/water-based hybrid nanofluid. *Journal of Indian Chemical Society, 97*(3), 414–419.
46. Sonawane, S., Thakur, P., & Paul, R. (2020). Study on thermal property enhancement of MWCNT based polypropylene (PP) nanocomposites. *Materials Today: Proceedings, 27*, 550–555.
47. Chen, C. H., & Yaji, K. (2025). Topology optimization for microchannel heat sinks with nanofluids using an Eulerian-Eulerian approach. *International Journal of Heat and Mass Transfer, 243*, 126870.
48. Sidik, N. A. C., Kean, T. H., Chow, H. K., Rajaandra, A., Rahman, S., & Kaur, J. (2018). Performance enhancement of cold thermal energy storage system using nanofluid phase change materials: A review. *International Communications in Heat and Mass Transfer, 94*, 85–95.
49. Siddiqui, O. K., Shams, A., & Al-Athel, K. (2025). A comprehensive review on the use of nanoparticles in nuclear power plants. *Arabian Journal for Science and Engineering, 50*(5), 2991–3015.

50. Li, J., Chen, D., Zhang, H., Jiang, Z., Sun, K., Fan, J., & Tang, Y. (2024). The tribological properties of nano-lubricants and their application on bearings: Recent research progress. *The International Journal of Advanced Manufacturing Technology, 134*(7), 3051–3082.
51. Malika, M. M., & Sonawane, S. S. (2019). Review on application of nanofluid/nano particle as water disinfectant. *Journal of Indian Association for Environmental Management (JIAEM), 39*(1–4), 21–24.
52. Malika, M., Bhad, R., & Sonawane, S. S. (2021). ANSYS simulation study of a low volume fraction CuO–ZnO/water hybrid nanofluid in a shell and tube heat exchanger. *Journal of the Indian Chemical Society, 98*(11), 100200.
53. Samykano, M. (2023). Hybrid photovoltaic thermal systems: Present and future feasibilities for industrial and building applications. *Buildings, 13*(8), 1950.
54. Mariraja, R., Harichandran, R., Vijayakumar, R., & Nichelson, A. (2024). Experimental analysis of solar desalination system performance with graphene and graphitic carbon nanopaint-coated solar absorbers. *Desalination, 592*, 118141.
55. Adio, S. A., Atofarati, E. O., Muritala, A. O., Huan, Z., & Veeredhi, V. R. (2025). Nanofluids flow boiling and convective heat transfer in microchannels: A systematic review and bibliometrics analysis. *Journal of Thermal Analysis and Calorimetry*, 1–33.
56. Thakur, P., Sonawane, S., Potoroko, I., & Sonawane, S. H. (2021). Recent advances in ultrasound-assisted synthesis of nano-emulsions and their industrial applications. *Current Pharmaceutical Biotechnology, 22*(13), 1748–1758.
57. Sonawane, A. V., Rikame, S., Sonawane, S. H., Gaikwad, M., Bhanvase, B., Sonawane, S. S., . . . & Gaikwad, R. (2024). A review of microbial fuel cell and its diversification in the development of green energy technology. *Chemosphere, 350*, 141127.
58. Ding, Y., Chen, H., Wang, L., Yang, C.-Y., He, Y., Yang, W., Lee, W. P., Zhang, L., & Huo, R. (2007). Heat transfer intensification using nanofluids. *KONA Powder and Particle Journal, 25*, 23–38.
59. Jiang, J. Z., Zhang, S., Fu, X. L., Liu, L., & Sun, B. M. (2019). Review of gas–liquid mass transfer enhancement by nanoparticles from macro to microscopic. *Heat and Mass Transfer, 55*(8), 2061–2072.
60. Xu, H., Xu, H., Yu, G., Xu, X., & Wang, F. (2025). Thermal integration of direct-indirect thermochemical reactors and charging-discharging thermal management strategies for solar thermal storage systems. *Solar Energy Materials and Solar Cells, 284*, 113485.
61. Malika, M., & Sonawane, S. S. (2021). The sono-photocatalytic performance of a novel water based Ti+ 4 coated Al (OH) 3-MWCNT's hybrid nanofluid for dye fragmentation. *International Journal of Chemical Reactor Engineering, 19*(9), 901–912.
62. Parate, V. R., Kawadkar, D. K., & Sonawane, S. S. (2011). Study of whey protein concentrate fortification in cookies variety biscuits. *International Journal of Food Engineering, 7*(2), 1–12.
63. Fan, X., Chen, H., Ding, Y., Plucinski, P. K., & Lapkin, A. A. (2008). Potential of 'nanofluids' to further intensify microreactors. *Green Chemistry, 10*(6), 670–677.
64. Ghamarpoor, R., Fallah, A., & Jamshidi, M. (2024). A review of synthesis methods, modifications, and mechanisms of ZnO/TiO2-based photocatalysts for photodegradation of contaminants. *ACS Omega, 9*(24), 25457–25492.
65. Chu, H., Yu, X., Jiang, H., Wang, D., & Xu, N. (2023). Progress in enhanced pool boiling heat transfer on macro-and micro-structured surfaces. *International Journal of Heat and Mass Transfer, 200*, 123530.
66. Upot, N. V., Rabbi, K. F., Khodakarami, S., Ho, J. Y., Mendizabal, J. K., & Miljkovic, N. (2023). Advances in micro and nanoengineered surfaces for enhancing boiling and condensation heat transfer: A review. *Nanoscale Advances, 5*(5), 1232–1270.

67. Singh, T., Atieh, M. A., Al-Ansari, T., Mohammad, A. W., & McKay, G. (2020). The role of nanofluids and renewable energy in the development of sustainable desalination systems: A review. *Water, 12*(7), 2002.
68. Raghavalu, K. V., & Rasu, N. G. (2018, March). Review on applications of nanoFluids used in vapour compression refrigeration system for cop enhancement. In *IOP conference series: Materials science and engineering* (Vol. 330, No. 1, p. 012112). IOP Publishing.
69. Aghel, B., Janati, S., Alobaid, F., Almoslh, A., & Epple, B. (2022). Application of nanofluids in CO2 absorption: A review. *Applied Sciences, 12*(6), 3200.
70. Parmar, H. B., Juybari, H. F., Yogi, Y. S., Nejati, S., Jacob, R. M., Menon, P. S., & Warsinger, D. M. (2021). Nanofluids improve energy efficiency of membrane distillation. *Nano Energy, 88*, 106235.
71. Bolto, B. A., & Xie, Z. (2018). Recent developments in fouling minimization of membranes modifed with silver nanoparticles. *Journal of Membrane Science and Research, 4*(3), 111–120.
72. Tong, Q., Fan, Z., Liu, Q., Qiao, S., Cai, L., Fu, Y., . . . & Sun, A. (2023). Research progress in nanofluid-enhanced oil recovery technology and mechanism. *Molecules, 28*(22), 7478.
73. Sidik, N. A. C., Jamil, M. M., Japar, W. M. A. A., & Adamu, I. M. (2017). A review on preparation methods, stability and applications of hybrid nanofluids. *Renewable and Sustainable Energy Reviews, 80*, 1112–1122.
74. Shankara, R. P., Banapurmath, N. R., Souza, A. D., & Dhaded, S. S. (2020, June). Experimental investigation of enhanced cooling performance with the use of hybrid nanofluid for automotive application. In *IOP conference series: Materials science and engineering* (Vol. 872, No. 1, p. 012074). IOP Publishing.
75. Fadaei, F., Shahrokhi, M., Dehkordi, A. M., & Abbasi, Z. (2017). Heat transfer enhancement of Fe3O4 ferrofluids in the presence of magnetic field. *Journal of Magnetism and Magnetic Materials, 429*, 314–323.
76. AbdulHussein, W. A., Abed, A. M., Mohammed, D. B., Smaisim, G. F., & Baghaei, S. (2022). Investigation of boiling process of different fluids in microchannels and nanochannels in the presence of external electric field and external magnetic field using molecular dynamics simulation. *Case Studies in Thermal Engineering, 35*, 102105.
77. Karimi, Y., Solaimany Nazar, A. R., & Motevasel, M. (2021). CFD simulation of nanofluid heat transfer considering the aggregation of nanoparticles in population balance model. *Journal of Thermal Analysis and Calorimetry, 143*(1), 671–684.
78. Kiani, M., Omiddezyani, S., Nejad, A. M., Ashjaee, M., & Houshfar, E. (2021). Novel hybrid thermal management for Li-ion batteries with nanofluid cooling in the presence of alternating magnetic field: An experimental study. *Case Studies in Thermal Engineering, 28*, 101539.
79. Ilyas, S. U., Narahari, M., Theng, J. T. Y., & Pendyala, R. (2019). Experimental evaluation of dispersion behavior, rheology and thermal analysis of functionalized zinc oxide-paraffin oil nanofluids. *Journal of Molecular Liquids, 294*, 111613.
80. Ramírez-Márquez, C., Al-Thubaiti, M. M., Martín, M., El-Halwagi, M. M., & Ponce-Ortega, J. M. (2023). Processes intensification for sustainability: Prospects and opportunities. *Industrial & Engineering Chemistry Research, 62*(6), 2428–2443.

2 Ionanofluids for Environmental Remediation

2.1 INTRODUCTION TO IONANOFLUIDS

Ionanofluids represent an emerging and highly promising category of advanced nanofluids, distinguished by their use of ionic liquids (ILs) as base fluids. Unlike traditional nanofluids that generally utilize water, ethylene glycol, oils, or other conventional solvents, ionanofluids incorporate ionic liquids, that is, salts that are liquid at temperatures below 100°C as dispersing media for nanoparticles. This innovative approach merges the advantageous properties of both ionic liquids and nanomaterials to create multifunctional fluids with exceptional thermal, physicochemical, and functional characteristics [1, 2].

2.1.1 What Are Ionic Liquids?

Ionic liquids are composed entirely of ions, typically consisting of bulky and asymmetric organic cations (e.g., imidazolium, pyridinium, ammonium, or phosphonium) paired with a wide range of organic or inorganic anions (e.g., tetrafluoroborate [BF_4^-], hexafluorophosphate [PF_6^-], bis(trifluoro methyl sulfonyl)imide [NTf_2^-]) [3]. The large size and structural asymmetry of these ions hinder crystallization, allowing these salts to remain liquid at or near room temperature.

Key attributes of ionic liquids include:

- *Negligible vapor pressure*, reducing the risk of evaporation and contributing to environmental safety.
- *High thermal stability*, enabling operation under extreme conditions.
- *Wide electrochemical windows*, supporting diverse redox applications.
- *Tunability*, as their properties can be customized by selecting different ion combinations [1, 4].

2.1.2 Concept of Ionanofluids

Ionanofluids are engineered by dispersing nanoparticles into ionic liquids to create stable colloidal suspensions. The resulting fluids inherit the distinctive properties of ionic liquids and gain additional functionalities from the nanoparticles [5]. Figure 2.1 represents the synthesis of ionanofluids. Depending on the choice of nanoparticles (e.g., metallic nanoparticles like Ag, Au; metal oxides like TiO_2, ZnO; carbon-based nanomaterials like graphene or CNTs), ionanofluids can be tailored for specific thermal, catalytic, or environmental applications [2, 6].

DOI: 10.1201/9781003628071-2

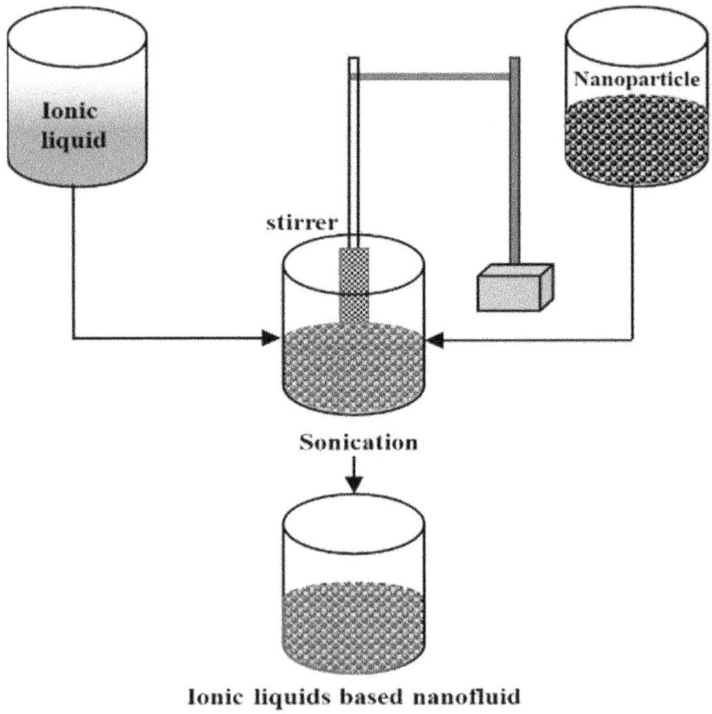

FIGURE 2.1 Synthesis of Ionanofluids [5].

The design of ionanofluids involves optimizing several parameters, such as:

- Nanoparticle type, shape, and size.
- Nanoparticle surface functionalization.
- Ionic liquid composition and purity.
- Interaction dynamics between nanoparticles and ionic liquid ions [1, 6].

This synergy between ILs and nanoparticles opens new possibilities for advanced heat transfer fluids, green solvents, lubricants, and environmental remediation agents.

2.1.3 Advantages of Ionanofluids

- **Enhanced thermal conductivity**: The inclusion of high-conductivity nanoparticles significantly improves heat transfer performance.
- **Superior stability**: Ionic liquids provide better electrostatic and steric stabilization compared to conventional base fluids.
- **Non-volatility**: Negligible vapor pressure minimizes environmental release and improves operational safety.
- **Flame retardancy**: Many ionic liquids exhibit inherent non-flammability.

- **Wide operating temperature range**: Suitable for both high-temperature and sub-ambient applications.
- **Customizability**: Molecular structure of ILs can be tailored to enhance compatibility and performance with specific nanoparticles.

2.1.4 Differences from Conventional Nanofluids and Hybrid Nanofluids

2.1.4.1 Base Fluid Composition

The most fundamental difference lies in the nature of the base fluid. Conventional nanofluids typically use water, ethylene glycol, propylene glycol, engine oils, or refrigerants. These fluids are widely available, low-cost, and easy to handle but come with limitations, such as limited thermal stability, higher vapor pressure, and corrosion concerns.

In contrast, ionanofluids use ionic liquids, which provide:

- Greater thermal and chemical stability
- Minimal evaporation losses
- Superior corrosion resistance in many cases [6, 3]

2.1.4.2 Thermal and Physicochemical Properties

Conventional nanofluids generally achieve modest enhancements in thermal conductivity (typically 10–50% higher than base fluids). However, they often experience increased viscosity and sedimentation issues [7].

Ionanofluids, due to the unique properties of ILs, show:

- Higher base thermal conductivities even before nanoparticle addition
- Better control over viscosity via molecular design of ILs
- Lower tendency toward sedimentation due to strong ion–nanoparticle interactions [5, 6]

2.1.5 Stability and Dispersion

Stability remains a major challenge for conventional nanofluids, as nanoparticles tend to agglomerate and settle over time, especially under static or cyclic thermal conditions. While surfactants and ultrasonication can improve short-term stability, long-term challenges persist [8].

Ionanofluids benefit from enhanced electrostatic and steric stabilization provided by the charged ionic environment. Furthermore, specific functional groups in ILs can form strong interactions with nanoparticles, reducing the need for additional surfactants and minimizing agglomeration [9].

2.1.6 Environmental and Safety Aspects

Conventional nanofluids, especially those based on organic solvents, may pose significant health and environmental risks due to volatility, flammability, and toxicity. Water-based nanofluids are safer but can lead to corrosion and biofouling in certain applications [10].

Ionic liquids used in ionanofluids often have negligible vapor pressure and high thermal stability, reducing volatile organic compound (VOC) emissions and operational hazards. Nevertheless, careful selection of ILs is necessary, as some anions and cations may exhibit toxicity or environmental persistence [11].

2.1.7 Application-Specific Advantages

2.1.7.1 Conventional Nanofluids

- Primarily focused on heat transfer enhancement in HVAC systems, automotive cooling, and electronics cooling
- Limited operational temperature range
- Moderate chemical inertness [12]

2.1.7.2 Hybrid Nanofluids

- Combine two or more types of nanoparticles for synergistic thermal or functional effects
- Improved thermal conductivity and sometimes enhanced specific heat
- Still limited by the base fluid characteristics [13]

2.1.7.3 Ionanofluids

- Extended temperature and chemical stability
- Tunable physicochemical properties for specific environmental remediation tasks, such as heavy metal adsorption or catalytic degradation of organic pollutants
- Potential for electrochemical and magnetic field-responsive applications [14]

2.1.7.4 Cost and Scalability

Conventional nanofluids generally have a lower cost due to inexpensive base fluids and relatively established production methods. Hybrid nanofluids add complexity and cost due to the combination of multiple nanoparticles and additional dispersion challenges [13].

Though Ionanofluids offer superior performance, they currently face high production costs attributed to the synthesis and purification of ionic liquids, as well as nanoparticle functionalization processes. However, ongoing research into green and scalable synthesis of ILs aims to overcome these barriers [15].

2.1.7.5 Sustainability and Green Chemistry Considerations

Growing emphasis on green chemistry has revealed some limitations in conventional nanofluids, especially regarding the environmental fate of nanoparticles and base fluid components. Ionic liquids, despite their synthetic origin, offer advantages such as:

- Recyclability and reusability due to low volatility.
- Potential for biodegradability through proper molecular design.
- Reduced secondary pollution during environmental remediation tasks [11, 16].

2.1.7.6 Research and Technological Integration

While both conventional and hybrid nanofluids have seen extensive laboratory and pilot-scale studies, ionanofluids remain relatively new. Their research currently focuses on fundamental property characterization, stability mechanisms, and proof-of-concept demonstrations in environmental and energy systems [17]. Advancements in computational modeling (e.g., molecular dynamics and quantum chemical calculations) are accelerating the design and prediction of ionanofluid behavior, bridging the gap between theoretical potential and experimental realization [18].

The emergence of ionanofluids marks a significant step forward in the evolution of advanced nanofluid systems. By harnessing the unique properties of ionic liquids combined with functional nanoparticles, ionanofluids overcome many of the challenges faced by traditional and hybrid nanofluids. Their superior thermal stability, tunable properties, and inherent safety features make them particularly promising for high-temperature, chemically aggressive, and environmentally sensitive applications [14, 19]. In the context of environmental remediation, ionanofluids offer tailored solutions for the adsorption and catalytic breakdown of a wide array of contaminants, including heavy metals, dyes, and persistent organic pollutants. Their ability to be designed at the molecular level offers unmatched flexibility and performance potential.

The differences from conventional and hybrid nanofluids highlight both the opportunities and challenges inherent to ionanofluid technology. While higher costs and scalability issues remain obstacles, the benefits in terms of performance, safety, and sustainability position ionanofluids as strong candidates for next-generation environmental and industrial applications. Continued research efforts, coupled with advances in green chemistry and nanotechnology, are expected to drive wider adoption and commercialization in the near future.

2.2 HYBRID AND FUNCTIONALIZED IONANOFLUIDS

2.2.1 Concept of Hybrid Ionanofluids

Hybrid ionanofluids are synthesized by using two or more than more than two types of nanoparticles in an ionic liquid. The goal is to achieve synergistic effects that surpass the performance of individual nanoparticle systems [20].

Advantages

- Enhanced thermal and electrical conductivities
- Improved stability due to multi-particle interactions
- Tailored functional properties for specific environmental remediation tasks

Examples

- Combining TiO_2 (photocatalytic) and CNTs (conductivity enhancement) in an imidazolium-based IL to simultaneously degrade organic pollutants and remove heavy metals [21].
- Incorporating Fe_3O_4 (magnetic) and ZnO (antibacterial) nanoparticles for multifunctional environmental fluids [22].

2.2.1.1 Functionalization of Nanoparticles

Functionalization refers to surface modification of nanoparticles to improve their compatibility with ionic liquids and to introduce specific reactive sites for targeted applications [23].

Methods of Functionalization

- **Chemical grafting:** Covalent bonding of functional groups (e.g., carboxyl, amino) onto nanoparticle surfaces
- **Surfactant coating:** Adsorption of organic molecules to improve dispersibility
- **Ionic liquid functionalization:** Direct attachment of ionic liquid moieties to nanoparticles to enhance compatibility and stability

Advantages of Functionalization

- Enhanced colloidal stability
- Improved interaction with target pollutants
- Increased selectivity and reactivity in environmental processes [23]

2.2.1.2 Applications of Hybrid and Functionalized Ionanofluids

Water and Wastewater Treatment

- Simultaneous adsorption of heavy metals and organic dyes
- Catalytic degradation of pharmaceutical contaminants [24]

Air and Gas Phase Remediation

- Capture and conversion of CO_2 using metal oxide and amine-functionalized nanoparticles
- Removal of volatile organic compounds (VOCs) through adsorption and catalytic breakdown [25]

Soil and Sediment Cleanup

- In situ remediation through targeted pollutant immobilization
- Magnetic separation for post-treatment recovery [26]

2.2.1.3 Challenges and Future Prospects

While hybrid and functionalized ionanofluids show immense promise, challenges such as high synthesis costs, complex functionalization procedures, and scalability remain. Future research should focus on:

- Developing green and cost-effective synthesis routes
- Exploring biodegradable and less toxic ionic liquids
- Establishing standardization for environmental applications [27]

The composition and types of ionanofluids are at the core of their multifunctional capabilities. The strategic combination of various ionic liquids with carefully selected nanoparticles enables the design of next-generation fluids tailored for diverse environmental remediation applications. Imidazolium, pyridinium, phosphonium, and ammonium-based ionic liquids each offer unique advantages and can be matched with metallic, metal oxide, or carbon-based nanoparticles to target specific contaminants or operational conditions [28].

2.3 PHYSICOCHEMICAL PROPERTIES OF IONANOFLUIDS

2.3.1 Thermal Conductivity

Ionanofluids have comparatively better thermal and electrical conductivity. This is due to the synergistic combination of ionic liquids (ILs) and dispersed nanoparticles.

Key Factors Contributing to High Thermal Conductivity in Ionanofluids

- **Ionic liquid contribution:** Certain ILs exhibit higher thermal conductivities than traditional organic solvents. For example, some imidazolium-based ILs can achieve values approaching or exceeding 0.2 W/m·K.
- **Nanoparticle addition:** Incorporation of highly conductive materials such as metallic (Ag, Cu) or carbon-based nanoparticles (CNTs, graphene) dramatically increases the overall conductivity of the fluid.
- **Brownian motion and micro-convection:** The random motion of nanoparticles within the ionic liquid promotes localized micro-convection, further enhancing heat transport.
- **Particle–ion interaction:** Strong electrostatic interactions between ions and nanoparticles improve particle distribution and reduce agglomeration, maintaining consistent thermal conductivity over time [14, 20, 28].

Applications

- High-performance cooling systems for electronics and reactors
- Thermal storage and heat exchanger fluids in renewable energy systems
- Enhanced solar thermal collectors

2.3.2 Electrical Conductivity

Electrical conductivity is equally crucial for certain advanced applications such as electrochemical devices, batteries, and electromagnetic shielding systems. Ionic liquids already possess intrinsic ionic conductivity due to the presence of mobile ions. When nanoparticles are added, especially conductive ones like graphene or metallic particles, the electrical conductivity is significantly boosted.

Mechanisms of Electrical Conductivity Enhancement

- **Charge carrier density:** The ionic nature of ILs provides an initial pool of charge carriers that facilitate conduction.
- **Percolation networks:** Dispersed conductive nanoparticles create percolation pathways, enabling continuous electron transport throughout the fluid.
- **Surface charge effects:** The modification of nanoparticle surfaces with IL moieties improves interface compatibility and promotes stable, long-range conductive networks [6].

Thus, it is possible to develop various nanofluids using different nanoparticles and base fluids. Figure 2.2 explores a few combinations of nanofluids for suitable applications.

Applications

- Electrochemical capacitors and batteries
- Smart thermal management systems with controllable electric fields
- Anti-static and electromagnetic interference (EMI) shielding fluids

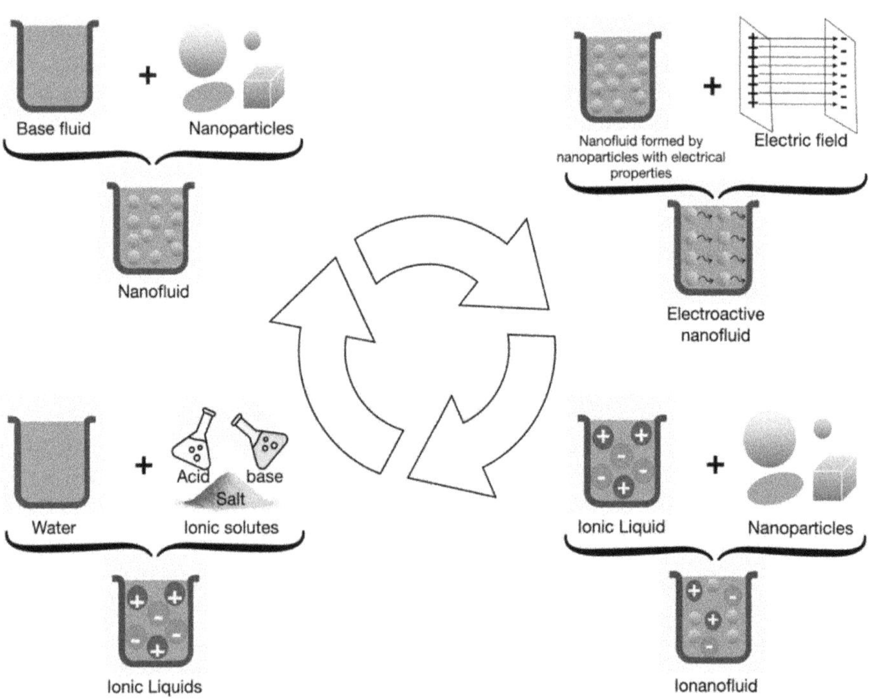

FIGURE 2.2 Possible combinations of nanoparticles and base fluids [29].

2.3.3 Viscosity

Viscosity plays a central role in determining the flow characteristics and heat transfer performance of fluids. Traditional nanofluids often suffer from significant increases in viscosity due to the presence of nanoparticles, which can hinder pumping and circulation, thereby offsetting thermal benefits.

Ionanofluids, however, offer tunable viscosity by leveraging the structural diversity of ionic liquids. The molecular design flexibility allows researchers to fine-tune cation and anion combinations to achieve desired viscosity ranges.

Key Aspects Influencing Viscosity in Ionanofluids

- **Ion structure:** The size, shape, and symmetry of cations and anions greatly influence fluid viscosity. For example, ILs with bulky or long-chain substituents generally exhibit higher viscosities.
- **Nanoparticle type and concentration:** Higher nanoparticle loadings increase viscosity, but functionalization strategies and proper dispersion can mitigate this effect.
- **Temperature dependence:** Like most fluids, ionanofluids show a decrease in viscosity with increasing temperature, enabling optimized performance across a range of operational conditions [4, 26].

Advantages of Tunable Viscosity

- Enables adaptation to various pumping and flow conditions
- Reduces energy consumption in circulating systems
- Balances trade-offs between enhanced heat transfer and ease of handling

2.3.4 Surface Tension

Surface tension is another crucial property that impacts fluid spreading, wetting, and capillary-driven flows. Ionic liquids can have widely varying surface tension values, typically between 20 and 50 mN/m, which can be tailored via molecular design.

Strategies to Tune Surface Tension in Ionanofluids

- **Cation/anion selection:** By choosing hydrophilic or hydrophobic ions, one can control the affinity of the fluid for different surfaces [30].
- **Nanoparticle addition:** Surface-active nanoparticles can reduce or modify surface tension, enabling better wetting and spreading on solid surfaces.
- **Functional additives:** Surfactants or functional molecules can be introduced to further adjust interfacial behavior without compromising bulk properties.

Applications of Tunable Surface Tension

- Improved spreading and wetting in heat exchangers
- Enhanced penetration in porous media for environmental remediation
- Tailored interfacial interactions in membrane separations

2.3.5 Chemical Stability

A defining feature of ionanofluids is their superior chemical stability, inherited primarily from the ionic liquid component. Ionic liquids are known for their resistance to oxidation, hydrolysis, and most common chemical degradations [31].

Reasons for High Chemical Stability

- **Strong ionic bonds:** The electrostatic forces holding ions together in ILs confer robust resistance against chemical breakdown.
- **Inertness toward common contaminants:** ILs are generally unreactive with many acids, bases, and oxidizing agents commonly found in environmental or industrial settings.
- **Synergistic nanoparticle effects:** Certain nanoparticles (e.g., TiO_2, ZnO) add photocatalytic self-cleaning or self-healing capabilities, further enhancing fluid longevity [32].

Applications Leveraging Chemical Stability

- Use in harsh chemical environments for pollutant extraction
- Longevity in closed-loop heat transfer systems
- Resistance to fouling and biofilm formation

2.3.6 Thermal Stability

Ionanofluids exhibit outstanding thermal stability, enabling them to function effectively across a wide range of temperatures without decomposition or significant property degradation. Many ILs can remain stable up to 400°C or even higher, depending on their ionic composition.

Factors Contributing to Thermal Stability

- **High decomposition temperatures of ILs:** Many ILs decompose only above 300°C.
- **Low vapor pressure:** Unlike traditional solvents, ILs do not evaporate easily, preventing fluid loss and composition changes at elevated temperatures.
- **Stabilized nanoparticles:** Properly functionalized nanoparticles resist agglomeration and maintain dispersion even under thermal cycling [33].

Benefits of Thermal Stability

- Prolonged operational life in high-temperature applications
- Reduced maintenance and replacement costs
- Enhanced safety due to minimal risk of vaporization or flammability

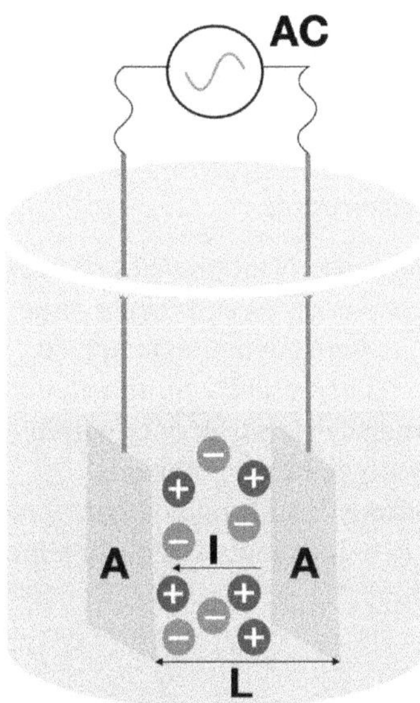

FIGURE 2.3 Representative scheme of an electrical conductivity measuring instrument composed of two electrodes, a cathode and an anode, immersed in an electrolyte solution and connected to a power supply. A represents the area of electrode surface, L is the distance between the electrodes, and I is the cell current [29].

Applications Requiring Thermal Stability

- Concentrated solar power systems
- High-temperature reactors and chemical processing equipment
- Advanced battery and fuel cell cooling

The unique physicochemical properties of ionanofluids—including their enhanced thermal and electrical conductivity, tunable viscosity and surface tension, and exceptional chemical and thermal stability—distinguish them from conventional and hybrid nanofluids. These attributes are critical to their success in demanding applications such as environmental remediation, advanced heat transfer, and energy storage.

The ability to precisely control and optimize these properties through careful selection and design of both the ionic liquid and nanoparticle components offers unprecedented flexibility and performance potential [34].

Figure 2.3 represents the electrical conductivity measurement apparatus.

2.4 MECHANISMS OF POLLUTANT REMOVAL BY IONANOFLUIDS

The following are a few mechanisms explaining the pollutant removal from the nanoparticles based on ionic liquids. The mechanisms may work individually or in combination with two or more mechanisms as per the process parameters.

2.4.1 Adsorption

As nanoparticles have a high surface area to volume ratio, the adsorption mechanism is generally found to be the most dominant mechanism for pollutant removal.

Key Factors Influencing Adsorption

- **Surface area and porosity:** Nanoparticles such as activated carbon, graphene oxide, and mesoporous silica provide large surface areas and pore structures that facilitate high contaminant uptake.
- **Functional groups:** The presence of functional groups (e.g., –OH, –COOH, $-NH_2$) on nanoparticle surfaces or within ionic liquid components improves binding affinity to target pollutants.
- **Ionic liquid interactions:** Ionic liquids can form ion-pair interactions, hydrogen bonds, or π–π interactions with organic molecules, enhancing selective adsorption [35].

Applications

- Removal of heavy metals (e.g., Pb^{2+}, Hg^{2+}, Cd^{2+})
- Adsorption of dyes (e.g., methylene blue, rhodamine B)
- Capture of pharmaceuticals and endocrine-disrupting chemicals (EDCs)

2.4.2 Photocatalysis

Photocatalysis involves the acceleration of photoreactions in the presence of a catalyst, typically under UV or visible light irradiation. In ionanofluids, the dispersed nanoparticles (e.g., TiO_2, ZnO, g-C_3N_4) act as photocatalysts, while ionic liquids can stabilize reactive intermediates and facilitate charge transfer.

Photocatalytic Mechanism

1. **Light absorption:** Photocatalyst nanoparticles absorb photons and generate electron-hole pairs.
2. **Charge separation:** Electrons and holes migrate to the surface, where they participate in redox reactions.
3. **Reactive species generation:** Holes can oxidize water or hydroxide ions to produce hydroxyl radicals (•OH), while electrons reduce oxygen to generate superoxide radicals ($O_2{\bullet}^-$).
4. **Pollutant degradation:** These reactive oxygen species (ROS) attack and mineralize organic pollutants into harmless byproducts such as CO_2 and H_2O.

Role of Ionic Liquids

- Enhance charge separation by acting as electron sinks
- Stabilize intermediate radicals
- Improve pollutant solubility and accessibility to catalytic sites

Applications

- Degradation of persistent organic pollutants (POPs)
- Breakdown of dyes and pharmaceutical residues
- Disinfection of microbial contaminants

2.4.3 Redox Reactions

Redox (reduction–oxidation) reactions are fundamental to the chemical transformation of pollutants. In ionanofluids, nanoparticles can act as electron donors or acceptors, while ionic liquids provide a unique electrochemical environment that supports and stabilizes redox processes.

Redox Mechanisms in Ionanofluids

- **Reduction of heavy metal ions:** Nanoparticles (e.g., zero-valent iron, noble metals) can reduce toxic metal ions (e.g., Cr^{6+} to Cr^{3+}), rendering them less mobile and toxic.
- **Oxidation of organic contaminants:** Nanoparticles can transfer electrons to oxidants, generating ROS that oxidize organic pollutants.
- **Synergistic effects:** Ionic liquids can modulate redox potentials and stabilize reduced or oxidized species, improving reaction efficiency.

Applications

- Detoxification of hexavalent chromium and arsenic compounds
- Oxidative decomposition of phenolic compounds
- Combined adsorption and redox treatment for industrial wastewater [36]

2.5 INTERACTION WITH HEAVY METALS, DYES, AND ORGANIC CONTAMINANTS

2.5.1 Heavy Metals

Heavy metals are among the most hazardous environmental pollutants due to their toxicity, persistence, and bioaccumulation potential. Ionanofluids demonstrate strong interactions with heavy metal ions through adsorption and redox mechanisms.

Mechanisms of Heavy Metal Removal

- **Electrostatic attraction:** Surface charges on functionalized nanoparticles attract oppositely charged metal ions.
- **Ion exchange:** Functional groups on nanoparticles and ionic liquid cations/anions facilitate exchange with metal ions.
- **Reductive precipitation:** Metal ions are reduced to insoluble forms and precipitated for easy separation [37, 38].

Advantages

- High selectivity for specific metal ions
- Simultaneous multi-metal removal
- Capability to operate in complex water matrices

Applications

- Industrial wastewater treatment (e.g., electroplating, mining effluents)
- Groundwater remediation
- Recovery of valuable metals from waste streams [37, 39]

2.5.2 Dyes

Synthetic dyes used in textiles, plastics, and paper industries are major water pollutants due to their complex structures and resistance to degradation.

Mechanisms of Dye Removal

- **Adsorption:** π–π interactions between aromatic dye molecules and nanoparticle surfaces (e.g., graphene oxide).
- **Photocatalysis:** Oxidative breakdown of chromophoric groups leading to complete mineralization.
- **Hydrophobic interactions:** Nonpolar dye molecules interact with hydrophobic moieties in ionic liquids [40–42].

Advantages

- Fast decolorization rates
- Effective even at low dye concentrations
- Reusability of ionanofluid systems after regeneration

Applications

- Treatment of textile wastewater
- Decolorization of printing and dyeing effluents
- Emergency spill response

2.5.3 Organic Contaminants

Organic contaminants encompass a wide range of chemicals, including pharmaceuticals, pesticides, personal care products, and endocrine-disrupting compounds.

Mechanisms of Organic Contaminant Removal

- **Adsorption:** Van der Waals forces, hydrogen bonding, and electrostatic interactions facilitate adsorption onto nanoparticle surfaces.
- **Photocatalytic degradation:** ROS generated during photocatalysis oxidize complex organic molecules into less harmful byproducts.

- **Redox transformations:** Conversion of toxic organics into less toxic or inert forms [43].

Advantages

- Broad-spectrum removal capability
- High degradation efficiency for recalcitrant compounds
- Reduction in toxicity and bioavailability of residual contaminants

Applications

- Treatment of hospital and pharmaceutical effluents
- Remediation of pesticide-laden agricultural runoff
- Cleanup of groundwater contaminated with organic solvents

The multifunctional mechanisms of pollutant removal offered by ionanofluids—including adsorption, photocatalysis, and redox reactions—position them as versatile and powerful tools for environmental remediation. By leveraging the unique interactions between nanoparticles and ionic liquids, ionanofluids achieve high selectivity, efficiency, and reusability in tackling a diverse range of pollutants such as heavy metals, dyes, and organic contaminants.

These advanced mechanisms not only address the immediate need for effective pollutant removal but also contribute to long-term environmental sustainability by enabling recovery, degradation, and detoxification processes in a single, integrated system. Continued research and development of ionanofluid formulations, along with optimization of operational parameters, hold the promise of transforming conventional water and wastewater treatment paradigms into more efficient, adaptable, and eco-friendly solutions.

2.6 APPLICATIONS OF IONANOFLUIDS IN WATER TREATMENT

Figure 2.4 shows various applications of ionanofluids.

2.6.1 Removal of Heavy Metals and Toxic Ions

Heavy metals and toxic ions such as lead (Pb^{2+}), cadmium (Cd^{2+}), mercury (Hg^{2+}), arsenic (As^{3+}), and chromium (Cr^{6+}) pose severe risks to ecosystems and human health. Ionanofluids have demonstrated great potential in efficiently removing these contaminants from water through a combination of adsorption, ion exchange, and redox reactions.

Mechanisms for Heavy Metal Removal

- **Adsorption and ion exchange:** Nanoparticles functionalized with specific ligands or groups (e.g., carboxyl, thiol, amine) can selectively bind heavy metal ions, while ionic liquids provide additional electrostatic interactions to enhance uptake.
- **Reductive precipitation:** Certain metal ions, such as Cr^{6+}, can be reduced to less toxic forms (Cr^{3+}) and subsequently precipitated out of the solution.
- **Synergistic effects:** The ionic liquid matrix not only stabilizes nanoparticles but also prevents agglomeration, maintaining high active surface areas [44, 45].

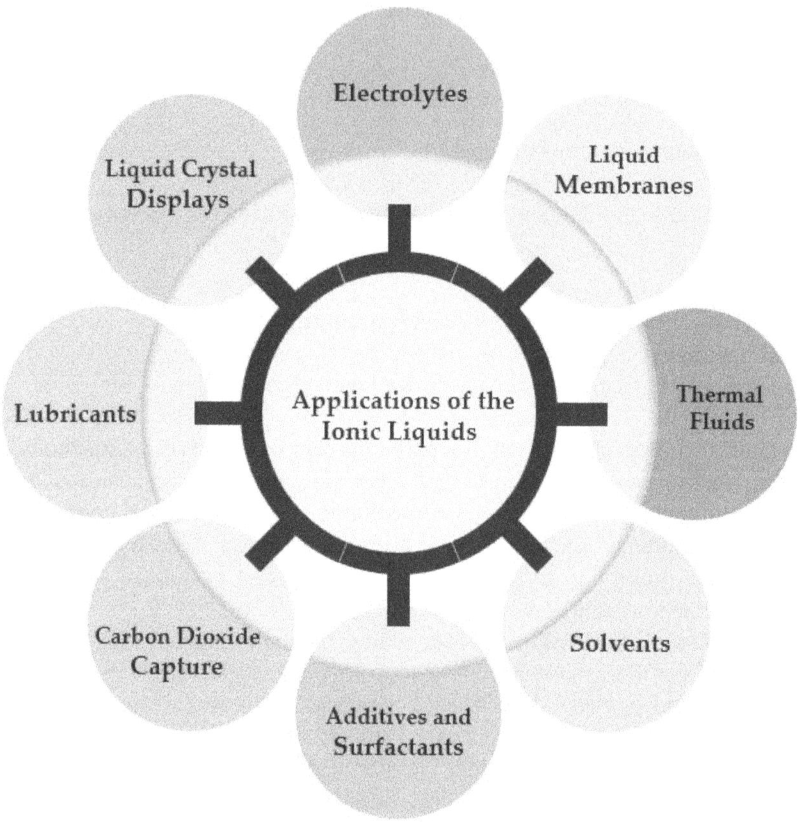

FIGURE 2.4 Various applications of ionanofluids [29].

Advantages

- High selectivity and efficiency even at trace concentrations
- Ability to treat multi-metal ion solutions simultaneously
- Reduced sludge generation compared to traditional precipitation methods [46]

Applications

- Industrial wastewater from mining, electroplating, and battery manufacturing
- Drinking water purification in areas with groundwater contamination
- Recovery of valuable metals from electronic waste leachates

2.6.2 Degradation of Organic Dyes and Pharmaceuticals

Organic dyes from textile, leather, and paper industries, along with pharmaceutical residues, represent major water pollutants. They are typically resistant to biodegradation and can persist in aquatic environments, affecting both wildlife and human health.

Mechanisms for Degradation

- **Photocatalysis:** Nanoparticles like TiO_2, ZnO, and g-C_3N_4 dispersed in ionanofluids act as effective photocatalysts under UV or visible light, generating reactive oxygen species (ROS) that degrade complex dye molecules and pharmaceutical compounds into harmless byproducts.
- **Adsorption:** Ionic liquids enhance the solubilization of hydrophobic contaminants, bringing them closer to the active catalytic sites on nanoparticles.
- **Combined mechanisms:** The simultaneous adsorption and photocatalytic degradation lead to faster and more complete removal compared to conventional systems [47, 48].

Advantages

- High efficiency in decolorization and mineralization
- Capability to degrade a broad spectrum of organic pollutants
- Reusability and regeneration of the ionanofluid systems [49]

Applications

- Treatment of effluents from textile and dyeing industries
- Removal of residual antibiotics and hormones in municipal wastewater
- Emergency treatment of spills and accidental releases of organic chemicals

2.6.3 Antimicrobial Activity and Biofilm Control

Water treatment systems frequently encounter issues related to microbial contamination and biofilm formation, which can reduce system efficiency and pose health risks. Ionanofluids have shown strong antimicrobial properties and biofilm control capabilities due to their unique composition.

Mechanisms for Antimicrobial Action

- **Disruption of microbial membranes:** The ionic liquid component interacts with lipid bilayers, leading to cell membrane destabilization and cell death.
- **Nanoparticle-mediated effects:** Metallic nanoparticles like silver (Ag) and copper (Cu) release ions that exhibit strong bactericidal activity.
- **Inhibition of biofilm formation:** The combined effects prevent the initial adhesion of microorganisms to surfaces and disrupt existing biofilms [50, 51].

Advantages

- Broad-spectrum antimicrobial activity against bacteria, fungi, and algae
- Reduction of chemical biocide usage, minimizing environmental toxicity
- Prevention of biofouling in filtration and membrane systems

Applications

- Disinfection of drinking water and wastewater
- Maintenance of clean surfaces in distribution and storage systems
- Integration into membrane modules to enhance antifouling properties

Ionanofluids present a versatile and efficient solution for various challenges in water treatment. Their ability to remove heavy metals and toxic ions, degrade organic dyes and pharmaceuticals, and inhibit microbial growth positions them as next-generation multifunctional water treatment agents. The combination of nanoparticles and ionic liquids allows for synergistic actions that go beyond the capabilities of traditional methods, offering higher efficiency, selectivity, and sustainability.

By integrating ionanofluids into existing treatment infrastructures, significant advancements can be made toward safer, cleaner, and more resilient water resources. Future research and scale-up efforts will further enhance their practical applicability and economic feasibility, paving the way for widespread adoption in industrial and municipal water treatment systems.

2.7 SOIL AND SEDIMENT REMEDIATION USING IONANOFLUIDS

2.7.1 Immobilization of Pollutants

Mechanisms of Pollutant Immobilization

- **Adsorption and binding:** Nanoparticles within ionanofluids can adsorb heavy metal ions and toxic compounds, preventing their migration and bioavailability. Functionalized nanoparticles (e.g., thiol-modified silica or graphene oxide) provide high surface area and active sites for strong binding interactions [52–55].
- **Ion exchange:** Ionic liquids contribute to the ion exchange process by facilitating the replacement of toxic ions with less harmful or inert ions in the soil matrix [56, 57].
- **Chemical stabilization:** The strong electrostatic interactions and redox capabilities of ionanofluids can convert mobile pollutant species into stable, insoluble forms [58].

Advantages

- Effective in complex, heterogeneous soil and sediment environments
- Minimizes leaching and secondary contamination risks [59–61]
- Potential to restore land for agricultural or residential use

Applications

- Remediation of mining-impacted soils with elevated heavy metal concentrations
- Stabilization of contaminated sediments in riverbeds and coastal areas
- In situ treatment of industrial spill sites to prevent pollutant spread

2.7.2 Detoxification of Persistent Organic Pollutants (POPs)

Persistent organic pollutants (POPs), such as polychlorinated biphenyls (PCBs), dioxins, and certain pesticides, are characterized by their long-term stability, bioaccumulation potential, and toxicity [62]. Conventional remediation techniques often struggle with the effective degradation or detoxification of these compounds.

Mechanisms of POP Detoxification

- **Photocatalytic degradation:** Photocatalytic nanoparticles (e.g., TiO_2, ZnO) dispersed in ionanofluids can generate reactive oxygen species (ROS) under light irradiation, breaking down complex POP molecules into less harmful or fully mineralized products [63].
- **Redox reactions:** Ionanofluids can facilitate redox transformations, converting POPs into less toxic or more biodegradable intermediates. For example, chlorinated compounds can be dechlorinated through reductive pathways [64].
- **Enhanced solubilization:** Ionic liquids improve the solubility and mobilization of hydrophobic POPs from soil and sediment matrices, making them more accessible to reactive sites on nanoparticles [65–67].

Advantages

- High efficiency even at low pollutant concentrations
- Capability to handle mixed contaminant scenarios
- Reduced need for excavation and off-site treatment, lowering costs and environmental impact

Applications

- In situ treatment of agricultural lands contaminated with organochlorine pesticides
- Remediation of industrial sites polluted with PCBs and polycyclic aromatic hydrocarbons (PAHs)
- Detoxification of sediments in harbors and water bodies impacted by historical industrial discharges

Ionanofluids represent a cutting-edge approach to soil and sediment remediation by providing a versatile platform for pollutant immobilization and detoxification. The synergistic effects of nanoparticles and ionic liquids enable the effective treatment of heavy metals, toxic ions, and persistent organic pollutants, addressing both mobility and toxicity.

Through targeted immobilization and chemical transformation mechanisms, ionanofluids not only mitigate immediate environmental risks but also contribute to the long-term restoration and sustainability of contaminated sites. As research progresses and field-scale demonstrations increase, ionanofluids are poised to become integral tools in global soil and sediment remediation strategies.

2.8 GAS PHASE AND AIR POLLUTION CONTROL USING IONANOFLUIDS

2.8.1 Capture of CO_2 and Other Greenhouse Gases

The alarming rise in atmospheric CO_2 and other greenhouse gases (GHGs) is a leading driver of climate change and global warming. Traditional gas capture methods, such as amine-based absorption and physical adsorption, suffer from limitations including high energy penalties, corrosiveness, and operational costs. Ionanofluids have emerged as a highly promising alternative for the selective and efficient capture of CO_2 and other GHGs due to their unique physicochemical properties.

Mechanisms of CO_2 Capture

- **Physical absorption:** The high solubility of CO_2 in certain ionic liquids allows for efficient physical absorption without chemical degradation of the solvent. The presence of nanoparticles enhances gas–liquid contact and increases the overall absorption rate [68–71].
- **Chemical absorption:** Functionalized ionic liquids with amine or carboxylate groups can chemically react with CO_2 to form stable carbamate or carbonate species. The addition of nanoparticles provides additional reactive sites and improves the kinetics of these reactions [72].
- **Synergistic effects:** The combination of ionic liquids and nanoparticles improves gas diffusivity, mass transfer rates, and CO_2 selectivity compared to conventional solvents [67].

Advantages

- High CO_2 capture efficiency with lower regeneration energy requirements
- Reduced corrosion and solvent degradation
- Flexibility to operate under various pressure and temperature conditions

Applications

- Post-combustion CO_2 capture from power plants and industrial flue gases
- Pre-combustion CO_2 removal in integrated gasification combined cycle (IGCC) systems
- Direct air capture systems for atmospheric CO_2 removal

2.8.2 Removal of Volatile Organic Compounds (VOCs)

Volatile organic compounds (VOCs) are a diverse group of carbon-based chemicals emitted from industrial processes, vehicle exhaust, solvents, and various consumer products. VOCs contribute to air pollution, smog formation, and pose serious health hazards to humans.

Mechanisms of VOC Removal

- **Adsorption:** Nanoparticles with large surface areas (e.g., activated carbon, metal-organic frameworks) dispersed in ionic liquids can adsorb VOC molecules effectively. The ionic liquid matrix facilitates uniform distribution and stabilization of nanoparticles, enhancing adsorption capacity [73–75].

- **Absorption and dissolution:** Ionic liquids can dissolve many VOCs due to their unique solvation capabilities, effectively removing them from the gas phase.
- **Catalytic oxidation:** Metal oxide nanoparticles (e.g., TiO_2, CeO_2) within ionanofluids can catalytically oxidize VOCs into less harmful products, especially under light irradiation (photocatalysis) [76].

Advantages

- High removal efficiency for a wide range of VOCs
- Reusability of ionanofluid systems through regeneration
- Reduced secondary pollution compared to thermal- or combustion-based VOC removal methods

Applications

- Air purification systems in industrial facilities
- Indoor air quality improvement in commercial and residential buildings
- VOC emission control in paint and coating manufacturing

2.8.3 Integration and Operational Benefits

2.8.3.1 Operational Flexibility

Ionanofluids can operate under a broad range of temperatures and pressures, enabling seamless integration into existing gas treatment and air purification systems without major retrofitting.

2.8.3.2 Reduced Energy Consumption

Compared to traditional solvent-based systems, ionanofluids require less energy for regeneration due to their lower volatility and stable chemical nature, leading to significant operational cost savings.

2.8.3.3 Environmental Safety

The non-volatile and non-flammable nature of ionic liquids reduces the risk of solvent loss and fire hazards, improving overall safety during gas treatment processes.

2.9 FUTURE PROSPECTS AND RESEARCH DIRECTIONS

2.9.1 Tailored Design of Ionanofluids

Future research is expected to focus on the molecular-level design of ionic liquids and the precise selection of nanoparticle types and functionalization to optimize selectivity, capacity, and kinetics for specific gas pollutants [77].

2.9.2 Hybrid Capture Systems

Combining ionanofluid systems with other technologies such as membrane separation, adsorption columns, or cryogenic distillation could create hybrid solutions with enhanced efficiency and broader applicability [78].

2.9.3 Large-Scale Implementation and Commercialization

Pilot-scale demonstrations and techno-economic assessments are necessary to move ionanofluid-based air pollution control technologies from laboratory research to commercial deployment. Partnerships between academia, industry, and government agencies will be crucial to overcome scalability challenges [79].

2.9.4 Environmental and Life Cycle Assessment

Assessing the full environmental impact and life cycle of ionanofluids, including their synthesis, usage, regeneration, and disposal, will help ensure sustainable development and adoption of these advanced materials [80].

2.10 STABILITY, RECYCLABILITY, AND ENVIRONMENTAL SAFETY OF IONANOFLUIDS

2.10.1 Strategies for Improving Stability

Stability of nanofluids is a very important parameter for the development of highly efficient nanofluids with best performance.

Key Strategies to Improve Stability

- **Surface modification of nanoparticles:** Functionalization with ligands, polymers, or surfactants can improve the dispersion stability of nanoparticles within the ionic liquid matrix, reducing the tendency for aggregation and sedimentation.
- **Covalent bonding approaches:** Covalently attaching nanoparticles to ionic liquid molecules or polymer chains creates a stable hybrid system, minimizing nanoparticle release and maintaining uniformity [81].
- **Optimized nanoparticle concentration:** Carefully controlling the concentration of nanoparticles prevents excessive clustering and viscosity changes that could compromise stability [82].
- **Controlled pH and ionic strength:** Manipulating the pH and controlling ionic strength of nanofluid helps to maintain the charge balance within ionanofluids, stabilizing dispersions against flocculation.

2.10.2 Reusability and Regeneration of Ionanofluids

Mechanisms Enabling Reusability

- **Reversible adsorption and desorption:** Pollutants adsorbed onto nanoparticle surfaces can be desorbed using mild thermal, chemical, or solvent treatments, regenerating the ionanofluid for subsequent cycles.

- **Chemical regeneration:** Ionic liquid components and nanoparticles can be chemically regenerated by washing with suitable solvents or by adjusting pH and ionic strength to restore original reactivity.
- **Thermal treatment:** Moderate heating can drive off adsorbed volatile pollutants without degrading the ionic liquid or nanoparticles, enabling multiple reuse cycles.

Factors Affecting Regeneration Efficiency

- Nature and strength of the interactions between pollutants and ionanofluid components
- Operating conditions such as temperature, pH, and solvent type during regeneration
- Presence of co-contaminants or fouling agents that may affect performance

Examples of Reusability Performance

- Studies have shown that ionanofluids used for heavy metal adsorption can be regenerated and reused for up to 10–20 cycles with minimal loss in efficiency.
- Photocatalytic ionanofluids maintain high degradation rates over multiple cycles when subjected to light-assisted regeneration techniques.

Economic and Operational Benefits

- Reduced chemical costs due to fewer replacement requirements
- Lower environmental footprint from decreased waste generation
- Consistent performance in long-term applications, enhancing process reliability [83]

2.10.3 Environmental Impact and Toxicity Assessment

Assessing the environmental impact and toxicity of ionanofluids is crucial for safe deployment in environmental and industrial applications.

Potential Environmental Risks

- **Ecotoxicity of ionic liquids:** Some ionic liquids can be toxic to aquatic organisms due to their strong surface activity and persistence. The choice of ionic liquid must consider biodegradability and minimal bioaccumulation potential.
- **Nanoparticle toxicity:** Released nanoparticles may exhibit cytotoxic, genotoxic, or ecotoxic effects, depending on their composition, size, and surface properties.
- **Secondary pollution:** Improper disposal or leakage could lead to secondary contamination of soil and water bodies [84].

Toxicity Assessment Methodologies

- **In vitro studies:** Assessing cytotoxicity using cell cultures to understand potential effects on human and animal health

- **Aquatic toxicity tests:** Using model organisms such as algae, daphnia, and fish larvae to evaluate impacts on aquatic ecosystems
- **Soil toxicity assays:** Evaluating effects on soil microorganisms and plant growth to understand terrestrial ecosystem impacts

Strategies to Minimize Environmental Risks

- **Design of greener ionic liquids:** Developing ionic liquids with inherently low toxicity, high biodegradability, and reduced bioaccumulation risk (e.g., cholinium-based and amino acid-based ILs).
- **Nanoparticle surface modification:** Using biocompatible coatings and surface functionalizations to reduce toxicity and improve environmental compatibility.
- **Life cycle assessment (LCA):** Comprehensive LCA studies help quantify the total environmental burden of ionanofluids from synthesis to disposal, guiding design improvements.
- **Strict regulation and disposal protocols:** Implementing guidelines for safe handling, use, and disposal to prevent accidental release and environmental contamination [85].

Regulatory Considerations

International and national regulatory bodies are beginning to develop frameworks addressing nanomaterial and ionic liquid safety. Adherence to guidelines set by agencies such as the US Environmental Protection Agency (EPA), European Chemicals Agency (ECHA), and national environmental protection authorities is essential for responsible ionanofluid deployment [86].

2.11 CHALLENGES AND LIMITATIONS OF IONANOFLUIDS

2.11.1 High Cost of Ionic Liquids

2.11.1.1 Factors Contributing to High Cost

- **Complex synthesis routes:** Many ILs require advanced organic synthesis techniques, including quaternization, anion exchange, and purification steps, each adding to the overall cost.
- **Limited large-scale production:** Most ILs are still produced in small batches or at laboratory scales, leading to high per-unit production costs.
- **Specialized raw materials:** Some ILs require rare or specialized chemical precursors, increasing both cost and supply chain vulnerability [87].

2.11.1.2 Economic Implications

- The high price of ILs significantly impacts the economic viability of ionanofluids, particularly in cost-sensitive sectors such as municipal water treatment or industrial air pollution control.

- The use of expensive ILs can make the total treatment cost prohibitive when compared to traditional adsorbents and solvents.

2.11.2 Scalability and Large-Scale Application Barriers

Translating laboratory-scale ionanofluid technologies to large-scale industrial and environmental applications poses considerable challenges. While ionanofluids have demonstrated excellent performance in controlled lab conditions, real-world scenarios involve complex matrices, variable operating conditions, and economic constraints.

2.11.2.1 Technical Barriers to Scalability

- **Mass transfer limitations:** Ensuring uniform mixing and dispersion of nanoparticles in large volumes is difficult, potentially affecting performance consistency.
- **Equipment compatibility:** Specialized materials and reactor designs are often needed to handle the unique properties of ionanofluids, such as higher viscosities and potential corrosiveness.
- **Thermal and mechanical stability:** Large-scale systems experience temperature gradients, shear forces, and pressure fluctuations, which can compromise ionanofluid stability and efficiency.

2.11.2.2 Operational Challenges

- **Monitoring and control:** Real-time monitoring of ionanofluid composition, pollutant loading, and performance requires sophisticated analytical tools and control systems.
- **Maintenance complexity:** Handling large volumes of ionanofluids may increase maintenance requirements, particularly concerning fouling, precipitation, or unexpected phase separation.

Economic and Logistical Issues

- Scaling up production and deployment involves substantial initial capital investment in specialized equipment and infrastructure.
- Transporting and storing large quantities of ionanofluids require careful consideration of safety, handling, and regulatory compliance.

Strategies for Overcoming Scalability Challenges

- **Modular system design:** Developing modular treatment units can facilitate gradual scaling while maintaining operational flexibility.
- **Hybrid approaches:** Combining ionanofluid systems with conventional technologies (e.g., membranes, adsorption columns) can reduce reliance on large single systems and optimize overall performance.
- **Pilot and demonstration projects:** Extensive pilot-scale studies are essential to gather data on operational stability, maintenance needs, and economic feasibility before full-scale implementation.

2.12 FUTURE RESEARCH DIRECTIONS AND OPPORTUNITIES FOR IONANOFLUIDS

2.12.1 Development of Green and Bio-Based Ionic Liquids

A critical future direction in the field of ionanofluids is the development of green and bio-based ionic liquids (ILs). Traditional ILs, while effective in various applications, often raise concerns regarding their environmental persistence, toxicity, and high production costs. The move toward greener ILs aligns with the broader principles of sustainable chemistry and offers a pathway to minimize the environmental footprint of ionanofluid technologies [88, 89].

Advantages of Green and Bio-Based ILs

- **Biodegradability:** Bio-based ILs are designed to be more readily degradable in natural environments, reducing long-term ecological impacts.
- **Lower toxicity:** Many bio-derived ILs utilize naturally occurring components, which are less toxic to aquatic and terrestrial organisms.
- **Renewable feedstocks:** Green ILs are synthesized from renewable resources such as amino acids, sugars, and choline derivatives, supporting circular economy principles.

Research Opportunities

- **New molecular designs:** Development of novel IL structures that combine functional performance with enhanced environmental compatibility.
- **Life cycle assessment (LCA):** Comprehensive LCA studies to compare the environmental impacts of green ILs with conventional ones, guiding design choices.
- **Cost reduction:** Exploration of scalable synthesis routes using agricultural and industrial waste as feedstocks to lower production costs.

2.12.2 Hybrid Systems with Synergistic Effects

Another promising research avenue is the integration of ionanofluids into hybrid systems designed to exploit synergistic effects. Combining ionanofluids with other advanced materials and technologies can enhance overall system performance and broaden application areas.

Potential Hybrid Systems

- **Nanocomposite membranes:** Incorporating ionanofluids into polymeric or ceramic membranes can create advanced filtration systems with improved selectivity, fouling resistance, and pollutant removal efficiency.
- **Hybrid adsorbent-absorbent systems:** Utilizing both adsorption (nanoparticles) and absorption (ionic liquids) mechanisms in a single system can maximize contaminant capture, especially in complex environmental matrices.

- **Photocatalytic reactors:** Integrating photocatalytic nanoparticles within ionanofluids can enable combined chemical degradation and capture of pollutants, making them highly versatile for water and air treatment applications.

Research Opportunities

- **Material compatibility:** Studying the interactions between ionanofluids and different matrix materials to ensure long-term stability and optimal performance.
- **System optimization:** Developing predictive models to design hybrid systems with tailored properties for specific pollutants and operational conditions.
- **Process intensification:** Exploring how hybrid systems can reduce energy consumption and footprint while increasing treatment capacity.

REFERENCES

1. Paul, T. C., Mahamud, R., & Khan, J. A. (2019). Multiphase modeling approach for ionic liquids (ILs) based nanofluids: Improving the performance of heat transfer fluids (HTFs). *Applied Thermal Engineering*, *149*, 165–172.
2. Minea, A. A., & Murshed, S. S. (2018). A review on development of ionic liquid based nanofluids and their heat transfer behavior. *Renewable and Sustainable Energy Reviews*, *91*, 584–599.
3. Plechkova, N. V., & Seddon, K. R. (2008). Applications of ionic liquids in the chemical industry. *Chemical Society Reviews*, *37*(1), 123–150.
4. MacFarlane, D. R., Kar, M., & Pringle, J. M. (2017). *Fundamentals of ionic liquids: From chemistry to applications*. John Wiley & Sons.
5. Paul, T. C., Tikadar, A., Mahamud, R., Salman, A. S., Morshed, A. M., & Khan, J. A. (2021). A critical review on the development of ionic liquids-based nanofluids as heat transfer fluids for solar thermal energy. *Processes*, *9*(5), 858.
6. Zhou, T., Gui, C., Sun, L., Hu, Y., Lyu, H., Wang, Z., . . . & Yu, G. (2023). Energy applications of ionic liquids: Recent developments and future prospects. *Chemical Reviews*, *123*(21), 12170–12253.
7. Das, S. K., Choi, S. U., Yu, W., & Pradeep, T. (2007). *Nanofluids: Science and technology*. John Wiley & Sons.
8. Behi, M., & Mirmohammadi, S. A. (2012). *Investigation on thermal conductivity, viscosity and stability of nanofluids* (Master of Science Thesis EGI-2012, Royal Institute of Technology (KTH), School of Industrial Engineering and Management, Department of Energy Technology, Division of Applied Thermodynamics and Refrigeration, Stockholm, Sweden).
9. Lingala, S. S. (2023). Ionic-liquid-based nanofluids and their heat-transfer applications: A comprehensive review. *ChemPhysChem*, *24*(22), e202300191.
10. Wang, X. Q., & Mujumdar, A. S. (2007). Heat transfer characteristics of nanofluids: A review. *International Journal of Thermal Sciences*, *46*(1), 1–19.
11. Pham, T. P. T., Cho, C. W., & Yun, Y. S. (2010). Environmental fate and toxicity of ionic liquids: A review. *Water Research*, *44*(2), 352–372.
12. Michaelides, E. E. (2014). *Nanofluidics*. Cham: Springer International Publishing.
13. Rubaiee, S., Yahya, S. M., Fazal, M. A., & Danish, M. (2023). Characterization of Al2O3, TiO2, hybrid Al2O3-TiO2 and graphene oxide nanofluids and their

performance evaluations in photovoltaic thermal system. *Journal of Thermal Analysis and Calorimetry, 148*(21), 11467–11477.

14. Lingala, S. S. (2023). Ionic-liquid-based nanofluids and their heat-transfer applications: A comprehensive review. *ChemPhysChem, 24*(22), e202300191.
15. Sabbaghan, M., & Shahvelayati, A. S. (2016). Green synthesis of symmetrical imidazolium based ionic liquids and their application in the preparation of ZnO nanostructures. *Ceramics International, 42*(3), 3820–3825.
16. Zhang, Q., Zhang, S., & Deng, Y. (2011). Recent advances in ionic liquid catalysis. *Green Chemistry, 13*(10), 2619–2637.
17. E Zhou, T., Gui, C., Sun, L., Hu, Y., Lyu, H., Wang, Z., . . . & Yu, G. (2023). Energy applications of ionic liquids: Recent developments and future prospects. *Chemical Reviews, 123*(21), 12170–12253.
18. Bedrov, D., Piquemal, J. P., Borodin, O., MacKerell, A. D., Jr., Roux, B., & Schröder, C. (2019). Molecular dynamics simulations of ionic liquids and electrolytes using polarizable force fields. *Chemical Reviews, 119*(13), 7940–7995.
19. Nancarrow, P., & Mohammed, H. (2017). Ionic liquids in space technology–current and future trends. *ChemBioEng Reviews, 4*(2), 106–119.
20. De Castro, C. A. N., Murshed, S. S., Lourenço, M. J., Santos, F. J., Lopes, M. L., & França, J. M. (2012). Ionanofluids: New heat transfer fluids for green processes development. In *Green solvents I: Properties and applications in chemistry* (pp. 233–249). Dordrecht: Springer Netherlands.
21. Wang, F., Li, S., Liang, J., Wang, Y., Song, H., Yang, J., . . . & Li, C. (2023). Removal and reuse of heavy metal ions on mildly oxidized Ti3C2@ BF membrane via synergistic photocatalytic-photothermal approach. *Journal of Hazardous Materials, 458*, 131954.
22. Safajou, H., Mizwari, Z. M., Rostaminia, A., Khojasteh, H., Aspoukeh, P., & Mazhari, M. P. (2024). Green synthesis and enhanced photocatalytic performance of rGO/ZnO/Fe3O4 nanocomposites: A sustainable approach to environmental remediation. *Journal of Fluorescence*, 1–20.
23. Wegner, S., & Janiak, C. (2017). Metal nanoparticles in ionic liquids. *Topics in Current Chemistry, 375*(4), 65.
24. Siddeeg, S. M., Tahoon, M. A., Alsaiari, N. S., Shabbir, M., & Rebah, F. B. (2021). Application of functionalized nanomaterials as effective adsorbents for the removal of heavy metals from wastewater: A review. *Current Analytical Chemistry, 17*(1), 4–22.
25. Shah, K. W., & Li, W. (2019). A review on catalytic nanomaterials for volatile organic compounds VOC removal and their applications for healthy buildings. *Nanomaterials, 9*(6), 910.
26. Woldeamanuel, M. M., Mohapatra, S., Senapati, S., Bastia, T. K., Panda, A. K., & Rath, P. (2024). Role of magnetic nanomaterials in environmental remediation. In *Iron oxide-based nanocomposites and nanoenzymes: Fundamentals and applications* (pp. 185–208). Cham: Springer International Publishing.
27. Pešić, J., Watson, M., Papović, S., & Vraneš, M. (2021). Ionic liquids: Review of their current and future industrial applications and their potential environmental impact. *Recent Patents on Nanotechnology, 15*(3), 225–244.
28. Greaves, T. L., & Drummond, C. J. (2008). Protic ionic liquids: Properties and applications. *Chemical Reviews, 108*(1), 206–237.
29. Pereira, J., Souza, R., Moita, A., & Moreira, A. (2023). Nanofluids and ionic fluids as liquid electrodes: An overview on their properties and potential applications. *Processes, 11*(11), 3189.
30. Wang, H., & Chen, X. (2022). A comprehensive review of predicting the thermophysical properties of nanofluids using machine learning methods. *Industrial & Engineering Chemistry Research, 61*(40), 14711–14730.

31. Porgar, S., Oztop, H. F., & Salehfekr, S. (2023). A comprehensive review on thermal conductivity and viscosity of nanofluids and their application in heat exchangers. *Journal of Molecular Liquids*, *386*, 122213.
32. Su, G., Yang, L., Liu, S., Song, J., Jiang, W., & Jin, X. (2024). Review on factors affecting nanofluids surface tension and mechanism analysis. *Journal of Molecular Liquids*, *407*, 125159.
33. Jóźwiak, B., & Boncel, S. (2020). Rheology of ionanofluids—a review. *Journal of Molecular Liquids*, *302*, 112568.
34. Garoz, D., Bueno, C., Larriba, C., Castro, S., Romero-Sanz, I., Fernandez de La Mora, J., . . . & Saito, G. (2007). Taylor cones of ionic liquids from capillary tubes as sources of pure ions: The role of surface tension and electrical conductivity. *Journal of Applied Physics*, *102*(6).
35. Paredes, X., Lourenço, M. J., Castro, C. N. D., & Wakeham, W. (2021). Thermal conductivity of ionic liquids and ionanofluids. Can molecular theory help?. *Fluids*, *6*(3), 116.
36. Oster, K., Hardacre, C., Jacquemin, J., Ribeiro, A. P., & Elsinawi, A. (2019). Ionic liquid-based nanofluids (ionanofluids) for thermal applications: An experimental thermophysical characterization. *Pure and Applied Chemistry*, *91*(8), 1309–1340.
37. García, M. J. S., Sánchez-Segado, S., Ortiz-Martínez, V. M., Godinez-Seoane, C., & Lozano-Blanco, L. J. (2024). Ionic liquid membrane technology for heavy metal removal from aqueous effluents. In *Membrane technologies for heavy metal removal from water* (pp. 242–255). CRC Press.
38. Pospiech, B., & Kujawski, W. (2015). Ionic liquids as selective extractants and ion carriers of heavy metal ions from aqueous solutions utilized in extraction and membrane separation. *Reviews in Chemical Engineering*, *31*(2), 179–191.
39. Fu, F., & Wang, Q. (2011). Removal of heavy metal ions from wastewaters: A review. *Journal of Environmental Management*, *92*(3), 407–418.
40. Badawi, A. K., Abd Elkodous, M., & Ali, G. A. (2021). Recent advances in dye and metal ion removal using efficient adsorbents and novel nano-based materials: An overview. *RSC Advances*, *11*(58), 36528–36553.
41. Yang, H., Zhang, J., Liu, Y., Wang, L., Bai, L., Yang, L., . . . & Chen, H. (2019). Rapid removal of anionic dye from water by poly (ionic liquid)-modified magnetic nanoparticles. *Journal of Molecular Liquids*, *284*, 383–392.
42. Ruan, W., Hu, J., Qi, J., Hou, Y., Zhou, C., & Wei, X. (2019). Removal of dyes from wastewater by nanomaterials: A review. *Advanced Materials Letters*, *10*(1), 9–20.
43. Al Hassan, M. K., Alfarsi, A., Nasser, M. S., Hussein, I. A., & Khan, I. (2023). Ionic liquids and NADES for removal of organic pollutants and heavy metals in wastewater: A comprehensive review. *Journal of Molecular Liquids*, *391*, 123163.
44. Rajadurai, V., & Anguraj, B. L. (2021). Ionic liquids to remove toxic metal pollution. *Environmental Chemistry Letters*, *19*(2), 1173–1203.
45. Wang, J., & Guo, X. (2023). Adsorption kinetics and isotherm models of heavy metals by various adsorbents: An overview. *Critical Reviews in Environmental Science and Technology*, *53*(21), 1837–1865.
46. De Silva, M., Cao, G., & Tam, K. C. (2025). Nanomaterials for the removal and detection of heavy metals: A review. *Environmental Science: Nano*, *12*(4), 2154–2176.
47. Muñoz-Bonilla, A., & Fernández-García, M. (2018). Poly (ionic liquid) s as antimicrobial materials. *European Polymer Journal*, *105*, 135–149.
48. Hussain, M. H., & Pozan Soylu, G. S. (2025). Synthesis of ionic liquid-assisted nanoparticles: High activity, fast removal for photodegradation of methylene blue in water. *Water, Air, & Soil Pollution*, *236*(2), 97.

49. Ayub, A., Wani, A. K., Chopra, C., Sharma, D. K., Amin, O., Wani, A. W., . . . & Singh, R. (2025). Advancing dye degradation: Integrating microbial metabolism, photocatalysis, and nanotechnology for eco-friendly solutions. *Bacteria*, *4*(1), 15.
50. Sharma, V. K., Gupta, J., Mitra, J. B., Srinivasan, H., Sakai, V. G., Ghosh, S. K., & Mitra, S. (2024). The physics of antimicrobial activity of ionic liquids. *The Journal of Physical Chemistry Letters*, *15*(27), 7075–7083.
51. Fang, Z., Zheng, X., Li, L., Qi, J., Wu, W., & Lu, Y. (2022). Ionic liquids: Emerging antimicrobial agents. *Pharmaceutical Research*, *39*(10).
52. Thakur, P., Sonawane, S. S., Sonawane, S. H., & Bhanvase, B. A. (2020). Nanofluids-based delivery system, encapsulation of nanoparticles for stability to make stable nanofluids. In *Encapsulation of active molecules and their delivery system* (pp. 141–152). Elsevier.
53. Bethi, B., Sonawane, S. H., Bhanvase, B. A., & Sonawane, S. S. (2021). Textile industry wastewater treatment by cavitation combined with Fenton and ceramic nanofiltration membrane. *Chemical Engineering and Processing-Process Intensification*, *168*, 108540.
54. Sen, J., Prakash, P., & De, N. (2015). Nano-clay composite and phyto-nanotechnology: A new horizon to food security issue in Indian agriculture. *Journal of Global Biosciences*, *4*(5), 2187–2198.
55. Samanta, S., Raja, M. A., Daniyal, M., Bilal, M., & Sheob, M. (2025). Advancing soil remediation and sustainable environmental engineering solutions through nanotechnology. In *Applications of nanomaterials in civil and environmental engineering: Towards sustainable development and environment remediation* (pp. 165–179). Singapore: Springer Nature Singapore.
56. Zhang, W. X., Cao, J., & Elliott, D. (2005). Iron nanoparticles for site remediation. *Nanotechnology and the Environment* (Chapter 33, pp. 248–255). ACS Publications.
57. Rothee, S. R., Heidari, H., Fortier, M. O., & Khan, E. (2024). Applications of ionic liquids in soil remediation: Mechanisms, efficiency and life cycle assessment. *Soil & Environmental Health*, *2*(3), 100097.
58. Fan, Y., Dastyar, W., Wang, X., Zhao, S., & Yuan, W. (2025). The study on a cost-effective ionic liquid for sustainable heavy metal-contaminated soil remediation. *Discover Soil*, *2*(1), 41.
59. Malika, M., & Sonawane, S. S. (2020). Effect of nanoparticle mixed ratio on stability and thermo-physical properties of CuO-ZnO/water-based hybrid nanofluid. *Journal of the Indian Chemical Society*, *97*(3), 414–419.
60. Sonawane, S., Thakur, P., & Paul, R. (2020). Study on thermal property enhancement of MWCNT based polypropylene (PP) nanocomposites. *Materials Today: Proceedings*, *27*, 550–555.
61. Ali, Q., Zia, M. A., Kamran, M., Shabaan, M., Zulfiqar, U., Ahmad, M., . . . & Maqsood, M. F. (2023). Nanoremediation for heavy metal contamination: A review. *Hybrid Advances*, *4*, 100091.
62. Bhandari, G. (2018). Environmental nanotechnology: Applications of nanoparticles for bioremediation. In *Approaches in bioremediation: The new era of environmental microbiology and nanobiotechnology* (pp. 301–315). Cham: Springer International Publishing.
63. Claudio, C., Iorio, E. D., Liu, Q., Jiang, Z., & Barron, V. (2017). Iron oxide nanoparticles in soils: Environmental and agronomic importance. *Journal of Nanoscience and Nanotechnology*, *17*(7), 4449–4460.
64. Devendrapandi, G., Liu, X., Balu, R., Ayyamperumal, R., Arasu, M. V., Lavanya, M., . . . & Karthika, P. C. (2024). Innovative remediation strategies for persistent organic pollutants in soil and water: A comprehensive review. *Environmental Research*, *249*, 118404.

65. Malika, M. M., & Sonawane, S. S. (2019). Review on application of nanofluid/nano particle as water disinfectant. *Journal of Indian Association for Environmental Management (JIAEM)*, *39*(1–4), 21–24.
66. Malika, M., Bhad, R., & Sonawane, S. S. (2021). ANSYS simulation study of a low volume fraction CuO–ZnO/water hybrid nanofluid in a shell and tube heat exchanger. *Journal of the Indian Chemical Society*, *98*(11), 100200.
67. Da Silva, W. L., & dos Santos, J. H. Z. (2021). Applications of ionic liquids in environmental remediation. In *Green sustainable process for chemical and environmental engineering and science* (pp. 15–21). Elsevier.
68. Zafar, A., Matuszek, K., MacFarlane, D. R., & Zhang, X. (2025). Recent progress and future perspectives of ionic liquid-based carbon dioxide capture and conversion. *Green Energy & Environment*, *10*(6), 1097–1129.
69. Thakur, P., Sonawane, S., Potoroko, I., & Sonawane, S. H. (2021). Recent advances in ultrasound-assisted synthesis of nano-emulsions and their industrial applications. *Current Pharmaceutical Biotechnology*, *22*(13), 1748–1758.
70. Sonawane, A. V., Rikame, S., Sonawane, S. H., Gaikwad, M., Bhanvase, B., Sonawane, S. S., . . . & Gaikwad, R. (2024). A review of microbial fuel cell and its diversification in the development of green energy technology. *Chemosphere*, *350*, 141127.
71. Soo, X. Y. D., Lee, J. J. C., Wu, W. Y., Tao, L., Wang, C., Zhu, Q., & Bu, J. (2024). Advancements in CO2 capture by absorption and adsorption: A comprehensive review. *Journal of CO2 Utilization*, *81*, 102727.
72. Numpilai, T., Pham, L. K. H., & Witoon, T. (2024). Advances in ionic liquid technologies for CO2 capture and conversion: A comprehensive review. *Industrial & Engineering Chemistry Research*, *63*(46), 19865–19915.
73. Malika, M., & Sonawane, S. S. (2021). The sono-photocatalytic performance of a novel water based Ti+ 4 coated Al (OH) 3-MWCNT's hybrid nanofluid for dye fragmentation. *International Journal of Chemical Reactor Engineering*, *19*(9), 901–912.
74. Parate, V. R., Kawadkar, D. K., & Sonawane, S. S. (2011). Study of whey protein concentrate fortification in cookies variety biscuits. *International Journal of Food Engineering*, *7*(2), 1–12.
75. Li, J., Li, B., Sui, G., Du, L., Zhuang, Y., Zhang, Y., & Zou, Y. (2021). Removal of volatile organic compounds from air using supported ionic liquid membrane containing ultraviolet-visible light-driven Nd-TiO2 nanoparticles. *Journal of Molecular Structure*, *1231*, 130023.
76. Drdova, S. (2023). *Integrated photocatalytic systems for VOCs control: Innovative light-driven catalyst-embedded membrane pump, enhanced activity of spark-ablated nanoparticles and insight into catalyst immobilization* (Doctoral dissertation, ETH Zurich).
77. Zia ul Mustafa, M., bin Mukhtar, H., Md Nordin, N. A. H., Mannan, H. A., Nasir, R., & Fazil, N. (2019). Recent developments and applications of ionic liquids in gas separation membranes. *Chemical Engineering & Technology*, *42*(12), 2580–2593.
78. Hospido, A., & Rodríguez, H. (2019). Life cycle assessment (LCA) of ionic liquids. In *Encyclopedia of ionic liquids* (pp. 1–9). Singapore: Springer.
79. Maciel, V. G., Wales, D. J., Seferin, M., Ugaya, C. M. L., & Sans, V. (2019). State-of-the-art and limitations in the life cycle assessment of ionic liquids. *Journal of Cleaner Production*, *217*, 844–858.
80. Bo, L., Zhang, X., Luo, Z., Saboori, T., Dehghan, M., Ghasemizadeh, M., . . . & Mahian, O. (2022). An overview of the applications of ionic fluids and deep eutectic solvents enhanced by nanoparticles. *Journal of Thermal Analysis and Calorimetry*, *147*(14), 7589–7601.

81. Wei, P., Pan, X., Chen, C. Y., Li, H. Y., Yan, X., Li, C., . . . & Yan, B. (2021). Emerging impacts of ionic liquids on eco-environmental safety and human health. *Chemical Society Reviews*, *50*(24), 13609–13627.
82. Righi, S., Morfino, A., Galletti, P., Samorì, C., Tugnoli, A., & Stramigioli, C. (2011). Comparative cradle-to-gate life cycle assessments of cellulose dissolution with 1-butyl-3-methylimidazolium chloride and N-methyl-morpholine-N-oxide. *Green Chemistry*, *13*(2), 367–375.
83. Zhang, J., Lu, C., Li, B., Juhasz, A., Du, Z., Zhu, L., . . . & Wang, J. (2022). Persistence and ecological impact of ionic liquids in soil. *Current Pollution Reports*, *8*(4), 556–568.
84. Buzea, C., Pacheco, I. I., & Robbie, K. (2007). Nanomaterials and nanoparticles: Sources and toxicity. *Biointerphases*, *2*(4), MR17–MR71.
85. Cuéllar-Franca, R. M., García-Gutiérrez, P., Taylor, S. R., Hardacre, C., & Azapagic, A. (2016). A novel methodology for assessing the environmental sustainability of ionic liquids used for CO 2 capture. *Faraday Discussions*, *192*, 283–301.
86. Zafar, A., Matuszek, K., MacFarlane, D. R., & Zhang, X. (2025). Recent progress and future perspectives of ionic liquid-based carbon dioxide capture and conversion. *Green Energy & Environment*, *10*(6), 1097–1129.
87. Jordan, A., & Gathergood, N. (2015). Biodegradation of ionic liquids—a critical review. *Chemical Society Reviews*, *44*(22), 8200–8237.
88. Gaida, B., & Brzęczek-Szafran, A. (2020). Insights into the properties and potential applications of renewable carbohydrate-based ionic liquids: A review. *Molecules*, *25*(14), 3285.
89. He, Z., & Alexandridis, P. (2017). Ionic liquid and nanoparticle hybrid systems: Emerging applications. *Advances in Colloid and Interface Science*, *244*, 54–70.

3 Hybrid Nanofluids for Carbon Capture

3.1 INTRODUCTION TO CO_2 EMISSIONS AND CLIMATE CHANGE

3.1.1 Global Impact of CO_2 Emissions

The most common greenhouse gas that has been released into the atmosphere since the beginning of human technology is carbon dioxide (CO_2). It mainly occurs due to fossil fuels like coal, oil, and natural gas, which are used to fuel other assets like homes, cars, and businesses. The CO_2 emissions in the world started exponentially growing since the beginning of the industrial revolution, which led to a significant alteration of the climate system in the world [1]. The report of the Intergovernmental Panel on Climate Change (IPCC) revealed that CO_2 generates approximately three-quarters of the total human-made greenhouse gases [2].

The buildup of CO_2 in the atmosphere comprises the greenhouse effect, whereby the radiated heat emitted on the surface of the earth is trapped, and the global temperature rises. These are the reasons that have resulted in the average global temperature rising about 1.2 degrees Celsius in comparison to the pre-industrial temperatures [3]. Although it might sound like a modest warming, effects of such warming are already dramatic and wide sweeping on other ecosystems, weather patterns, and human communities.

3.1.2 Key Consequences of Elevated CO_2 Levels Include

- **Sea-level rise:** The coastal physical infrastructure is at risk due to the increase in sea level as a result of the warming climate, and sea-level rise is taking place worldwide due to the rise in temperature and also the expansion of the seawater.
- **Extreme weather:** There have been more and stronger hurricanes, droughts, heatwaves, and prolonged heavy precipitation, meaning a dramatic loss of economic resources coupled with human lives.
- **Ocean acidification:** The seawater absorption of CO_2 has reduced its pH, which has bad impacts on marine life, especially shell-forming life forms and coral reefs.
- **Biodiversity loss:** The losses that have to be faced are due to rapid climate changes, which disrupt the habitats, causing extinction of species as well as loss of biodiversity, which goes against the resilience of the ecosystems and functions of services.

Furthermore, CO_2 emissions exacerbate public health challenges. Increased temperatures contribute to the spread of vector-borne diseases, heat-related illnesses, and

DOI: 10.1201/9781003628071-3

air pollution-related respiratory problems. Economically, climate-related disasters strain national budgets and hinder sustainable development, particularly in vulnerable developing nations.

3.1.3 Need for Advanced CO_2 Capture Technologies

Recognizing the urgent need to limit global warming to well below 2°C, preferably to 1.5°C, as stipulated in the Paris Agreement, nations and scientific communities are increasingly focusing on reducing CO_2 emissions [4]. While transitioning to renewable energy sources and improving energy efficiency are essential, these measures alone are insufficient to meet global climate goals. Thus, there is a growing consensus on the critical role of carbon capture and storage (CCS) technologies in achieving net-zero emissions.

CO_2 capture technologies are designed to remove CO_2 from large point sources such as power plants, industrial facilities, and even directly from ambient air (known as direct air capture or DAC). The captured CO_2 can then be stored underground in geological formations or utilized in various industrial applications, including enhanced oil recovery and the production of synthetic fuels and chemicals.

3.1.4 Importance of CO_2 Capture Technologies

- **Bridging the transition:** While the global shift to renewable energy is underway, fossil fuels still account for a significant portion of global energy consumption. CCS offers a transitional solution to reduce emissions from existing infrastructure.
- **Mitigating industrial emissions:** Certain industries, such as cement, steel, and chemical manufacturing, inherently produce CO_2 during production processes. CCS enables decarbonization of these hard-to-abate sectors.
- **Negative emissions potential:** Emerging approaches like bioenergy with carbon capture and storage (BECCS) and DAC are essential for achieving net-negative emissions scenarios.

Despite their promise, existing CCS technologies face several challenges, including high energy requirements, cost concerns, and public acceptance. This has driven intense research and innovation aimed at developing more efficient, cost-effective, and environmentally friendly CO_2 capture solutions.

3.1.5 Brief Overview of Current Capture Methods and Their Limitations

Current CO_2 capture techniques are broadly classified into three main categories: pre-combustion, post-combustion, and oxy-fuel combustion capture. Each has its own advantages, disadvantages, and specific applications.

3.1.5.1 Pre-Combustion Capture

In pre-combustion capture, carbon is removed from fuels before combustion. This process involves gasifying fuel (such as coal or biomass) to produce a synthesis gas (syngas) composed mainly of hydrogen and carbon monoxide. The CO is then converted to CO_2 via the water-gas shift reaction, and the resulting CO_2 is separated, leaving hydrogen as a clean fuel.

Advantages

- Higher CO_2 concentration in the gas stream, simplifying separation
- Potential for hydrogen production as a valuable byproduct

Limitations

- High capital investment and complex infrastructure requirements
- Suitable primarily for new plants rather than retrofitting existing facilities

3.1.5.2 Post-Combustion Capture

Post-combustion capture is the most mature and widely studied technology, involving the removal of CO_2 from flue gases after fossil fuels have been burned. The flue gas typically contains low concentrations of CO_2 (10–15% for coal-fired power plants), nitrogen, and other impurities [5].

Most Common Methods

- **Chemical absorption:** Using amine-based solvents (e.g., monoethanolamine, MEA) to chemically bind CO_2
- **Physical absorption:** Using solvents like Selexol or Rectisol under high pressures
- **Adsorption:** Utilizing solid sorbents such as zeolites, activated carbons, and metal-organic frameworks (MOFs)
- **Membrane separation:** Employing selective membranes to separate CO_2 from other gases

Advantages

- Can be retrofitted to existing power plants and industrial facilities
- Relatively well-understood and commercially available

Limitations

- Significant energy penalty associated with solvent regeneration and compression
- Solvent degradation and corrosion issues
- High operational and maintenance costs

3.1.5.3 Oxy-Fuel Combustion Capture

Oxy-fuel combustion involves burning fuel in a mixture of pure oxygen and recycled flue gas instead of air. This produces a flue gas that is primarily composed of CO_2 and water vapor, simplifying CO_2 separation after condensation.

Advantages

- High CO_2 concentrations in the flue gas, making separation more straightforward
- Potential for higher energy efficiency in power cycles

Limitations

- High cost and energy demand for oxygen production
- Need for new combustion system designs and extensive retrofitting

3.1.5.4 Direct Air Capture (DAC)

DAC technologies capture CO_2 directly from ambient air using chemical sorbents or solid adsorbents. While still in early stages of commercialization, DAC holds potential for addressing dispersed emissions and generating negative emissions.

Advantages

- Can be deployed anywhere, independent of emission sources
- Essential for offsetting residual emissions and achieving net-zero targets

Limitations

- Very low CO_2 concentration in air (~0.04%), making the process energy-intensive and costly
- Scalability and economic feasibility being significant hurdles

3.1.6 General Limitations and Challenges

Despite progress, current CO_2 capture methods share common challenges:

- **Energy intensity:** There is a high consumption of energy, which is used in capturing, regeneration, as well as compression, and this decreases the efficiency and increases the cost [6].
- **Economic feasibility:** The high cost of capital investment and operation costs makes it impossible to be implemented widely, unless policies are incentives or prices are charged on carbon [7].
- **Durability of materials:** Solvents and sorbents have the tendency to degrade with time, which causes reduction in performance and increases the maintenance requirements [8].
- **On the environment:** The probable environmental hazards of solvent leakages, solvent by-products, and the wastes.

The growing danger of climatic changes increases the importance of creating and introducing efficient CO_2 capture methods. Even though these methods have been the basis of conventional capture methods, the shortcomings of these systems have raised the need to have new systems that are more efficient, economical, and environmentally friendly. Nanotechnology and, more specifically, nanofluids are one such novel frontier and hold promise as a solution to address the barriers of conventional CO_2 capture schemes. This chapter will focus on the background and essential characteristics of nanofluids, their peculiarities, and the role they may play in transforming climate mitigation and CO_2 capture activities in the following sections.

3.2 FUNDAMENTALS OF NANOFLUIDS: DEFINITION AND BASIC CONCEPT

Nanofluids exhibit several unique thermophysical properties that make them attractive for gas absorption processes such as CO_2 capture. These properties contribute to enhancing mass transfer rates, absorption capacity, and overall efficiency of gas–liquid systems.

3.2.1 Enhanced Thermal Conductivity

Among the potential modifications that nanofluids can bring, enhanced thermal conductivity is considered one of the greatest. Nanoparticles, and in particular those with high intrinsic thermal conductivity (e.g., metals and metal oxides), have the capacity to dramatically increase the amount of heat that can be transferred within the fluid. The gas absorption processes are very critical, such that effective heat management is crucial in a gas absorption system because gas absorption can be exothermic. Increased thermal conductivity with increased speed favors the removal of heat and hence increases the rate of absorption and eliminates deterioration of performance caused by a rise in temperature [9].

3.2.2 Increased Specific Heat Capacity

At least some nanofluids have a larger specific heat capacity than the base fluid. This characteristic causes the fluid to take up and hold more of the thermal energy and without a large increase in temperature, stabilizing the absorption process further.

3.2.3 Improved Mass Transfer Rates

The presence of nanoparticles disrupts boundary layers and induces micro-convection effects within the liquid phase. These effects increase the mass transfer coefficient, allowing for faster CO_2 diffusion into the liquid bulk. This enhanced mass transfer directly contributes to higher absorption rates and reduced equipment size.

3.2.4 Modulated Viscosity

There may be an increase or a decrease in viscosity of nanofluids that relies on the concentration of nanoparticles, the size, form, and surface treatment of nanoparticles. The viscosity must be carefully controlled to ensure that it does not impede pumping and mixing processes while still providing the desired enhancements in heat and mass transfer.

3.2.5 Tunable Surface Tension

Nanoparticles are able to change surface tension of the base fluid to enhance the gas-liquid contact for high efficiency. Low surface tension, enhancing bubble breakup and reducing the amount of interfacial area occupied by liquid, increases gas absorption interfacial area [10].

3.2.6 Stability and Dispersion Behavior

A stable dispersion of nanoparticles is crucial to maintain the enhanced properties of nanofluids. Advances in surfactant technology, surface functionalization, and optimized mixing techniques (such as ultrasonication) have enabled the creation of nanofluids with high stability and uniform particle distribution.

3.2.7 Advantages of Nanofluids Compared to Conventional Solvents

Applications of nanofluids in gas absorption, especially those involving CO_2 capture, have many benefits compared to conventional solvents, like aqueous amines or physical absorbents.

3.2.8 Enhanced Absorption Capacity and Rate

The rate and capacity of absorption of conventional solvents do not usually meet the expectations, particularly with low-concentration CO_2 streams. These limitations could be suppressed by adding nanoparticles, due to mass transfer rates increased through micro-mixing and Brownian motion-driven convection, and hence a more rapid capture of CO_2.

3.2.9 Reduced Energy Penalties

One of the main disadvantages of the traditional chemical solvents is the high energy cost of regeneration of solvents. Nanofluids, particularly the ones targeted to physical absorption, may provide a reduced requirement of regeneration energy, because of less tight binding interactions with the CO_2 and better thermal characteristics that enable recovering heat.

3.2.10 Improved Process Efficiency

The combined enhancements in thermal conductivity and mass transfer translate to overall process efficiency gains. Smaller absorber and stripper units, lower operational costs, and improved throughput can be achieved with nanofluid systems.

3.2.11 Lower Solvent Degradation and Corrosion

Traditional amine-based solvents suffer from degradation over time, leading to byproduct formation and equipment corrosion. Nanofluids, especially those with the nanoparticles of inert nature and the base fluids of stable ones, show greater chemical stability and a significantly reduced corrosion tendency, which increases the system lifetime and maintenance expenses.

3.2.12 Flexibility in Design and Tuning

It is possible to finely control the properties of nanofluids by choosing suitable types, sizes, shapes, and surface functionalization of the nanoparticles. It makes it possible to design nanofluids to absorb the requirements of any operation conditions and concentrations of CO_2.

3.2.13 Potential for Hybrid and Multifunctional Systems

Nanofluids allow the creation of a multifunctional capture system through the participation of photocatalytic, magnetic, or catalytic nanoparticles. These types of hybrid nanofluids have the capability to take up CO_2 and degrade pollutants in a combined process, combine with renewable energy systems, or be magnetically removed and recycled.

3.2.14 Environmental Benefits

Nanofluids also have more benefits because they guarantee an overall decline in greenhouse gas emissions and resource consumption through the consumption of less energy and prolonged solvent life. Moreover, the study of green nanoparticles and biodegradable base fluids is more promising to achieve sustainability in the future.

3.2.15 Adaptability to Existing Infrastructure

In many cases, nanofluids could be incorporated into slots in basic absorption systems that require no retrofit. Their ability to be dropped in and operate as a replacer or to boost the performance of the regular solvents offers the opportunity to shift the use in a small step without significant capital outlay.

Nanofluids are a revolution in the sphere of gas absorption and CO_2 capturing. This makes their particular thermophysical characteristics, namely, the high thermal conductivity, greater rates of mass transfer, the possibility to change the viscosity, and surface tension highly propitious to replace basic solvents. The adaptability of nanofluid allows various possibilities to make the new-generation, energy-conserving, and ecologically supported CO_2 capture apparatus.

Later sections of the chapter will address particular nanoparticle types, processes of CO_2 entrapment using nanofluids, experimental and computational methods, along with existing challenges and opportunities for future research that should be conducted to maximally realize the power of nanofluids in industrial applications. Figure 3.1 shows the stability improvement methods.

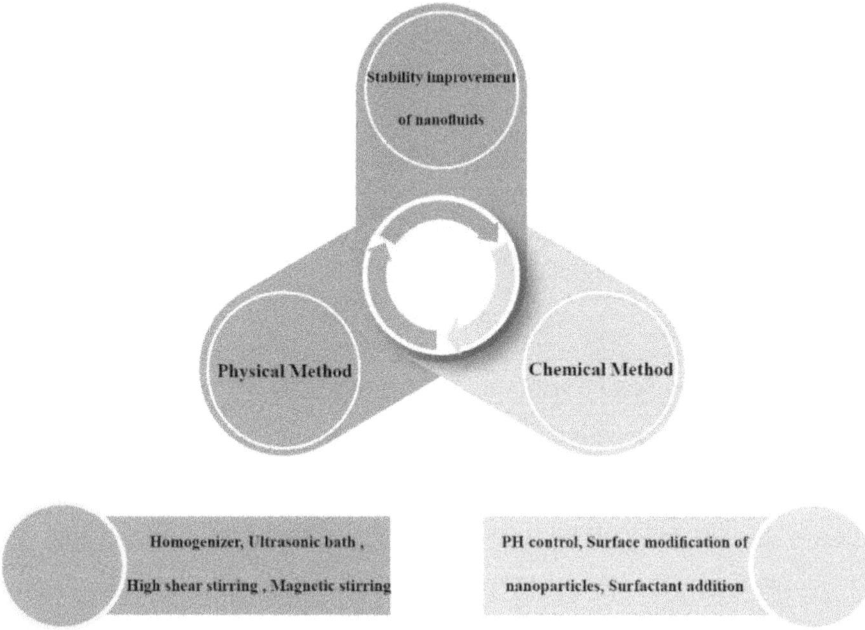

FIGURE 3.1 Stability improvement methods [9].

3.3 MECHANISMS OF CO_2 CAPTURE WITH NANOFLUIDS

3.3.1 Physical Absorption Enhancement

The fact that nanofluids can augment the physical process of absorption becomes one of the main pros of utilizing nanofluids in the CO_2 capture process. Unlike chemical absorption, physical absorption does not depend on chemical bonding, rather on solvent physical absorption of the gas in the soluble state. Capturing of CO_2 through high-pressure or low-temperature conditions involves the application of traditional physical solvents like water, methanol, or Selexol. Nonetheless, limited mass transfer rate and overall absorption capacities are among the drawbacks of these solvents.

The use of nanofluids has considerable benefits in the enhancement of physical absorption in a number of ways.

3.3.1.1 Increased Mass Transfer Coefficient

The micro-convection in the fluid is increased due to the dispersion of nanoparticles in the solvent. The Brownian motion of nanoparticles disrupts the boundary layer of gas bubbles, giving way to thinner diffusion layers and increasing the mass transfer coefficients. It enhances the speed of the inclusion of CO_2 into the solvent, improving general absorption [11].

3.3.1.2 Enhanced Interfacial Area

The dynamics of the bubble can be changed by the use of nanoparticles, which enhance a bubble breakup and inhibit bubble coalescence. Consequently, this massively enhances the area of contact between the gas and liquid phases, which provides an even greater improvement in mass transfer. An increased area of the interface enables increased CO_2 to be in contact with the liquid phase and enhances rapid absorption.

3.3.1.3 Improved Thermal Conductivity

The exothermic nature of the gas absorption processes means there is a need to remove the heat in order to ensure the driving forces of the mass transfer are maintained. Nanofluids have more thermal conductivity than conventional solvents, and this enables a large amount of heat produced in the process of absorption to be dissipated. The absorption is not hindered by localized hotspots, thus keeping the rate high, and degradation of the solvents cannot occur on a large scale [12].

3.3.1.4 Tunable Viscosity and Surface Tension

Surface tension can be decreased by adding nanoparticles, which improves the dispersion of bubbles as well as gas–liquid contact efficiency. Despite the fact that viscosity may rise marginally based on concentration and type of nanoparticles, the considered design leads to the absence of complications in the flow and mixing.

The combination of these innovations leads to greatly improved physical absorption efficiencies, which means that nanofluids present an opportunity to achieve the next generation of CO_2 capture systems, in particular, in environments where chemical absorption is impractical or undesired.

3.3.2 Chemical Absorption and Reaction Kinetics

When CO_2 is absorbed in a chemical process, the molecule reacts with a solvent and is chemically bound to form a compound. The popular route is the use of amine-type solvents such as monoethanolamine (MEA). The most efficient ones have sluggish kinetics, large energy barriers to regeneration, and solvent degradation [13].

Nanofluids also improve significantly on the mechanisms of chemical absorption:

3.3.2.1 Accelerated Reaction Rates

Nanoparticles may serve as nano-catalysts, which quicken the reaction process of CO_2 with chemical absorbents. For example, nanoparticles with high surface activity, such as metal oxides (e.g., TiO_2, MgO), can lower activation energies, allowing faster formation of carbamates or bicarbonates in amine solutions. Faster kinetics translate to higher capture rates and reduced absorber size [14].

3.3.2.2 Enhanced Diffusivity and Mass Transfer

As with physical absorption, nanoparticles enhance micro-convection, which improves CO_2 transport from the gas phase to the bulk liquid. In chemical absorption, this means a quicker supply of CO_2 to reactive sites within the solvent. Combined with accelerated reaction rates, this leads to superior overall absorption performance.

3.3.2.3 Reduction in Solvent Degradation

Nanoparticles can help stabilize chemical solvents by inhibiting oxidative degradation pathways. Certain nanoparticles can scavenge free radicals or stabilize intermediate compounds, extending solvent lifetimes and reducing operational costs.

3.3.2.4 Energy Savings in Regeneration

Improved heat transfer properties of nanofluids allow more efficient thermal regeneration of chemical solvents. By facilitating faster and more uniform heating, the energy required to strip CO_2 from the solvent is reduced, improving overall process economics [15].

Such mechanisms make nanofluid an appealing addition to chemical absorption systems, and this could lead to retrofitting of existing amine-based plants with nanofluid formulations to achieve better results.

3.3.3 Role of Nanoparticle Surface Area and Reactivity

Surface area and reactivity of nanoparticles are very important points that relate to their effectiveness as nanofluids in CO_2 capture. The nanoparticles exhibit large values of specific surface area, which can be as big as hundreds of square meters per gram. This immense surface area ensures that there are many active sites that can interact with CO_2 molecules either physically, chemically, or catalytically [16].

3.3.4 Physical Adsorption on Nanoparticle Surfaces

Nanoparticles comprising a highly porous or high surface area structure, for example, mesoporous silica or activated carbon, can physically adsorb CO_2 to their surface. When dispersed in a fluid, these nanoparticles are used as dispersed adsorbents, which can capture CO_2 even before it is able to reach the liquid phase inside the liquid. This dual absorption–adsorption mechanism significantly boosts CO_2 uptake capacity.

3.4 CATALYTIC SURFACE REACTIONS

Reactive nanoparticles, especially metal oxides, are effective in promoting surface reactions with CO_2 and transforming it into solid or dissolved forms of carbonate species. As an example, the nanoparticle of magnesium oxide is able to react directly with CO_2 to produce magnesium carbonate. This is a reactive absorption process that supplements the use of chemical solvents and has the potential of lowering the concentration of solvent to be used [17].

3.4.1 Enhanced Interfacial Phenomena

The presence of nanoparticles at the gas–liquid interface can modify interfacial tension and stabilize intermediate reaction complexes. Surface-modified nanoparticles can be designed to preferentially localize at the interface, enhancing the local CO_2 concentration and reaction probability [18].

3.4.2 Synergistic Effects with Solvents

Functionalized nanoparticles can introduce specific chemical groups that promote CO_2 binding, creating hybrid capture mechanisms that combine physical absorption, chemical reaction, and surface adsorption. Such synergistic designs provide flexibility in tailoring nanofluids to specific CO_2 capture scenarios, including high-pressure systems, low-concentration flue gases, or ambient air capture [19].

3.4.3 Thermodynamic and Kinetic Considerations

The effectiveness of nanoparticles in CO_2 capture also depends on their surface energy, morphology, and electronic structure. Research shows that nanoscale materials can lower activation barriers for CO_2 hydration and reaction steps. Higher surface curvature and high-defect surfaces enhance further reactivity and cause nanofluids to be more efficacious even at low nanoparticle loadings.

The processes described in the CO_2 absorption of nanofluids entail the advantages of improved physical absorption, rapid chemical reactions, and large surface areas of interactions of nanoparticles. Nanofluids are capable of surpassing most of the limitations that are characterized by conventional solvents by breaking boundary layers, enhancing heat transfer, and creating a catalytic effect, respectively. The opportunity of high reactive surface area and pore structure of nanoparticles provides the way toward highly customized nanofluids that are application specific with optimized traits to capture CO_2.

Future studies are expected to address how to optimize nanoparticle features and investigate multifunctional, hybrid structures, and how long-term stability and regeneration processes work. These innovations already show promise in the future to make CO_2 capture technologies more efficient and scalable and less harmful to the environment as a way to mitigate climate change.

3.5 TYPES OF NANOPARTICLES USED IN CO_2 CAPTURE

3.5.1 Metal Oxides (e.g., MgO, CaO)

Metal oxide nanoparticles have attracted immense interest from researchers investigating the CO_2 capture process based on their high reactivity, chemical stability, and low cost of technological usage. The most notable among them are magnesium oxide (MgO) and calcium oxide (CaO).

3.5.1.1 Magnesium Oxide (MgO)

MgO nanoparticles are highly basic and of high surface area, which is equally superior in CO_2 adsorption. The result when MgO and CO_2 interact is magnesium carbonate ($MgCO_3$), which is a solid-stable product [20].

Advantages of MgO Nanoparticles

- **High surface reactivity:** The large surface area and porosity of MgO nanoparticles enhance CO_2 adsorption capacity.
- **Thermal stability:** MgO retains its structure at high temperatures, enabling regeneration and reuse.
- **Cost-effective:** Magnesium is abundant and relatively inexpensive compared to other materials [21].

Applications in Nanofluids

When suspended in a base fluid, MgO nanoparticles act as dispersed solid sorbents. This makes their presence beneficial due to the enhancement of physical and chemical CO_2 absorption, providing more reactive sites in the fluid and increasing the overall system capture efficiency [22].

3.5.1.2 Calcium Oxide (CaO)

The CaO nanoparticles were previously applied in the process of chemical looping and carbonation-calcination cycle in the CO_2 capture. When CaO is reacted with CO_2, it forms calcium carbonate ($CaCO_3$) [23].

Advantages of CaO Nanoparticles

- **High capture capacity:** CaO can capture large amounts of CO_2 per unit mass.
- **Rapid reaction kinetics:** This means that carbonation reaction is speedy even in moderate temperatures.
- **Regenerability:** CaO can be regenerated by calcination at high temperatures, releasing pure CO_2 for storage or utilization.

Challenges

- **Sintering:** If the particle size is calcination-carbonated repeatedly, then sintering may take place and result in a decline in the surface area and trapping capacity.
- **Agglomeration:** Nanofluids need high-level stabilization to make sure that the dispersion is stable.

3.5.1.3 Other Metal Oxides

CO_2 is also captured in other metal oxides: titanium dioxide (TiO_2), aluminum oxide (Al_2O_3), and zinc oxide (ZnO). The materials are commonly used as enhancers or supplements to the performance of chemical solvents and hybrid capture systems [24].

3.5.2 Carbon-Based Nanoparticles (Graphene, CNTs)

The use of carbon-based nanoparticles such as graphene, carbon nanotubes (CNTs), and activated carbon is advantageous due to the outstanding qualities that they present in relation to CO_2 capture [25].

3.5.2.1 Graphene and Graphene Oxide

Graphene is a single-layer sheet of carbon atoms arranged in a two-dimensional honeycomb lattice. Its derivative, graphene oxide (GO), contains oxygen functional groups that improve dispersibility in aqueous and organic solvents.

Advantages of Graphene-Based Nanoparticles

- **High surface area:** Graphene's theoretical surface area exceeds 2,600 m^2/g, providing abundant adsorption sites.
- **Functionalization flexibility:** Oxygen-containing groups on GO allow further chemical modifications to enhance CO_2 affinity.
- **Mechanical strength and stability:** Graphene exhibits excellent mechanical properties, contributing to long-term operational stability.

Applications

When incorporated into nanofluids, graphene and GO increase CO_2 solubility and facilitate faster mass transfer. Functionalized graphene can also act as a catalyst for CO_2 reduction reactions, broadening the scope of potential applications.

3.5.2.2 Carbon Nanotubes (CNTs)

CNTs are cylindrical structures composed of rolled graphene sheets. They exist as single-walled (SWCNTs) or multi-walled (MWCNTs) forms.

Advantages of CNTs

- **Exceptional surface area:** High specific surface areas provide numerous adsorption sites.
- **Unique adsorption behavior:** CO_2 molecules can be adsorbed inside nanotube cavities and on external surfaces.
- **Chemical tunability:** CNTs can be functionalized with amine or other reactive groups to improve CO_2 binding.

Applications

CNT-based nanofluids demonstrate improved CO_2 absorption kinetics due to enhanced micro-convection and higher gas–liquid contact efficiency. They are particularly promising in hybrid systems combining adsorption and absorption mechanisms.

3.5.2.3 Activated Carbon Nanoparticles

Activated carbon is known for its porous structure and high adsorption capacity. In nanoparticle form, it provides even greater surface area and more accessible micropores.

Advantages

- **High adsorption capacity:** Effective for capturing CO_2 at various pressures and temperatures.

- **Economic viability:** Produced from inexpensive raw materials like biomass or coal.
- **Versatility:** Suitable for both physical adsorption and hybrid chemical processes.

Challenges

- **Agglomeration tendency:** Stabilizing activated carbon nanoparticles in fluids can be challenging.
- **Limited chemical reactivity:** Primarily used for physical adsorption rather than chemical absorption.

3.5.3 Functionalized and Hybrid Nanoparticles

Researchers have gone a step further in improving the CO_2 capture efficiency; the development of functionalized and hybrid nanoparticles was carried out. These particles combine different materials or introduce reactive functional groups, enabling synergistic effects.

3.5.3.1 Functionalized Nanoparticles

Functionalization involves chemically modifying nanoparticles to introduce specific groups that improve CO_2 interaction. For example:

- **Amine-functionalized silica nanoparticles:** Amines have a strong affinity for CO_2 and form carbamate or bicarbonate species upon reaction.
- **Amine-grafted CNTs or graphene:** Enhance chemical binding and improve dispersion in solvents.
- **Ionic liquid-coated nanoparticles:** Provide dual benefits of high CO_2 solubility and chemical selectivity.

Advantages

- **Better selectivity:** Functional groups attack CO_2 and do not attack other gases.
- **Improved dispersion:** The aggregation is lowered, and stability is enhanced in nanofluids through surface modifications.
- **Less regeneration energy:** Reducing desorption temperatures through functionalization of systems saves on energy used.

3.5.3.2 Hybrid Nanoparticles

Hybrid nanoparticles integrate multiple materials to harness complementary properties. Examples include:

- **Metal oxide-carbon composites:** Combine the high reactivity of metal oxides with the high surface area of carbon materials.
- **Metal-organic frameworks (MOFs):** Hybrid structures combining metal ions and organic linkers, known for their tunable porosity and high CO_2 uptake.
- **Magnetic nanoparticles with functional coatings:** Facilitate easy separation and regeneration using external magnetic fields.

Advantages

- **Synergistic performance:** Hybrid structures exhibit enhanced CO_2 capture efficiency due to combined adsorption and catalytic properties.
- **Multifunctionality:** Some hybrid nanoparticles enable simultaneous capture and conversion of CO_2 into value-added products.
- **Operational flexibility:** Hybrid designs allow adaptation to different process conditions and capture scenarios.

Challenges and Future Directions

Although functionalized and hybrid nanoparticles show high potential when it comes to boosting performance, there are limitations:

- **Synthesis complexity:** Functionalization and hybridization processes can be technically demanding and costly.
- **Stability concerns:** Long-term operational stability, including resistance to fouling and chemical degradation, needs thorough evaluation.
- **Environmental impact:** Potential toxicity and environmental effects of novel nanoparticles require careful assessment.

Work proceeds to make the synthesis greener, offer better scalability, and safety to the environment. Research in the future also aims at developing next-generation nanoparticle designs that will combine CO_2 capture with catalytic conversion, integration of renewable energy, and monitoring of a smart system.

The nanofluid efficiency and viability largely depend on the nanoparticles used and are a key factor in CO_2 capture applications. Metal oxides such as MgO and CaO are highly reactive and provide sufficient capture capacity. Carbon-based nanoparticles have extraordinarily high surface areas as well as tunability, and functionalized and hybrid nanoparticles allow custom and multifunctional capture opportunities.

The CO_2 capture technologies are continuing to progress with the engineering of nanoparticles, surface chemistry, and hybrid designs. With the help of such innovative materials, nanofluids have enormous potential in changing typical gas absorption systems to highly efficient, sustainable, and multifaceted solutions to overcome global climatic problems.

3.6 BASE FLUIDS AND ADDITIVES

3.6.1 Choice of Base Liquids (e.g., Water, Amine Solutions, Ionic Liquids)

The choice of base fluids is an essential part of designing a nanofluid CO_2 capture. Besides providing the suspension of the nanoparticle, the base fluid is also a principal determinant of the bulk absorption ability and the mass transfer performance, along with operating stability of the nanofluid system.

3.6.1.1 Water

The base fluid most commonly used is water because it has a low cost, high heating capacity, and is tolerant of the environment. The CO_2 absorption performance can be enhanced using nanoparticles so that a desired outcome can be achieved based on improved thermal conductivity and mass transfer when the nanoparticles are dispersed in water. However, pure water has a relatively low CO_2 solubility and weak chemical affinity for CO_2, which limits its practical use in large-scale capture applications unless combined with other chemical agents [26].

Advantages of Water as a Base Fluid

- Abundant and inexpensive
- Environmentally benign
- High specific heat capacity supports thermal management

Limitations

- Low intrinsic CO_2 absorption capacity
- Tendency for nanoparticle aggregation without proper stabilization [27]

3.6.1.2 Amine Solutions

Aqueous amine solutions, such as monoethanolamine (MEA), diethanolamine (DEA), and methyldiethanolamine (MDEA), are widely used in industrial CO_2 capture due to their high chemical reactivity with CO_2. Incorporating nanoparticles into amine solutions can further enhance the absorption rate and reduce the energy needed for regeneration.

Advantages

- High CO_2 absorption capacity and selectivity
- Established operational experience and commercial use

Enhancements with Nanoparticles

- Increased mass transfer rates
- Improved thermal conductivity aids in heat recovery during regeneration
- Reduced solvent degradation when nanoparticles act as radical scavengers

Challenges

- Corrosiveness of amine solutions
- High energy demand for solvent regeneration
- Need for precise nanoparticle stabilization to avoid precipitation [28]

3.6.1.3 Ionic Liquids

Ionic liquids (ILs) are salts that remain liquid at or near room temperature and have emerged as promising solvents for CO_2 capture due to their negligible vapor pressure, high thermal stability, and tunable properties.

Advantages of ILs as Base Fluids

- High CO_2 solubility due to specific chemical interactions
- Customizable cations and anions enable property tuning
- Low volatility reduces solvent loss and environmental release

Synergy with Nanoparticles

- Nanoparticles enhance thermal and mass transfer properties of ILs
- Enable multifunctional capture processes, including catalytic conversion of CO_2 [29]

Challenges

- High viscosity, which can impede mass transfer and fluid handling
- High cost of ILs limits large-scale deployment

3.6.2 Role of Surfactants and Stabilizers

Stability is a major challenge when designing nanofluids, as nanoparticles naturally tend to aggregate due to van der Waals forces and surface energy minimization. Surfactants and stabilizers play crucial roles in maintaining a uniform dispersion of nanoparticles in base fluids [30].

3.6.2.1 Surfactants

Surfactants are amphiphilic molecules that adsorb at interfaces and reduce surface or interfacial tension. In nanofluids, surfactants adsorb on nanoparticle surfaces, providing steric or electrostatic stabilization [31].

Types of Surfactants Used

- **Anionic surfactants:** Sodium dodecyl sulfate (SDS), which provides strong electrostatic repulsion
- **Cationic surfactants:** Cetyltrimethylammonium bromide (CTAB), commonly used with negatively charged particles
- **Nonionic surfactants:** Tween 80 and Triton X-100, offering steric stabilization without altering surface charge significantly

Advantages

- Prevent nanoparticle aggregation and sedimentation
- Improve nanofluid homogeneity
- Enhance gas–liquid interfacial contact

Challenges

- Possible interference with chemical absorption reactions
- Potential foaming issues in gas–liquid systems
- Environmental and operational impacts of surfactant residues

3.6.2.2 Stabilizers

Beyond surfactants, stabilizers include polymers and functionalization agents that can improve long-term dispersion stability [32–34].

Examples

- Polyvinylpyrrolidone (PVP): Provides steric stabilization
- Polyacrylic acid: Enhances electrostatic repulsion in aqueous systems
- Silane coupling agents: Used to modify nanoparticle surfaces for better compatibility with non-aqueous solvents

Roles in Nanofluid Systems

- Maintain consistent thermal and mass transfer properties over time
- Facilitate easier regeneration and reusability of nanofluids
- Minimize risks of fouling and equipment blockage [35]

3.6.3 Influence on CO_2 Solubility and Mass Transfer

The interplay between base fluids, nanoparticles, surfactants, and stabilizers directly influences CO_2 solubility and mass transfer performance.

3.6.3.1 Enhancement of CO_2 Solubility

Incorporating nanoparticles and choosing appropriate base fluids can enhance CO_2 solubility through various mechanisms:

- **Chemical interaction:** Ionic liquids and functionalized nanoparticles can chemically bind CO_2, increasing the overall solubility.
- **Surface adsorption:** Nanoparticles provide additional surfaces where CO_2 can be physically adsorbed, supplementing solubility in the base fluid [36].

This results in higher overall capture capacities and more efficient operation, especially under low CO_2 concentration conditions such as flue gas or ambient air capture.

3.6.3.2 Improvement of Mass Transfer Rates

Mass transfer in gas–liquid absorption involves several resistance layers: the gas film, the liquid film, and the bulk liquid. Nanoparticles in nanofluids help mitigate liquid-side mass transfer resistance by inducing micro-convection through Brownian motion and altering hydrodynamic conditions at the interface [37].

Key Mechanisms

- **Boundary layer disruption:** Nanoparticles disturb the laminar sublayer near the gas–liquid interface, thinning the mass transfer boundary layer.
- **Enhanced mixing:** Increased micro-convection promotes rapid redistribution of CO_2 from the interface into the bulk fluid.

- **Interfacial area increase:** Nanoparticles can promote finer gas bubble formation, increasing the effective interfacial area for absorption.

3.6.3.3 Combined Effects on Overall Performance

This combination of high CO_2 solubility and significantly increased mass transfer leads to a number of operating advantages:

- **Higher absorption rates:** Faster CO_2 uptake allows for smaller absorber columns and higher throughput.
- **Reduced equipment size and cost:** Improved efficiency can translate into more compact process designs, lowering capital expenditure.
- **Energy savings:** Enhanced mass transfer reduces the need for excessive solvent flow rates or elevated pressures, leading to energy-efficient capture systems [38–40].

Optimizing nanofluids to suit CO_2 capture demands that the right base fluids be selected and correct additives in the form of surfactants and stabilizers be included. Water is reasonably sustainable and economical, even though the intrinsic CO_2 solubility is low, whereas more reactive amine solutions are sensitive to handling, to ensure that energy costs are not vastly hindered and corrosion limited. Despite a greater expense of production and working complexities, the ionic liquids have offered considerable alternatives with greater chemical adjustability along with CO_2 affinity.

The long-term stability, uniformity of nanofluids, and subsequent performance metrics—CO_2 solubility, mass transfer rate, and operational reliability—rely directly on surfactants and stabilizers, and inevitably prop up the rate of production and output. Their preciseness of selection and loading in nanofluid systems also allow the best possible mass or heat transfer operation [41].

Next-generation nanofluids can transform CO_2 capture technologies by using advanced base fluids, strategic additives, and unique nanoparticle designs to make technologies more efficient, scalable, and sustainable. The next step of research should be concentrated on the synergistic blending of base fluids and additives, the search for green surfactants, and profound research on long-term stability and environmental effects that would lead to the implementation in industry on large scale.

3.7 PREDICTIVE MODELS FOR ABSORPTION KINETICS AND MASS TRANSFER

3.7.1 Fundamentals of CO_2 Absorption Kinetics

CO_2 absorption in liquid solvents is governed by physical and chemical mechanisms. The two-film theory, which divides the gas–liquid interface into thin stagnant films, provides a conceptual basis for understanding mass transfer. In hybrid nanofluids, the presence of nanoparticles modifies both film resistance and reaction kinetics [42].

3.7.2 Enhancement Factors and Sherwood Number Correlations

The introduction of nanoparticles enhances mass transfer through micro-convection and Brownian motion. Predictive models often incorporate enhancement factors, E, to account for this improvement:

$$k_{L,eff} = E \times k_L$$

where k_L is the liquid-side mass transfer coefficient in the absence of nanoparticles. Sherwood number (Sh) correlations for hybrid nanofluids have been developed, generally expressed as:

$$Sh = f(Re, Sc, \phi)$$

where Re is the Reynolds number, Sc is the Schmidt number, and ϕ is the nanoparticle volume fraction [43–45].

3.7.3 Reaction–Diffusion Models

In reactive absorption (e.g., with MEA), the interplay between diffusion and reaction kinetics determines overall absorption performance. Reaction–diffusion models can simulate nanoparticle effects by modifying the effective diffusivity and reaction rate constants. For instance, nanoparticle-induced micro-mixing reduces local concentration gradients, leading to higher reaction rates [46].

3.7.4 Machine Learning and Data-Driven Models

With the increasing availability of experimental and simulation data, machine learning approaches have been employed to predict CO_2 absorption rates in nanofluids. Algorithms such as artificial neural networks (ANNs) and support vector machines (SVMs) can model nonlinear relationships between operational parameters and performance metrics, offering rapid and accurate predictions that complement mechanistic models [47].

3.7.5 Multiscale Modeling Frameworks

Hybrid approaches combining molecular dynamics (MD) simulations with continuum-scale models are gaining popularity. MD simulations elucidate nanoparticle interactions and local transport phenomena, which are then integrated into CFD or reactor-scale models for more accurate performance predictions. For example, MD studies show that surface-modified nanoparticles increase CO_2 solubility, an effect that can be embedded into larger-scale models [48–50].

3.7.6 Optimization of Process Parameters

Optimizing process parameters such as nanoparticle concentration, solvent composition, flow rates, temperature, and column design is critical for achieving maximum

CO_2 capture efficiency while minimizing operational costs. Modeling and simulation provide a powerful toolset for systematic optimization without extensive and expensive experimental trials [51].

3.7.7 Challenges and Future Perspectives

3.7.7.1 Data Availability and Validation

Model validation relies heavily on high-quality experimental data. Currently, systematic datasets on hybrid nanofluids under varied operational conditions are limited. Collaborative efforts between experimentalists and modelers are crucial to bridge this gap.

3.7.7.2 Integration with Carbon Capture and Utilization (CCU)

Future models should also consider downstream integration of captured CO_2, such as conversion to value-added products. Coupled models that simulate absorption and subsequent catalytic conversion processes will be critical for assessing overall system viability.

3.7.7.3 Advances in AI and Hybrid Modeling

The integration of AI and hybrid modeling approaches holds promise for accelerating model development and improving predictive capabilities. Future research may focus on developing digital twins of CO_2 capture plants using hybrid nanofluids, enabling real-time monitoring and optimization [52].

Modeling and simulation have emerged as indispensable tools in advancing the application of hybrid nanofluids for CO_2 capture. CFD approaches provide detailed insights into flow dynamics and mass transfer, while predictive kinetic models elucidate complex reaction and diffusion phenomena. Optimization techniques further enable the design of efficient and sustainable CO_2 capture processes, balancing technical performance with economic and environmental considerations.

As research progresses, continued refinement of multiscale models, integration with AI techniques, and collaborative validation with experimental data will pave the way for large-scale deployment of hybrid nanofluids in industrial CO_2 capture systems. These advances represent a crucial step toward achieving global carbon neutrality goals and addressing the pressing challenges of climate change [53–55].

3.8 CHALLENGES AND LIMITATIONS

3.8.1 Stability and Agglomeration Issues

3.8.1.1 Fundamentals of Nanofluid Stability

Nanoparticles dispersed in a base fluid tend to aggregate due to van der Waals forces, magnetic interactions (in magnetic nanoparticles), or other inter-particle attractions. This leads to sedimentation, clogging, and a significant reduction in performance over time.

3.8.1.2 Agglomeration Mechanisms

Several mechanisms contribute to nanoparticle agglomeration:

- **Brownian motion-induced collision:** At nanoscales, random particle movement promotes frequent collisions, encouraging agglomeration.
- **van der Waals forces:** These weak attractive forces become significant at the nanoscale, pulling particles together.
- **Electrostatic interactions:** In cases where particles have surface charges, double-layer compression can cause particles to come closer and stick.
- **Shear-induced aggregation:** High shear rates in absorption columns or pumps can cause nanoparticles to collide and form aggregates.

The formation of larger clusters affects the effective surface area available for CO_2 capture, directly reducing the enhancement effects initially offered by nanoparticles.

3.8.1.3 Impact on Thermophysical Properties

Agglomeration not only reduces active surface area but also alters thermophysical properties. Viscosity may increase unexpectedly due to larger aggregates, leading to higher pumping energy requirements. The thermal conductivity and specific heat capacity of the nanofluid can also become non-uniform, impacting heat transfer efficiency in absorbers and regenerators.

3.8.1.4 Strategies for Enhancing Stability

Several methods have been developed to improve the stability of nanofluids:

- **Surface modification:** Functionalization of nanoparticles with hydrophilic or hydrophobic groups improves dispersion by enhancing compatibility with the base fluid.
- **Use of surfactants and dispersants:** Adding chemical agents can reduce surface tension and electrostatic attractions. However, they may introduce additional complications such as foaming or chemical reactivity.
- **pH control:** Adjusting the pH can increase the surface charge of particles, improving electrostatic repulsion.
- **Ultrasonication:** Ultrasonic treatment helps to break up agglomerates, though it is not always feasible for continuous operations.

3.8.1.5 Long-Term Performance and Monitoring

A major concern is the lack of comprehensive data on the long-term stability of hybrid nanofluids under realistic operating conditions. While short-term laboratory experiments often show promising results, extended tests over hundreds or thousands of operational hours are necessary to validate stability claims. Real-time monitoring technologies for detecting aggregation and sedimentation in operating systems are still under development and are critical for future commercialization.

3.8.2 Economic Considerations and Scale-Up Barriers

3.8.2.1 High Cost of Nanoparticle Production

One of the most significant economic barriers to the widespread use of hybrid nanofluids in CO_2 capture is the high cost associated with nanoparticle production. Synthesis methods such as sol-gel, hydrothermal, and chemical vapor deposition are often expensive and energy-intensive. The price of high-purity nanoparticles (especially hybrid or doped types) can be several orders of magnitude higher than conventional absorbents.

For example, the cost of graphene-based nanomaterials remains prohibitive, primarily due to the complexity of manufacturing processes and low production yields. Similarly, metal oxide nanoparticles (e.g., TiO_2, Al_2O_3) require strict control of particle size, morphology, and surface properties, further driving up costs.

3.8.2.2 Process Design Complexity

Hybrid nanofluid systems necessitate significant modifications in process design. Existing CO_2 capture facilities are optimized for traditional solvents like monoethanolamine (MEA).

3.8.2.3 Cost–Benefit Analysis and Techno-Economic Assessments

Comprehensive techno-economic assessments (TEAs) are crucial for evaluating the feasibility of hybrid nanofluids. These analyses must account for:

- Nanoparticle production and supply chain costs.
- Additional capital and operational expenditures.
- Expected improvements in capture efficiency and energy savings.
- Maintenance and nanoparticle replenishment expenses.
- End-of-life management and disposal costs.

Recent TEAs suggest that unless nanoparticle costs can be substantially reduced and operational reliability ensured, hybrid nanofluids may struggle to compete with improved traditional solvents or emerging solid sorbents in large-scale applications.

3.8.3 Environmental and Health Safety Concerns

3.8.3.1 Nanotoxicology and Occupational Exposure

Nanoparticles exhibit unique physicochemical properties that also render them potentially hazardous to human health and the environment. Due to their small size, nanoparticles can penetrate biological barriers, enter cells, and cause oxidative stress, inflammation, or cytotoxicity.

During manufacturing, handling, and maintenance of nanofluid systems, there is a risk of occupational exposure through inhalation, dermal contact, or accidental ingestion. Workers in facilities employing hybrid nanofluids may require specialized personal protective equipment (PPE), rigorous training, and continuous air quality monitoring.

3.8.3.2 Environmental Release and Ecological Impact

Unintended release of nanoparticles into the environment poses significant ecological risks. Nanoparticles can persist in water bodies, accumulate in sediments, and enter the food chain, potentially affecting aquatic and terrestrial organisms.

3.8.4 Future Research Directions and Possible Solutions

3.8.4.1 Development of Smart Nanofluids

Smart nanofluids with stimuli-responsive behavior (e.g., pH-sensitive, temperature-sensitive) offer the potential to dynamically adjust properties during operation, enhancing performance and stability while reducing risks of agglomeration.

3.8.4.2 Hybrid Modeling for Predictive Maintenance

Advanced modeling and digital twin approaches can predict stability issues and guide proactive maintenance. Machine learning models trained on operational data can help optimize nanoparticle dosing, flow conditions, and regeneration strategies, thereby extending nanofluid life and reducing costs.

3.8.4.3 Standardized Safety Protocols

Establishing standardized protocols for toxicity testing, exposure monitoring, and environmental impact assessment will facilitate safer deployment. International collaboration on regulatory harmonization and certification schemes will also help mitigate risks.

3.8.4.4 Circular Economy and Resource Recovery

Incorporating principles of circular economy into nanofluid systems, such as designing for reuse, establishing take-back programs for nanoparticles, and developing closed-loop systems, can alleviate economic and environmental concerns.

Hybrid nanofluids represent an exciting and potentially transformative technology for carbon dioxide capture, offering enhanced mass transfer and improved energy efficiency. However, realizing their full potential requires addressing significant challenges related to stability, economic feasibility, and environmental safety. Agglomeration and sedimentation undermine performance and operational reliability, while high nanoparticle costs and complex process integration hinder economic viability. Concurrently, potential risks to human health and the environment necessitate robust safety measures, regulatory frameworks, and public engagement.

Future progress will depend on multidisciplinary innovation, including advances in nanoparticle design, process engineering, safety science, and economic analysis. Through integrated research and collaborative action, it is possible to overcome these challenges and pave the way for the sustainable deployment of hybrid nanofluid systems in large-scale CO_2 capture and beyond. As climate change intensifies and global carbon emissions continue to rise, innovative technologies such as hybrid nanofluids are emerging as potential game-changers for CO_2 capture. While current research has demonstrated considerable promise in laboratory and pilot-scale studies, realizing the full potential of hybrid nanofluids requires a concerted effort in advancing materials science, process engineering, and policy frameworks.

3.8.4.5 Development of Novel Nanoparticles and Functionalized Systems

3.8.4.5.1 The Need for Advanced Nanoparticles

The core advantage of nanofluids lies in their ability to enhance heat and mass transfer characteristics through the inclusion of nanoparticles. However, traditional nanoparticles such as alumina (Al_2O_3), silica (SiO_2), and titania (TiO_2) are not always optimized for CO_2 capture processes, especially when stability, reactivity, and cost are considered. Thus, developing novel nanoparticles with tailored properties is critical.

3.8.4.5.2 Functionalized Nanoparticles for Enhanced Reactivity

Functionalization involves modifying the surface of nanoparticles with specific chemical groups or coatings to impart desired properties. For CO_2 capture applications, functionalized nanoparticles can serve as active sites for chemical absorption, thereby complementing the solvent phase.

3.8.4.6 Multifunctional and Hybrid Nanoparticles

Research into stimuli-responsive (smart) nanoparticles is also gaining traction. These nanoparticles can change their behavior in response to external stimuli such as temperature, pH, or electric fields, enabling dynamic control over capture processes and potentially reducing energy consumption in regeneration steps.

3.8.4.7 Computational Design and High-Throughput Screening

Advances in computational chemistry and materials informatics allow for the rapid design and screening of new nanoparticles. By combining density functional theory (DFT), molecular dynamics (MD) simulations, and machine learning algorithms, researchers can predict nanoparticle properties, interactions with CO_2, and stability in different solvents before experimental synthesis.

High-throughput screening accelerates the identification of optimal materials, thereby significantly shortening the development cycle for new nanofluids.

3.8.4.8 Stability-Engineered Systems

As discussed in earlier sections, nanoparticle agglomeration remains a major limitation. Future efforts will likely focus on designing stability-engineered nanoparticles that incorporate anti-agglomeration strategies at the molecular level. Techniques such as self-assembled monolayers, steric stabilization through polymer grafting, and electrostatic tuning are expected to play a critical role.

3.8.4.9 Integration with Hybrid and Smart Capture Systems

3.8.4.9.1 Synergistic Capture Processes

The future of CO_2 capture will not rely on a single technology but rather on integrated hybrid systems that combine the strengths of different approaches. Hybrid nanofluids can be integrated with existing chemical, physical, and membrane-based capture technologies to maximize efficiency and minimize energy penalties.

For example, combining nanofluid-based absorption with membrane separation can leverage the high mass transfer rates of nanofluids and the selective permeation advantages of membranes. Such integration could enable staged capture processes with improved overall performance.

3.8.4.9.2 Smart Capture Systems with Real-Time Adaptability

Smart systems equipped with sensors and real-time feedback mechanisms can dynamically adjust operational parameters (e.g., flow rates, temperatures, nanoparticle concentrations) to optimize capture performance under varying flue gas compositions and flow conditions. Nanofluids can be designed to work within such adaptive frameworks, enabling highly responsive and energy-efficient capture operations.

For instance, stimuli-responsive nanofluids that alter their viscosity or reactivity based on CO_2 concentration could lead to self-regulating capture processes, reducing operator intervention and energy usage.

3.8.4.9.3 Integration with Carbon Utilization and Storage

Beyond capture, integrating nanofluid-based systems with downstream carbon utilization (e.g., conversion to fuels or chemicals) or storage processes is crucial for developing closed-loop carbon management strategies. Hybrid nanofluids could be engineered to facilitate direct conversion of captured CO_2 through catalytic sites embedded on nanoparticles, transforming the fluid from a passive absorbent to an active reactor medium.

For example, nanoparticles functionalized with metal catalysts (e.g., nickel, copper) can enable CO_2 hydrogenation to methane or methanol within the capture medium itself. This approach, known as reactive absorption-conversion, represents a frontier area in process intensification.

3.8.4.9.4 Process Intensification and Compact Designs

Future capture systems will prioritize compact, modular designs to enable deployment in decentralized settings such as small power plants, marine vessels, and mobile units. Hybrid nanofluids, with their superior mass and heat transfer properties, align well with these goals.

REFERENCES

1. Boden, T. A. (2010). *Global, regional, and national fossil-fuel CO2 emissions.* http://cdiac.ornl.gov/trends/emis/tre_glob.html.
2. Change, I. P. O. C. (2007). Climate change 2007: The physical science basis. *Agenda*, *6*(7), 333.
3. Shaftel, H., Jackson, R., & Callery, S. (2018). *Global climate change: Vital signs of the planet.* NASA's Jet Propuls. Lab. Calif. Inst. Technol. https://climate. nasa. gov/causes/. 8.
4. Gao, Y., Gao, X., & Zhang, X. (2017). The 2 C global temperature target and the evolution of the long-term goal of addressing climate change—from the United Nations framework convention on climate change to the Paris Agreement. *Engineering*, *3*(2), 272–278.
5. Boot-Handford, M. E., Abanades, J. C., Anthony, E. J., Blunt, M. J., Brandani, S., Mac Dowell, N., . . . & Fennell, P. S. (2014). Carbon capture and storage update. *Energy & Environmental Science*, 7(1), 130–189.

6. Zheng, Y., Gao, L., He, S., & Jin, H. (2023). Reduction potential of the energy penalty for CO_2 capture in CCS. *Frontiers in Energy*, *17*, 390–399.
7. Campbell, R. J. (2013). Increasing the efficiency of existing coal--fired power plants-. *Congressional Research Service Reports*.
8. House, K. Z., Harvey, C. F., Aziz, M. J., & Schrag, D. P. (2009). The energy penalty of post--combustion CO_2 capture & storage and its implications for retrofitting the U-.-S-. -installed base-. *Energy & Environmental Science*, *2*(2), 193–205.
9. Aghel, B., Janati, S., Alobaid, F., Almoslh, A., & Epple, B. (2022). Application of nanofluids in CO_2 absorption: A review. *Applied Sciences*, *12*(6), 3200. https://doi.org/10.3390/app12063200
10. Tawfik, M. M. (2017). Experimental studies of nanofluid thermal conductivity enhancement and applications: A review. *Renewable and Sustainable Energy Reviews*, *75*, 1239–1253.
11. Zhang, Z., Cai, J., Chen, F., Li, H., Zhang, W., & Qi, W. (2018). Progress in enhancement of CO2 absorption by nanofluids: A mini review of mechanisms and current status. *Renewable Energy*, *118*, 527–535.
12. Zhang, N., Zhang, X., Pan, Z., & Zhang, Z. (2018). A brief review of enhanced CO 2 absorption by nanoparticles. *International Journal of Energy for a Clean Environment*, *19*(3–4).
13. Zhang, H., Wang, B., Xiong, M., Gao, C., Ren, H., & Ma, L. (2022). Process intensification in gas-liquid mass transfer by nanofluids: Mechanism and current status. *Journal of Molecular Liquids*, *346*, 118268.
14. Yu, W., Wang, T., Park, A. H. A., & Fang, M. (2019). Review of liquid nano-absorbents for enhanced CO 2 capture. *Nanoscale*, *11*(37), 17137–17156.
15. Bolourchian Tabrizi, S. Z., Shahhosseini, S., & Ghaemi, A. (2021). Insights into the mass transfer mechanisms of nanofluids: A CO2 absorption study. *Energy & Fuels*, *35*(24), 20172–20184.
16. Shahbaz, F., Singh, I., Krishnan, P., & Celik, K. (2022). Life cycle assessment of brucite and synthetic MgO produced from reject brine using different alkalis. *Journal of Cleaner Production*, *380*, 135071.
17. Sarikhani, K., Jeddi, K., Thompson, R. B., Park, C. B., & Chen, P. (2015). Adsorption of surface-modified silica nanoparticles to the interface of melt poly (lactic acid) and supercritical carbon dioxide. *Langmuir*, *31*(20), 5571–5579.
18. Gargiulo, N., Pepe, F., & Caputo, D. (2014). CO2 adsorption by functionalized nanoporous materials: A review. *Journal of Nanoscience and Nanotechnology*, *14*(2), 1811–1822.
19. Costentin, C., & Savéant, J. M. (2016). Catalysis at the nanoscale may change selectivity. *Proceedings of the National Academy of Sciences*, *113*(42), 11756–11758.
20. Wang, X., He, T., Hu, J., & Liu, M. (2021). The progress of nanomaterials for carbon dioxide capture via the adsorption process. *Environmental Science: Nano*, *8*(4), 890–912.
21. Alkadhem, A. M., Elgzoly, M. A., Alshami, A., & Onaizi, S. A. (2021). Kinetics of CO2 capture by novel amine-functionalized magnesium oxide adsorbents. *Colloids and Surfaces A: Physicochemical and Engineering Aspects*, *616*, 126258.
22. Liu, W., An, H., Qin, C., Yin, J., Wang, G., Feng, B., & Xu, M. (2012). Performance enhancement of calcium oxide sorbents for cyclic CO2 capture: A review. *Energy & Fuels*, *26*(5), 2751–2767.
23. Stankovich, S., Dikin, D. A., Dommett, G. H., Kohlhaas, K. M., Zimney, E. J., Stach, E. A., . . . & Ruoff, R. S. (2006). Graphene-based composite materials. *Nature*, *442*(7100), 282–286.

24. Lourenço, M. A., Fontana, M., Jagdale, P., Pirri, C. F., & Bocchini, S. (2021). Improved CO2 adsorption properties through amine functionalization of multi-walled carbon nanotubes. *Chemical Engineering Journal, 414*, 128763.
25. Fan, Y., & Jia, X. (2022). Progress in amine-functionalized silica for CO2 capture: Important roles of support and amine structure. *Energy & Fuels, 36*(3), 1252–1270.
26. Lee, H., Calvin, K., Dasgupta, D., Krinner, G., Mukherji, A., Thorne, P., . . . & Ruane, A. C. (2024). Climate change 2023 synthesis report summary for policymakers. *Climate Change 2023 Synthesis Report: Summary for Policymakers.*
27. Rochelle, G. T. (2009). Amine scrubbing for CO2 capture. *Science, 325*(5948), 1652–1654.
28. Babamohammadi, S., Shamiri, A., & Aroua, M. K. (2015). A review of CO2 capture by absorption in ionic liquid-based solvents. *Reviews in Chemical Engineering, 31*(4), 383–412.
29. Bernardo, P., Jansen, J. C., Bazzarelli, F., Tasselli, F., Fuoco, A., Friess, K., . . . & Clarizia, G. (2012). Gas transport properties of Pebax®/room temperature ionic liquid gel membranes. *Separation and Purification Technology, 97*, 73–82.
30. Wang, C., Luo, X., Zhu, X., Cui, G., Jiang, D. E., Deng, D., . . . & Dai, S. (2013). The strategies for improving carbon dioxide chemisorption by functionalized ionic liquids. *Rsc Advances, 3*(36), 15518–15527.
31. Yasmeen, I., Ilyas, A., Shamair, Z., Gilani, M. A., Rafiq, S., Bilad, M. R., & Khan, A. L. (2020). Synergistic effects of highly selective ionic liquid confined in nanocages: Exploiting the three component mixed matrix membranes for CO2 capture. *Chemical Engineering Research and Design, 155*, 123–132.
32. Malika, M., & Sonawane, S. S. (2021). The sono-photocatalytic performance of a novel water based Ti+ 4 coated Al (OH) 3-MWCNT's hybrid nanofluid for dye fragmentation. *International Journal of Chemical Reactor Engineering, 19*(9), 901–912.
33. Parate, V. R., Kawadkar, D. K., & Sonawane, S. S. (2011). Study of whey protein concentrate fortification in cookies variety biscuits. *International Journal of Food Engineering, 7*(2), 1–12.
34. Harrison, R., Saborit, J. D., Dor, F., Henderson, R., Penney, D., Benignus, V., & Verrier, A. (2010). *WHO guidelines for indoor air quality: Selected pollutants.* Geneva: World Health Organization. Available from: https://www.ncbi.nlm.nih.gov/books/NBK138705/
35. Zhu, M. (2025). Ionic-liquid/metal–organic-framework composites: Synthesis and emerging sustainable applications. *Inorganic Chemistry Frontiers, 12*(1), 39–84.
36. Kanakaraju, D., Natashya, P. P., Lim, Y. C., & Tan, I. A. W. (2025). Functionalized TiO2-waste-derived photocatalytic materials for emerging pollutant degradation: Synthesis and optimization. *Environmental Monitoring and Assessment, 197*(9), 1–22.
37. Welton, T. (2015). Solvents and sustainable chemistry. *Proceedings of the Royal Society A: Mathematical, Physical and Engineering Sciences, 471*(2183), 20150502.
38. Thakur, P., Sonawane, S., Potoroko, I., & Sonawane, S. H. (2021). Recent advances in ultrasound-assisted synthesis of nano-emulsions and their industrial applications. *Current Pharmaceutical Biotechnology, 22*(13), 1748–1758.
39. Sonawane, A. V., Rikame, S., Sonawane, S. H., Gaikwad, M., Bhanvase, B., Sonawane, S. S., . . . & Gaikwad, R. (2024). A review of microbial fuel cell and its diversification in the development of green energy technology. *Chemosphere, 350*, 141127.
40. Ma, C., Wang, N., Ye, N., & Ji, X. (2021). CO2 capture using ionic liquid-based hybrid solvents from experiment to process evaluation. *Applied Energy, 304*, 117767.
41. Shiflett, M. B., & Yokozeki, A. (2005). Solubilities and diffusivities of carbon dioxide in ionic liquids:[bmim][PF6] and [bmim][BF4]. *Industrial & Engineering Chemistry Research, 44*(12), 4453–4464.

42. Zhao, D., Wu, M., Kou, Y., & Min, E. (2002). Ionic liquids: Applications in catalysis. *Catalysis Today*, *74*(1–2), 157–189.
43. Malika, M. M., & Sonawane, S. S. (2019). Review on application of nanofluid/nano particle as water disinfectant. *Journal of Indian Association for Environmental Management (JIAEM)*, *39*(1–4), 21–24.
44. Malika, M., Bhad, R., & Sonawane, S. S. (2021). ANSYS simulation study of a low volume fraction CuO–ZnO/water hybrid nanofluid in a shell and tube heat exchanger. *Journal of the Indian Chemical Society*, *98*(11), 100200.
45. Li, J., Liu, H., Deng, Y., Liu, G., Chen, Y., & Yang, J. (2016). Emerging nanostructured materials for the catalytic removal of volatile organic compounds. *Nanotechnology Reviews*, *5*(1), 147–181.
46. Xu, Y., Chiam, S. L., Leo, C. P., & Hu, Z. (2025). Recent advances in photocatalytic TiO2-based membranes for eliminating water pollutants. *Separation & Purification Reviews*, 1–18.
47. Sonawane, S. S., Thakur, P. P., Malika, M., & Ali, H. M. (2023). Recent advances in the applications of green synthesized nanoparticle based nanofluids for the environmental remediation. *Current Pharmaceutical Biotechnology*, *24*(1), 188–198.
48. Malika, M., & Sonawane, S. S. (2020). Effect of nanoparticle mixed ratio on stability and thermo-physical properties of CuO-ZnO/water-based hybrid nanofluid. *Journal of the Indian Chemical Society*, *97*(3), 414–419.
49. Sonawane, S., Thakur, P., & Paul, R. (2020). Study on thermal property enhancement of MWCNT based polypropylene (PP) nanocomposites. *Materials Today: Proceedings*, *27*, 550–555.
50. Kshirsagar, D. P., & Venkatesh, M. A. (2021). A review on hybrid nanofluids for engineering applications. *Materials Today: Proceedings*, *44*, 744–755.
51. Gu, Y., & Jérôme, F. (2013). Bio-based solvents: An emerging generation of fluids for the design of eco-efficient processes in catalysis and organic chemistry. *Chemical Society Reviews*, *42*(24), 9550–9570.
52. Shang, D., Liu, X., Bai, L., Zeng, S., Xu, Q., Gao, H., & Zhang, X. (2017). Ionic liquids in gas separation processing. *Current Opinion in Green and Sustainable Chemistry*, *5*, 74–81.
53. Thakur, P., Sonawane, S. S., Sonawane, S. H., & Bhanvase, B. A. (2020). Nanofluids-based delivery system, encapsulation of nanoparticles for stability to make stable nanofluids. In *Encapsulation of active molecules and their delivery system* (pp. 141–152). Elsevier.
54. Bethi, B., Sonawane, S. H., Bhanvase, B. A., & Sonawane, S. S. (2021). Textile industry wastewater treatment by cavitation combined with Fenton and ceramic nanofiltration membrane. *Chemical Engineering and Processing-Process Intensification*, *168*, 108540.
55. Hasib-ur-Rahman, M., Siaj, M., & Larachi, F. (2010). Ionic liquids for CO2 capture—development and progress. *Chemical Engineering and Processing: Process Intensification*, *49*(4), 313–322.

4 Hybrid Nanofluids for Renewable and Sustainable Energy

4.1 INTRODUCTION TO RENEWABLE AND SUSTAINABLE ENERGY NEEDS

However, in the 21st century, the world community encounters energy production, consumption, and environmental sustainability issues like never before. As the global population continues to increase, mainly due to its current population of 7.5 billion people, with a projected figure of over 9 billion people by 2050, the energy demand will be at an alarming level [1]. This growth in demand is further catalyzed by the current rapid rates of urbanization, industrialization, and the increasing infiltration of energy-intensive electric vehicles, data centers, and advanced manufacturing systems.

In the historical past, the energy sector in the world has been dominated by fossil fuels, which are coal, oil, and natural gas. The International Energy Agency (IEA) estimates that fossil fuel is currently the primary source of energy in the world (about 80%) [2]. Although they have been instrumental in enhancing economic development and boosting standards of living, intensive use of these has had serious environmental consequences, which include climate change, air pollution, and depletion of resources, among others.

The biggest contributors to global warming are greenhouse gas (GHG) emissions caused mainly by the combustion of fossil fuels. The levels of carbon dioxide (CO_2), methane (CH_4), and nitrous oxide (N_2O) have increased rapidly, causing significant global warming, accelerated polar ice melt, and more frequent extreme weather events. [3]. The Intergovernmental Panel on Climate Change (IPCC) states that achieving net-zero emissions by 2050 is essential to limit global warming to 1.5°C. It also emphasizes that global carbon dioxide emissions must be reduced by approximately 50% in 2030.

Besides climate change, fossil fuels are the cause of extreme levels of air pollution that lead to the early deaths of millions of people all over the globe annually [4]. Energy security is also at risk in the long run due to the depletion of fossil fuels that are non-renewable and are more serious to the energy security of the countries that rely on imported sources.

Consequent to these looming concerns, the world has been aggressively trying to look into the practice of renewable sources of energy and the use of sustainable energy sources. Alternatives to fossil fuels are the renewable sources of energy in the form of solar, wind, geothermal, biomass, and hydropower, and these are clean and work without any end. In the past 20 years, innovations in technology, encouraging

DOI: 10.1201/9781003628071-4

policies, and declining prices have driven the use of renewables. As such, the price of solar photovoltaic (PV) modules has declined by almost 90% since 2010, with the levelized cost of wind energy also falling by more than 70% [5].

Although the aforementioned trends are encouraging, there are numerous challenges related to incorporating renewable energy into the current energy systems. Renewable sources are also unreliable as they vary due to time and place. Thus, there is an imbalance in the power supply. An example is that solar energy cannot be used at night and may be affected by bad weather, such as clouds, and wind energy is dependent on the fluctuating wind directions. To solve these problems of intermittency challenges, high-quality energy storage technologies that provide consistency and reliability of energy output must be put in place, as well as improved heat management.

Also, the efficiency of the conversion and utilization of energy is of great importance in reducing the loss of energy and increasing the sustainability. This will demand constant advancement in the fields of material science, thermal engineering, and system integration, all of which have enormous potential in emerging technology such as nanotechnology, and, even narrower, hybrid nanofluids.

4.2 ROLE OF ADVANCED THERMAL MANAGEMENT AND ENERGY STORAGE

Two of the most critical areas in the modern energy system include thermal management and energy storage, especially in such renewable resource-based energy systems [6]. Successful thermal management encompasses the regulation and riddance of heat in a system that keeps close attention to the ideal operating circumstances, improves performance, and averts harm or loss. Energy storage, in contrast, is the capture of energy generated at one time so that it can be used at another and therefore decouples supply and demand and leads to constant availability of energy [7]. The thermal management is essential in increasing the efficacy of the system when viewed on renewable energy applications. As an example, in photovoltaic cells, high temperatures lower the electrical conversion efficiency and may destroy sensitive parts. Proper cooling systems have the potential of extending the lifespan and output of solar panels one notch higher [8]. This is also true for concentrated solar power (CSP) systems, which use heat-transfer fluids to transport and store thermal energy, enabling power generation even during cloudy periods or after sunset.

Wind turbines also benefit from effective thermal management. The gearbox and the generator parts have to be maintained at certain temperature boundaries in order to remain efficient and not fail. Similarly, biomass and geothermal technologies in most cases imply high temperature applications that demand sophisticated cooling/heat transfer systems in order to achieve maximum energy recovery and utilization [9].

The complementary role that energy storage performs is to break the intermittency of renewable sources. An example is thermal energy storage (or TES) systems that are employed in solar thermal systems in which the surplus heat is stored during the peak. Various phase change materials (PCMs) and sensible heat storage systems have received attention toward this end and have been installed. In the interim, electrochemical storage technologies (such as batteries) absorb electrical energy directly, which can stabilize and balance grid loads and loads.

Advanced thermal fluids and materials are also essential to the integration of these systems to achieve a better level of effectiveness, reliability, and flexibility. Improved thermal-conversion/dissipation fluids will enable smaller, more efficient heat exchangers, higher rates of heat dissipation, and less power needed to pump the fluid. Furthermore, they make it possible to create new energy systems that are able to work under higher temperatures and pressure, which opens up the opportunities of new thermodynamic cycles and reduces operating expenses. But conventional heat transfer liquids like water, ethylene glycol, and mineral oils are intrinsically limited with regard to heat capacity and thermal conductivity [10]. Such shortcomings are a drawback to satisfying the tough needs of the present renewable systems of energy. This has led to new experiments on different heat transfer and storage mediums such as nanofluids and, lately, the hybrid nanofluids. Hybrid nanofluids, composed of more than one type of nanoparticle dispersed in a base fluid, offer synergistic enhancements in thermal and transport properties. They can be engineered to possess tailored properties, such as high thermal conductivity, optimal viscosity, improved specific heat, and enhanced stability. These attributes make hybrid nanofluids particularly attractive for applications in solar thermal collectors, PV cooling systems, TES, and many other renewable energy technologies.

In essence, advanced thermal management and energy storage are linchpins for realizing the full potential of renewable and sustainable energy systems. Hybrid nanofluids represent a transformative solution, offering a pathway to significantly improved heat transfer efficiency, reduced energy losses, and higher system reliability.

4.3 HYBRID NANOFLUIDS FOR RENEWABLE ENERGY

The concept of nanofluids, pioneered in the mid-1990s, involves dispersing nanoscale particles (typically less than 100 nm in size) into a conventional base fluid to enhance its thermal properties [11]. While single-component nanofluids have demonstrated substantial improvements in thermal conductivity and convective heat transfer, they often face challenges related to stability, agglomeration, and limited tunability of properties [12].

Hybrid nanofluids are an advanced iteration of this concept, incorporating two or more distinct types of nanoparticles into a single base fluid. This approach leverages the complementary properties of different nanoparticles to achieve superior thermal performance and operational stability. For example, combining metal oxide nanoparticles (which often exhibit high thermal conductivity) with carbon-based nanostructures (which offer excellent stability and specific heat) can yield a hybrid system with enhanced overall heat transfer capabilities [13].

Several key attributes make hybrid nanofluids particularly promising for renewable and sustainable energy applications:

4.3.1 Enhanced Thermal Conductivity

One of the most significant advantages of hybrid nanofluids is their superior thermal conductivity compared to both conventional fluids and mono nanofluids. Higher thermal conductivity enables more efficient heat absorption and dissipation, which is critical for solar thermal collectors, CSP systems, and PV cooling. Experimental studies have demonstrated that hybrid nanofluids can exhibit thermal conductivity improvements of up to 50% or more over traditional fluids [14].

4.3.2 Improved Convective Heat Transfer

The mixed nanoparticles produce synergistic effects on the heat transfer in hybrid nanofluids, resulting in increased heat transfer coefficients in the convective regime. As a result, smaller heat exchanger designs, lighter mass, and lower power requirements of the pump are possible. These properties are important in applications where space and efficiency are critical.

4.3.3 Tailored Thermophysical Properties

Hybrid nanofluids can be designed to have an optimal thermal conducting viscosity and specific heat capacity as compared to traditional fluids. The fluid can be optimized by changing the type, size, concentration, and shape of nanoparticles used so that the fluid characteristics meet the operation requirements. This flexibility creates an opportunity to deploy them in a broad variety of renewable applications, such as high-temperature solar thermal loops to low-temperature PV cooling.

4.3.4 Enhanced Stability

Nanofluids have much to fear of agglomeration and sedimentation, which could result in poor performance and clogging of the systems. Instead, hybrid nanofluids may be characterized by enhanced stability over longer periods of time, having in mind the synergistic stabilizing agents and the optimization of interactions between nanoparticles. The increased stability results in improved life cycles of the thermal systems and lowers maintenance needs.

4.3.5 Potential for Multifunctionality

In renewable energy applications, multifunctionality is highly advantageous. Simultaneous thermal management and functionalities, including optical absorption to harvest solar energy, anti-corrosiveness, and even magnetic or electric responsiveness, may be achieved using hybrid nanofluids. The resulting multifunctionality allows the possibility of incorporating optimized control approaches and multi-energy systems.

4.3.6 Applications across Diverse Renewable Systems

The hybrid nanofluids have proved to be applicable in many renewable energy areas. As an example, the operating temperatures of CSP plants could be increased by using hybrid nanofluids as an excellent heat transfer and storage medium, boosting power cycle efficiency. They have the potential to increase cooling in PV systems and, therefore, electric output and durability. During the use of geothermal systems and biomass, the hybrid nanofluids promote easier recovery of heat and energy.

4.3.7 Compatibility with Emerging Energy Storage Technologies

Hybrid nanofluids have been demonstrated as a suitable integration in thermal energy storage systems, especially in systems using phase change material storage or molten salt based. They allow faster response times and increased storage efficiency by increasing the rate at which heat can be transferred during both charging and discharging. This compatibility is essential as the intermittency of renewable sources of energy needs to be taken care of [15].

4.3.8 Alignment with Global Decarbonization Goals

The hybrid nanofluids lead to detouring the dependency on fossil fuels and abating the release of GHG by increasing the effectiveness and stability of the renewable energy systems. Having an enhanced system performance will enable additional renewable energy to be utilized and stored, so the shift toward a low-carbon energy future can occur more easily.

4.3.9 Economic and Operational Benefits

Though the early cost of research and development in hybrid nanofluids is likely to be expensive, their investment can lead to huge savings in the system operation in the future. Increased transfer of heat makes the pumps and fans require less energy consumption, reduces maintenance costs, as well as prolongs the lifespan of the systems, which makes them generally more economically viable.

4.3.10 Future Adaptability and Scalability

Due to the evolution in renewable energy technologies, the versatility of hybrid nanofluids promises them as a solution to future energy systems at scale. The further reduction of costs and the enhancement of performance that are envisioned through refinement of nanomaterial synthesis, technique of functionalization, and manufacturing processes make the use of hybrid nanofluids a feasible option in practical implementation.

The transition to renewable and sustainable energy sources is not merely a technical challenge but a societal imperative. Addressing climate change, ensuring energy security, and achieving sustainable economic growth all hinge on the ability to harness and utilize clean energy effectively.

Hybrid nanofluids, through their advanced thermal and transport properties, offer a transformative solution to many of the technical barriers facing renewable energy systems today. Their potential to improve thermal management and energy storage directly supports the integration and scalability of solar, wind, geothermal, biomass, and hydrogen energy systems.

Continued research and development, coupled with supportive policies and industrial collaboration, will be crucial for realizing the full potential of hybrid nanofluids.

As the field matures, we can anticipate hybrid nanofluids playing an increasingly central role in advancing the efficiency, reliability, and sustainability of next-generation energy systems.

4.4 FUNDAMENTALS OF HYBRID NANOFLUIDS

4.4.1 Fundamentals of Hybrid Nanofluids

4.4.1.1 Definition and Composition (Multi-Nanoparticle Systems)

Over the past few decades, rapid advancements in nanotechnology have significantly impacted the field of thermal fluids and heat transfer enhancement. A significant breakthrough was made with the introduction of nanofluids, engineered colloidal suspensions of nanoparticles in conventional base fluids, which provided a practical way of enhancing the thermal properties of a fluid to go beyond the constraints of conventional fluids [16]. Nevertheless, in spite of their encouraging results, single-component nanofluids possess some inevitable limitations, such as stability loss, a lack of tunability, and inadequate thermal properties at different work conditions [17]. In order to solve these drawbacks, scientists have established a superior type of fluid, which is referred to as hybrid nanofluids.

Hybrid nanofluids are progressive engineered fluids, which are a combination of more than two kinds of nanoparticles in a carrier fluid. Unlike conventional nanofluids that utilize a single nanoparticle type (e.g., Al_2O_3, CuO, TiO_2, SiO_2), hybrid nanofluids leverage the synergistic effects of multiple nanoparticles to achieve enhanced and customizable thermal and rheological properties [18].

4.4.1.2 Concept and Evolution

The concept of hybrid nanofluids emerged from the need to overcome the drawbacks of mono nanofluids, particularly in demanding thermal applications. By carefully selecting and combining different nanoparticles—each contributing unique properties—it is possible to create a fluid with superior thermal conductivity, enhanced specific heat capacity, reduced viscosity, and improved long-term stability [19].

For instance, metal nanoparticles (like copper or silver) are well-known for their high thermal conductivity, but they tend to suffer from high densities and sedimentation issues. On the other hand, carbon-based nanomaterials (such as graphene, carbon nanotubes, or nanodiamonds) offer excellent thermal conductivity along with low densities and good dispersion stability. Combining these in a hybrid system allows one to capitalize on the strengths of each type while mitigating their individual weaknesses.

4.4.1.3 Composition and Formulation

The formulation of hybrid nanofluids involves three primary components:

1. **Base fluid**: The traditional heat transfer medium (e.g., water, ethylene glycol, oil, refrigerants). The base fluid provides the primary medium for heat exchange and flow.

2. **Multiple nanoparticles**: At least two different nanoparticles, chosen based on desired properties such as thermal conductivity, chemical stability, optical behavior, or magnetic responsiveness. Common combinations include the following:
    - Metal/metal oxide hybrid (e.g., Cu-Al_2O_3, TiO_2-Al_2O_3)
    - Metal/carbon hybrid (e.g., Cu-graphene, Al_2O_3-CNT)
    - Metal oxide/carbon hybrid (e.g., TiO_2-graphene)
    - Triple or more complex systems (e.g., TiO_2-Al_2O_3-Cu)
3. **Surfactants or stabilizers**: Chemical agents added to prevent agglomeration and improve the colloidal stability of nanoparticles within the base fluid. Examples include sodium dodecyl sulfate (SDS), polyvinylpyrrolidone (PVP), or cetyltrimethylammonium bromide (CTAB).

4.4.2 Synthesis and Dispersion Methods

The preparation of hybrid nanofluids is a crucial factor determining their final properties. Common methods include:

- **Two-step method**: Individual nanoparticles are synthesized separately (often via chemical or physical routes), then mixed into the base fluid in the desired proportions using ultrasonication, high-shear mixing, or magnetic stirring [20].
- **One-step method**: Nanoparticles are directly synthesized within the base fluid, minimizing oxidation and improving dispersion stability. This method is more complex but offers superior control over particle distribution and size [21].
- **Hybridization techniques**: In some cases, hybrid nanoparticles are created first (e.g., core-shell structures or composite nanoparticles) before being dispersed, offering even more controlled synergistic properties.

The final concentration of nanoparticles is typically kept below 5 vol% to ensure a balance between improved thermal properties and acceptable viscosity levels for practical pumping and flow.

4.4.3 Comparison with Conventional Nanofluids

To fully appreciate the value of hybrid nanofluids, it is essential to compare them systematically with conventional nanofluids across key performance metrics and operational attributes.

4.4.3.1 Thermal Conductivity

Conventional nanofluids have demonstrated significant enhancements in thermal conductivity compared to pure base fluids. For example, the addition of 1 vol% Al_2O_3 nanoparticles can improve the thermal conductivity of water by approximately 10–20%. However, hybrid nanofluids have shown even greater enhancements due to the combined conductive pathways created by different nanoparticles.

Studies report that hybrid nanofluids can achieve thermal conductivity improvements of 30–60% at similar or even lower concentrations. For instance, a hybrid nanofluid with Al_2O_3 and Cu nanoparticles can exhibit a thermal conductivity up to 55% higher than water, outperforming both Al_2O_3-water and Cu-water nanofluids individually.

4.4.3.2 Specific Heat Capacity

While single nanoparticle nanofluids often show a slight reduction in specific heat capacity compared to base fluids, hybrid nanofluids can be tailored to maintain or even improve specific heat capacity by optimizing the nanoparticle mixture. This feature is particularly important in applications involving energy storage, such as in solar thermal systems or phase change materials (PCMs).

4.4.3.3 Viscosity and Flow Behavior

A major challenge with conventional nanofluids is increased viscosity, which leads to higher pumping power requirements and pressure drops in heat exchanger systems. Hybrid nanofluids provide a way to mitigate this issue by selecting nanoparticles that introduce minimal viscosity penalties while enhancing thermal properties. For example, carbon-based nanoparticles can help lower the effective viscosity compared to metallic counterparts.

4.4.3.4 Stability and Sedimentation Resistance

Conventional nanofluids are prone to sedimentation and agglomeration, especially under long-term operation or high-temperature conditions. Hybrid nanofluids often show improved stability due to the synergistic interactions between different nanoparticles and better surface functionalization possibilities. In hybrid systems, certain nanoparticles can act as stabilizers for others, forming more stable dispersions.

4.4.4 Heat Transfer Performance

The combined effect of improved thermal conductivity, optimized viscosity, and enhanced convective heat transfer coefficients results in superior overall heat transfer performance for hybrid nanofluids. Experimental and numerical studies have shown that heat transfer coefficients of hybrid nanofluids can exceed those of conventional nanofluids by 10–40%, depending on the system configuration and flow regime.

4.4.4.1 Multifunctionality and Adaptability

Hybrid nanofluids can be designed for multifunctionality beyond just thermal management. For example, magnetic nanoparticles can impart magneto-responsive behavior, enabling controllable flow and targeted heat transfer. Similarly, optically active nanoparticles can improve solar absorption in solar thermal collectors. This multifunctionality is typically harder to achieve with single-component nanofluids.

4.4.4.2 Economic and Operational Considerations

While both conventional and hybrid nanofluids currently involve higher initial costs compared to traditional fluids, the long-term operational savings from reduced energy consumption, enhanced system longevity, and lower maintenance requirements make hybrid nanofluids increasingly attractive. Moreover, ongoing advancements in nanoparticle synthesis and scale-up are expected to reduce costs further.

4.4.4.3 Key Thermophysical and Chemical Advantages

The unique composition of hybrid nanofluids offers a range of thermophysical and chemical advantages, making them highly suited for next-generation energy and thermal applications.

4.4.4.3.1 Superior Thermal Conductivity

The most prominent advantage is the significantly higher thermal conductivity, which directly enhances heat transfer efficiency. The dual (or multi) nanoparticle system allows for the formation of highly conductive networks within the base fluid, facilitating rapid energy transport. This is particularly beneficial for applications such as solar collectors, electronic cooling, and industrial heat exchangers.

Theoretical models, such as the Maxwell-Garnett model and effective medium theories, have been extended to describe hybrid systems, often predicting higher conductivity than traditional single-nanoparticle suspensions.

4.4.4.3.2 Enhanced Specific Heat Capacity

In energy storage applications, fluids with higher specific heat capacity are desirable, as they can store more energy per unit mass or volume. By selecting nanoparticles with complementary thermal capacities, hybrid nanofluids can be engineered to maintain or enhance specific heat, optimizing them for thermal energy storage and solar applications.

4.4.4.3.3 Optimized Viscosity and Flow Properties

Viscosity plays a critical role in determining the required pumping power and flow dynamics of a heat transfer fluid. Hybrid nanofluids allow for fine-tuning of viscosity by balancing the particle shapes, sizes, and interaction forces. For example, spherical nanoparticles may increase viscosity more than tubular or platelet-like structures; by combining different geometries, a balance can be struck between enhanced thermal performance and manageable flow resistance.

4.4.4.3.4 Improved Convective Heat Transfer Coefficients

In forced or natural convection applications, hybrid nanofluids have demonstrated higher convective heat transfer coefficients than single-component nanofluids. This enhancement results from the combined effects of improved thermal conductivity, nanoparticle-induced turbulence, and micro-convection around dispersed particles.

4.4.4.4 Greater Dispersion Stability

One of the most significant operational challenges in nanofluid use is maintaining stable dispersion without particle sedimentation. Hybrid nanofluids often exhibit improved stability due to:

- The synergistic electrostatic and steric stabilization provided by mixed nanoparticles.

- Possible self-assembly behavior, where certain nanoparticles act as physical barriers preventing agglomeration.
- Enhanced surface functionalization techniques tailored for hybrid systems.

This stability translates to consistent thermal performance and reduces system maintenance needs.

4.4.4.5 Tunable Optical and Radiative Properties

Hybrid nanofluids can be engineered to possess specific optical absorption and scattering characteristics, making them ideal for direct absorption solar collectors (DASCs) and hybrid photovoltaic/thermal (PV/T) systems. By incorporating nanoparticles with strong plasmonic or photothermal properties (e.g., gold, silver, or carbon-based materials), hybrid nanofluids can convert a broader range of solar spectra into thermal energy efficiently.

4.4.4.6 Chemical and Corrosion Resistance

Some hybrid nanofluid formulations include nanoparticles that provide corrosion inhibition properties, protecting heat exchanger surfaces and pipelines. For example, adding certain metal oxides or rare earth elements can create protective layers on metal surfaces, enhancing the system's operational lifespan and reducing corrosion-related failures.

4.4.4.7 Multifunctional Capabilities

Beyond heat transfer, hybrid nanofluids can offer additional functionalities such as:

- Magnetic responsiveness (by including Fe_3O_4 or other magnetic nanoparticles), allowing flow control using external magnetic fields.
- Lubrication improvements, beneficial in moving mechanical assemblies.
- Antimicrobial or anti-biofouling properties when integrated with specific bioactive nanoparticles, helpful in water and bioenergy applications.

4.4.4.8 Reduced Environmental Footprint

Although nanoparticle production still involves certain environmental challenges, the improved efficiency of hybrid nanofluids can lead to substantial reductions in overall energy consumption and associated emissions. Improved thermal management and energy storage translate directly into lower fossil fuel usage, supporting decarbonization efforts.

4.5 THERMAL AND TRANSPORT PROPERTIES RELEVANT TO ENERGY APPLICATIONS

Hybrid nanofluids marked a thrilling advance in thermo-energy control and heat transfer applications that can produce fine-tuned properties much more advanced than the conventional fluids, as well as nanofluids. Their use on renewable and sustainable energy systems significantly depends on their exclusive thermal and transport properties that directly affect the performance of the devices and systems of energy, making them a compact, efficient, and reliable energy source.

4.5.1 Enhanced Thermal Conductivity and Heat Capacity

4.5.1.1 Thermal Conductivity: Fundamental Importance

Thermal conductivity (k) is a primary parameter determining the heat transfer capability of a fluid. It describes the rate at which heat can be conducted through a material due to a temperature gradient. In most conventional fluids, such as water, ethylene glycol, and oils, thermal conductivity is relatively low (e.g., water ≈ 0.6 W/m·K, ethylene glycol ≈ 0.25 W/m·K), which limits their heat removal capacity and necessitates larger surface areas or higher flow rates in heat exchangers.

Hybrid nanofluids address this limitation by incorporating highly conductive nanoparticles, such as metals (Cu, Ag), metal oxides (Al_2O_3, TiO_2), or carbon-based nanomaterials (graphene, CNTs). The inclusion of multiple nanoparticles allows the creation of interconnected thermal networks, effectively bridging gaps in the fluid and facilitating rapid energy transfer.

4.5.1.2 Mechanisms Contributing to Thermal Conductivity Enhancement

Several mechanisms contribute to the improved thermal conductivity of hybrid nanofluids:

1. **Brownian motion of nanoparticles**: The rapid random movement of nanoparticles at nanoscale contributes to micro-convection, facilitating additional heat transfer paths.
2. **Percolation networks**: At higher nanoparticle concentrations, continuous networks of conductive particles can form, further enhancing k.
3. **Liquid layering effect**: Near the nanoparticle surface, a structured liquid layer forms, with thermal properties different from the bulk fluid, assisting heat transfer.
4. **Synergistic effects in hybrid systems**: By combining different nanoparticles (e.g., metal and oxide), hybrid nanofluids leverage complementary attributes (e.g., high thermal conductivity from metals, stability from oxides), achieving better performance than single-type nanofluids.

4.5.1.3 Experimental Evidence of Enhanced Thermal Conductivity

Numerous studies have reported thermal conductivity enhancements ranging from 10% to over 100% in hybrid nanofluids, depending on nanoparticle type, concentration, and preparation method.

- Sundar et al. (2016) demonstrated that a hybrid nanofluid containing graphene nanoplatelets and Al_2O_3 in water exhibited a thermal conductivity increase of 66% at 0.1 vol% [22].
- Madhesh et al. (2014) reported a 34% enhancement in thermal conductivity for Cu-TiO_2 hybrid nanofluid in ethylene glycol-water mixtures [23].

- Akilu et al. (2018) found that a hybrid nanofluid with MgO-MWCNTs dispersed in ethylene glycol showed significant thermal conductivity improvements even at very low concentrations [24].

4.5.1.4 Heat Capacity: Balancing Storage and Transfer

Specific heat capacity (Cp) defines the amount of heat required to raise the temperature of a unit mass by one degree Celsius. High Cp is desirable for energy storage applications, while for rapid heat removal, thermal conductivity is more critical.

Nanoparticles typically have lower specific heat than base fluids; hence, adding nanoparticles generally reduces the overall Cp. However, by carefully choosing and combining nanoparticles in a hybrid nanofluid, it is possible to mitigate or even reverse this effect.

For example:

- Metal oxide nanoparticles (like Al_2O_3, TiO_2) have moderate Cp and contribute to maintaining higher heat capacity.
- Carbon-based nanoparticles (e.g., graphene) offer high thermal conductivity and can help maintain acceptable Cp levels.

By adjusting the ratio and types of nanoparticles, hybrid nanofluids can be tailored to achieve both enhanced thermal conductivity and an adequate Cp, crucial for systems involving both heat transfer and temporary heat storage, such as solar thermal collectors integrated with thermal energy storage tanks.

4.5.1.5 Implications for Energy Applications

Enhanced thermal conductivity and carefully tuned heat capacity are fundamental for:

- Solar collectors (faster and more efficient heat absorption).
- Geothermal heat extraction (improved subsurface heat capture).
- Electronics and PV cooling (efficient removal of heat hotspots).
- Thermal energy storage systems (optimized energy uptake and release cycles).

4.5.2 Improved Convective Heat Transfer

4.5.2.1 Role of Convection in Thermal Systems

While thermal conductivity primarily governs conduction, convective heat transfer is crucial in most practical energy systems where fluids are in motion. Convective heat transfer coefficient (h) is a measure of the fluid's ability to transfer heat to or from a surface, and it depends on fluid properties (thermal conductivity, viscosity, specific heat), flow conditions (velocity, turbulence), and geometry.

Hybrid nanofluids enhance convective heat transfer in both laminar and turbulent regimes, making them particularly attractive for applications requiring compact and high-efficiency heat exchangers.

4.5.2.2 Mechanisms behind Improved Convection

Several factors contribute to improved convective performance of hybrid nanofluids:

1. **Higher thermal conductivity**: Directly improves energy transfer from the solid boundary to the bulk fluid.
2. **Nanoparticle-induced turbulence**: Suspended nanoparticles disturb the thermal and velocity boundary layers, promoting micro-mixing and enhancing local turbulence.
3. **Reduced boundary layer thickness**: Enhanced energy transport near the wall leads to thinner thermal boundary layers, increasing heat transfer rates.
4. **Modified flow patterns**: Depending on particle shape and concentration, hybrid nanofluids can alter flow structures, enhancing secondary flows and mixing.

4.5.2.3 Experimental Evidence and Case Studies

Several studies have demonstrated the superior convective heat transfer of hybrid nanofluids:

- Suresh et al. (2011) used Al_2O_3-Cu/water hybrid nanofluids on a laminar flow to a tube and noted an epiglottis between 13% and 19% increment in Nusselt number than single-element nanofluids of a similar Reynolds number [25].
- Nagarajan et al. (2017) noted that TiO_2-SiO_2/water hybrid nanofluids enhance convective heat transfer coefficient by up to 32% in a circular tube with turbulent flow of fluid at an aqueous phase [26].
- As it was demonstrated by Esfe et al. (2016), due to the synergy impact, Cu-Al_2O_3/EG nanofluids demonstrated even higher heat transfer coefficients than those of pure Cu nanofluid or Al_2O_3/EG nanofluid [27].

4.5.2.4 Numerical Simulations and Modeling

The behavior of hybrid nanofluid in a complex flow offers more understanding in the field of computational fluid dynamics (CFD) simulation. Finite models that include the effects of interactions between the nanoparticles and non-Newtonian behavior (where present) have predicted very large increases in the convective performance.

Hybrid nanofluid examples include wavy channel/microchannel heat sink nanofluid flows exhibiting up to 40% greater heat removal capabilities and thus warrant use in advanced energy systems like PV/T collectors and battery thermal management systems.

4.6 PRACTICAL CONSIDERATIONS IN ENERGY SYSTEMS

The enhanced convective heat transfer of hybrid nanofluids can lead to:

- Smaller heat exchanger footprints.
- Lower required fluid flow rates and pumping power.
- Improved system response times.
- Higher overall thermal efficiency.

These advantages are particularly relevant in renewable energy systems where weight and space constraints, such as in rooftop solar collectors or electric vehicle battery packs, play a critical role [28].

4.6.1 Importance of Stability

Maintaining a stable suspension of nanoparticles is crucial for consistent thermal performance and preventing operational problems such as clogging, erosion, and sedimentation. Hybrid nanofluids often exhibit better stability than mono nanofluids due to the interaction between different nanoparticles, leading to more robust colloidal structures [29].

Key strategies and factors that contribute to stability include:

- **Surface functionalization**: Chemical treatment of nanoparticles (e.g., silanization) improves surface charge, enhancing electrostatic repulsion and reducing agglomeration [30].
- **Surfactant addition**: Agents such as SDS or CTAB adsorb on particle surfaces, providing steric stabilization [31].
- **pH control**: Adjusting pH can increase the zeta potential, improving suspension stability.
- **Hybrid synergy**: Certain nanoparticles in hybrid systems can act as "anchors," preventing others from settling.

Experimental stability evaluations typically involve visual sedimentation tests, zeta potential measurements, and particle size distribution analysis over time. A zeta potential above ±30 mV is generally considered indicative of good electrostatic stability. Figure 4.1 shows morphology of hybrid nanofluids using TEM.

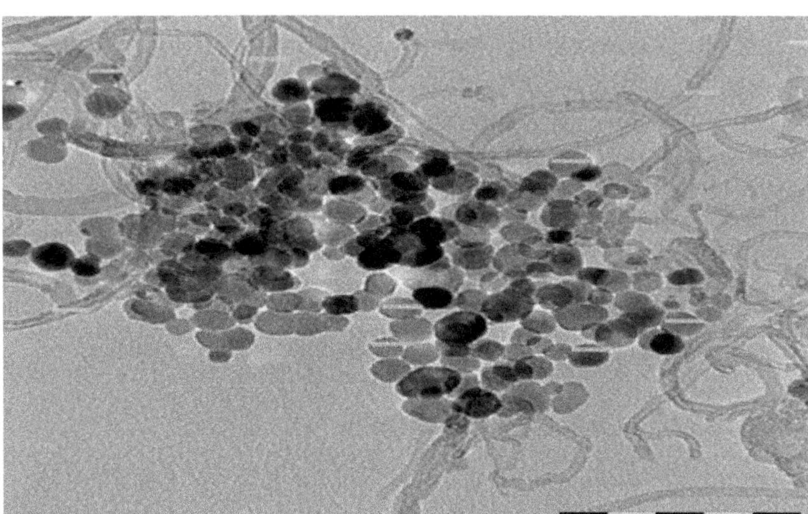

FIGURE 4.1 Morphological representation of hybrid nanofluids using transmission electron microscopy [31].

4.6.2 Rheological Properties: Viscosity and Flow Behavior

The rheological behavior of hybrid nanofluids is complex and influenced by nanoparticle concentration, shape, size, and temperature. Viscosity is a critical parameter as it affects pumping power, flow regimes, and heat transfer effectiveness.

Observations and Trends

- Low concentrations (≤0.5 vol%): Many hybrid nanofluids remain Newtonian, exhibiting constant viscosity independent of shear rate.
- Moderate to high concentrations (>0.5 vol%): Shear-thinning (pseudoplastic) behavior is often observed, where viscosity decreases with increasing shear rate. This behavior is advantageous in energy systems because it reduces pumping power at higher flow rates [32–34].
- Temperature effects: Viscosity generally decreases with increasing temperature, which can mitigate some viscosity-induced penalties in high-temperature applications [35].

Case Examples

- Sundar et al. (2018) studied the viscosity of MWCNT-Fe_3O_4 hybrid nanofluids and observed shear-thinning behavior at higher concentrations, with a significant viscosity reduction at higher temperatures, beneficial for solar collector applications [36–38].
- Sahoo et al. (2015) reported that Cu-Al_2O_3 hybrid nanofluids showed lower viscosity than pure Cu nanofluids at the same volume fraction, highlighting the benefit of hybrid formulations [39].

4.6.3 Pumping Power and Energy Penalty Considerations

While higher viscosity typically implies higher pumping power requirements, hybrid nanofluids can be optimized to provide enhanced heat transfer performance without disproportionately increasing flow resistance. The goal is to achieve a higher thermal performance index (TPI), which considers both heat transfer enhancement and viscosity increase.

In renewable energy systems, careful design and hybrid nanofluid formulation can result in net energy gains despite viscosity changes. For instance, solar thermal loops using hybrid nanofluids have demonstrated overall system efficiencies up to 20% higher than those using water, even after accounting for pumping energy.

4.6.4 Long-Term Operational Behavior

Under operational conditions, hybrid nanofluids must withstand the following:

- High thermal cycling (e.g., daily solar collector temperature swings)
- High shear environments (e.g., in microchannel flows)
- Possible exposure to UV radiation, oxidizing agents, or reactive surfaces

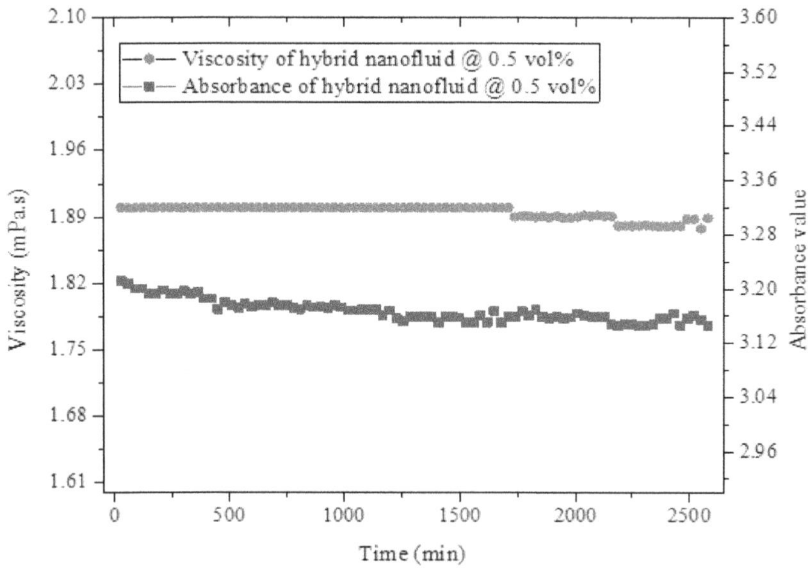

FIGURE 4.2 Stability analysis [31].

Studies involving extended cyclic tests and aging simulations have shown that hybrid nanofluids generally maintain better long-term stability than single-component nanofluids, with less agglomeration and minimal performance degradation. Figure 4.2 shows the stability data for the hybrid nanofluid systems.

4.6.5 Applications for Renewable and Sustainable Energy Systems

The combination of enhanced thermal conductivity, improved convective heat transfer, and robust stability/rheological behavior positions hybrid nanofluids as key enablers in next-generation renewable and sustainable energy systems.

4.6.5.1 Solar Thermal Applications

In solar collectors and concentrators, hybrid nanofluids can:

- Increase optical absorption efficiency (particularly in direct absorption configurations).
- Improve overall thermal conversion efficiency.
- Enable compact, high-power-density designs.

4.6.5.2 Photovoltaic Thermal (PV/T) Systems

Hybrid nanofluids used as cooling media can:

- Reduce PV cell temperatures, improving electrical efficiency.
- Enable simultaneous heat recovery, boosting total system efficiency.
- Enhance durability and reduce maintenance intervals.

4.6.5.3 Geothermal and Waste Heat Recovery

The superior heat transfer performance of hybrid nanofluids can lead to:

- Enhanced heat extraction rates from geothermal reservoirs.
- Improved economic viability of low-grade heat recovery systems.
- Better adaptability to varying subsurface thermal profiles.

4.6.5.4 Energy Storage Systems

Hybrid nanofluids integrated with thermal storage units can:

- Improve charging and discharging rates.
- Allow higher temperature operating ranges.
- Enable hybrid configurations with phase change materials for advanced storage.

Hybrid nanofluids have emerged as a transformative solution in the field of thermal management and energy transfer, driven by their remarkable thermophysical and chemical properties. Their enhanced thermal conductivity and carefully managed heat capacity enable more efficient heat transport and energy storage, while superior convective heat transfer capabilities facilitate smaller, more effective, and more flexible thermal systems. Importantly, excellent stability and tailored rheological behavior ensure reliable, long-term operation even under demanding conditions, supporting a wide range of renewable and sustainable energy applications.

The continued development of hybrid nanofluids through advanced nanoparticle engineering, optimized formulations, and application-specific customization promises to unlock new levels of performance in solar, geothermal, bioenergy, and advanced energy storage systems, driving forward the global transition toward a more sustainable and resilient energy future.

4.7 HYBRID NANOFLUIDS IN SOLAR THERMAL ENERGY SYSTEMS

Solar thermal energy systems have become a critical component in the global transition toward renewable energy. They convert solar radiation into thermal energy, which can then be used for electricity generation, industrial process heat, and residential or commercial heating applications. Despite their maturity and deployment worldwide, the efficiency of solar thermal systems is often limited by traditional working fluids (e.g., water, ethylene glycol, mineral oils) that exhibit poor thermal properties and limited optical absorption capacity.

Hybrid nanofluids, consisting of two or more types of nanoparticles suspended in a base fluid, offer a revolutionary opportunity to improve heat transfer performance, optical absorptivity, and overall system efficiency. This chapter explores the application of hybrid nanofluids in solar collectors and concentrators, their role in enhancing absorptivity and spectral selectivity, and supporting case studies and performance evaluations [40–42].

4.8 USE IN SOLAR COLLECTORS AND CONCENTRATORS

4.8.1 Background on Solar Thermal Collectors

Solar thermal collectors are devices designed to capture solar radiation and convert it into thermal energy. Broadly, they are classified into two categories:

1. **Non-concentrating collectors**: Such as flat-plate collectors (FPCs) and evacuated tube collectors (ETCs), which absorb both direct and diffuse radiation.
2. **Concentrating collectors**: Including parabolic trough collectors (PTCs), dish collectors, and linear Fresnel reflectors, which focus sunlight onto a small receiver area to achieve higher temperatures.

In both cases, the choice of working fluid is critical, as it determines the rate of heat extraction, system efficiency, and operating temperature limits. Traditional fluids have low thermal conductivity and poor optical absorption, which restrict performance [43].

4.8.1.1 Hybrid Nanofluids as Advanced Heat Transfer Fluids

Hybrid nanofluids address these limitations through:

- **Superior thermal conductivity**: Enhancing heat transfer from the collector surface to the fluid.
- **Enhanced convective heat transfer**: Promoting better heat removal even at lower flow rates.
- **Improved thermal stability**: Allowing operation at higher temperatures without significant degradation.
- Potential direct absorption capability: Acting simultaneously as a heat transfer medium and an optical absorber.

In direct absorption solar collectors (DASCs), where the fluid itself absorbs solar radiation instead of a solid surface, hybrid nanofluids can dramatically improve the solar-to-thermal conversion efficiency. By tuning the type, concentration, and optical properties of nanoparticles, they can be engineered to absorb a broad spectrum of sunlight, converting it efficiently into heat.

4.8.1.2 Hybrid Nanofluids in Concentrating Solar Systems

Concentrating solar power (CSP) systems operate at higher temperatures (up to 400–600°C) to drive steam turbines or other thermodynamic cycles. Hybrid nanofluids, with their improved thermal stability and heat transfer capabilities, are particularly well-suited for CSP applications.

Hybrid nanofluids have been explored as heat transfer fluids in:

- **Parabolic trough collectors (PTCs)**: Where they can increase heat removal rates and reduce required collector length.

- **Central receiver systems (solar towers)**: Offering potential for higher heat flux handling and improved heat storage integration.
- **Dish collectors**: Where they help maintain high thermal gradients efficiently, supporting high-concentration solar flux.

These improvements can enable higher system efficiencies, smaller collector footprints, and reduced operational costs, supporting the economic viability of CSP technology.

4.8.2 Absorptivity and Spectral Selectivity Enhancements

4.8.2.1 Fundamentals of Solar Absorptivity

Absorptivity refers to the fraction of incident solar radiation absorbed by a material or fluid. In solar thermal applications, maximizing absorptivity is crucial for efficient heat conversion. Spectral selectivity, meanwhile, refers to the ability of a material to absorb solar radiation within the solar spectrum (0.3–2.5 μm) while minimizing thermal re-radiation losses in the infrared range (>2.5 μm).

Conventional working fluids are largely transparent to solar radiation and rely on absorber plates to convert sunlight into heat. In contrast, hybrid nanofluids can directly absorb incoming solar energy, allowing them to serve dual roles as both heat transfer and absorber mediums [44–46].

4.8.2.2 Enhancing Optical Properties with Hybrid Nanofluids

Hybrid nanofluids achieve high absorptivity and tailored spectral selectivity through careful design of nanoparticle compositions:

1. **Metal nanoparticles (e.g., Cu, Ag)**: Exhibit strong surface plasmon resonance, enhancing absorption in the visible range.
2. **Carbon-based nanoparticles (e.g., graphene, carbon nanotubes)**: Provide broadband absorption across the solar spectrum, with additional benefits in infrared absorption.
3. **Metal oxides (e.g., TiO_2, Al_2O_3)**: Offer UV absorption capabilities and improve thermal stability.

By combining these nanoparticles, hybrid nanofluids can be designed for optimal absorption profiles, achieving close to blackbody-like absorption with minimal thermal re-radiation.

4.8.2.3 Plasmonic Enhancement and Solar Spectrum Tailoring

Metallic nanoparticles contribute plasmonic effects, which create localized electromagnetic field enhancements around the particles. These effects can be tuned by adjusting nanoparticle size, shape, and dielectric environment, allowing hybrid nanofluids to be customized for maximal solar absorption.

For example:

- Spherical silver nanoparticles can be tuned for maximum absorption in the 400–500 nm range.

- Graphene oxide can be used to extend absorption into the near-infrared range.
- TiO_2 nanoparticles complement this by capturing UV radiation.

This spectral tailoring is especially valuable in direct absorption solar collectors, where conventional fluids suffer from low absorption and high transmittance losses.

4.8.2.4 Reduction of Radiative Losses

Hybrid nanofluids can be engineered with low emissivity in the infrared range, reducing heat losses due to re-radiation. This spectral selectivity ensures more energy is retained within the system, further improving thermal efficiency.

4.8.2.5 Impact on Thermal Conversion Efficiency

Experimental studies have shown that using hybrid nanofluids in DASCs can increase optical efficiency by up to 20–30% compared to conventional fluids. Enhanced spectral selectivity minimizes thermal radiation losses, particularly under high-concentration sunlight conditions.

4.8.3 Case Studies and Performance Metrics

4.8.3.1 Case Study 1: Hybrid Nanofluid in Flat Plate Collectors

Study Overview

A hybrid nanofluid consisting of Al_2O_3 and Cu nanoparticles in water was tested in a flat plate collector in India. The study aimed to evaluate the thermal efficiency improvements over traditional water-based systems.

Key Findings

- The collector efficiency increased from 55% (with water) to 68% (with hybrid nanofluid).
- Outlet fluid temperature was higher by ~7–9°C, reducing required auxiliary heating.
- The hybrid nanofluid showed minimal sedimentation after 60 days of continuous operation, demonstrating excellent stability [47].

Implications

Such efficiency improvements translate to significant operational savings and reduced system sizes for given heating demands, enhancing feasibility for residential and commercial solar water heating.

4.8.3.2 Case Study 2: Direct Absorption Solar Collector Using Hybrid Nanofluids

Study Overview

A laboratory-scale DASC was developed using a hybrid nanofluid composed of graphite and silver nanoparticles dispersed in ethylene glycol-water mixture. The hybrid fluid was chosen for its broadband optical absorption.

Key Findings

- Absorptivity reached up to 95% across 0.3–2.5 μm spectral range.
- The maximum collector efficiency exceeded 80% at high flow rates.
- Significant reduction in thermal gradient across the collector, leading to more uniform temperature distribution.

Implications

Such performance enables higher operating temperatures and improved compatibility with high-efficiency thermal storage systems, crucial for off-grid and hybrid PV/T setups.

4.8.3.3 Case Study 3: Parabolic Trough Collector with Hybrid Nanofluid

Study Overview

A parabolic trough collector using a hybrid nanofluid made of MWCNT (multi-walled carbon nanotubes) and SiO_2 nanoparticles in therminol oil was evaluated under high solar flux conditions [48–50].

Key Findings

- Outlet fluid temperature reached 380°C, higher than with pure therminol oil.
- Thermal efficiency improved by 12–18% depending on flow rates and solar intensity.
- The hybrid nanofluid showed stable rheological behavior at high temperatures, with minimal viscosity increase.

Implications

These results support the viability of hybrid nanofluids in CSP applications, allowing for smaller collector fields and higher thermal-to-electric conversion efficiencies.

4.8.3.4 Performance Metrics: Comparative Analysis

To evaluate hybrid nanofluids' performance in solar thermal systems, several key metrics are considered:

1. **Collector efficiency (η)**: Defined as the ratio of useful heat gain to incident solar energy. Hybrid nanofluids generally improve η by 10–30% over water or oil-based systems.
2. **Effective absorptivity ($\alpha_{\_eff}$)**: Measures the fluid's capacity to absorb solar energy across the relevant spectrum. Hybrid nanofluids have demonstrated $\alpha_{\_eff} > 0.9$ in optimized formulations.
3. **Heat removal factor (FR)**: Indicates how effectively the heat is carried away from the collector surface. Higher FR values with hybrid nanofluids reduce stagnation temperatures and improve safety.
4. **Thermal efficiency under varying conditions**: Hybrid nanofluids maintain higher efficiency under fluctuating solar irradiance, wind speeds, and ambient temperatures due to their superior thermal inertia.

5. **Pumping power and thermal performance index (TPI)**: While nanoparticle addition increases viscosity slightly, the overall gain in heat transfer compensates, resulting in TPI values >1.2 in most studies.

4.8.4 Field Demonstrations and Pilot Installations

Pilot projects in various countries have started integrating hybrid nanofluids into solar thermal setups:

- **Spain**: Hybrid nanofluids in PTC fields reduced total collector area by 15% while maintaining the same thermal output.
- **China**: Hybrid nanofluid-based DASCs improved seasonal hot water supply performance by 22% in northern cold climates.
- **Middle East**: Hybrid nanofluids used in hybrid PV/T systems achieved combined electrical and thermal efficiencies up to 75%.

These field data confirm laboratory findings, highlighting practical operational benefits such as reduced system sizes, lower auxiliary energy needs, and longer collector lifespans due to lower thermal cycling stresses.

4.9 APPLICATIONS IN PHOTOVOLTAIC AND HYBRID PV/T SYSTEMS

The world energy system is undergoing a lightning-fast change to decarbonization and sustainability. Photovoltaic (PV) systems are among the renewable technologies that have proved very critical in green electricity production. Nonetheless, these classic PV designs have inherent efficiency limits because of thermal effects of performance degradation and wastage of the solar spectrum. In these, to overcome these difficulties, hybrid photovoltaic/thermal (PV/T) systems have been considered, which integrate the generation of electricity and heat in a combined system.

It has recently been understood that hybrid nanofluids introduce a revolutionary medium that can increase the thermal management and the energy capture efficiency of PV and PV/T systems. This chapter explores their application in cooling photovoltaic panels, efficiency improvements in hybrid PV/T systems, and supporting evidence from experimental and simulation studies.

4.9.1 Cooling of Photovoltaic Panels

4.9.1.1 Temperature Dependence of PV Efficiency

Photovoltaic cells, typically made from crystalline silicon, exhibit a negative temperature coefficient of power output, generally around −0.4 to −0.5% per °C increase in temperature. During peak solar irradiation, PV module temperatures can rise above 60–70°C, leading to significant reductions in electrical efficiency—sometimes by more than 20% compared to standard test conditions (25°C).

Effective cooling strategies are thus crucial to maintain or enhance PV efficiency, prolong module lifespan, and reduce thermal-induced degradation.

4.9.1.2 Traditional Cooling Approaches

Historically, PV module cooling has been attempted through:

- Passive cooling, such as using heat sinks, fins, or natural convection airflow.
- Active air cooling, employing forced air to remove excess heat.
- Active liquid cooling, where a coolant fluid circulates behind the PV panel to absorb and carry away heat.

Liquid cooling generally offers superior thermal performance compared to air-based methods, yet conventional fluids like water and glycol mixtures have limited thermal conductivities and heat capacities, restricting the potential benefits.

4.9.2 Advantages of Hybrid Nanofluids for PV Cooling

Hybrid nanofluids address these limitations through several mechanisms:

1. **Enhanced thermal conductivity**: Facilitates rapid heat extraction from PV back surfaces.
2. **Improved convective heat transfer coefficients**: Supports efficient heat removal even at moderate flow rates.
3. **Broad optical absorption**: Can also serve as spectral filters, selectively absorbing sub-bandgap photons that do not contribute to electrical generation but add thermal load.
4. **Reduced temperature gradients**: Promote more uniform cooling across the PV surface, minimizing thermal stresses and hotspots.

4.9.3 Case Study: Water-Based Al_2O_3–Cu Hybrid Nanofluid

In a comparative study conducted on a rooftop PV installation in Malaysia, a hybrid nanofluid composed of Al_2O_3 and Cu nanoparticles in water (volume fraction ~0.1%) was used for active liquid cooling.

Key Outcomes

- Average PV module temperature was reduced by 15°C compared to uncooled systems.
- Electrical efficiency improved by approximately 10–12%.
- The hybrid nanofluid maintained stable dispersion over 90 days, indicating operational viability.

Operational Benefits

- **Higher energy yield**: Enhanced cooling directly correlates to increased daily and seasonal energy outputs.
- **Increased module lifespan**: Reducing operating temperatures mitigates thermal degradation mechanisms, such as delamination and solder joint fatigue.

- **Reduced maintenance costs**: Hybrid nanofluids can potentially incorporate anti-biofouling and corrosion-resistant properties, further lowering maintenance requirements.

4.9.4 Efficiency Improvements in Hybrid PV/T Systems

4.9.4.1 Principle of PV/T Systems

PV/T systems are designed to harness the entire solar spectrum, converting part into electricity and the rest into usable heat. In a standard PV/T setup, a thermal collector is attached to the backside of the PV module, extracting heat and lowering PV temperatures, thereby increasing electrical efficiency while producing thermal energy for water or air heating.

4.9.4.2 Thermal and Electrical Performance Interplay

The total efficiency of a PV/T system is the sum of its electrical efficiency (η_e) and thermal efficiency (η_{th}). Effective cooling increases η_e while providing valuable thermal output, thereby significantly increasing overall energy utilization.

However, the effectiveness of PV/T systems is often limited by the poor thermal performance of conventional heat transfer fluids.

4.9.4.3 Hybrid Nanofluids: Unlocking New Performance Thresholds

Hybrid nanofluids have been identified as ideal candidates for PV/T systems due to their dual capacity to enhance thermal energy collection and improve electrical output by cooling the PV module.

Key Contributions

1. **Enhanced thermal conductivity and heat capacity**: Enable rapid heat absorption and transfer to storage or usage points, supporting higher thermal efficiencies.
2. **Improved optical properties**: Some hybrid nanofluids act as direct absorbers of infrared and ultraviolet radiation, effectively reducing thermal load on the PV layer.
3. **Lower operating temperatures**: Maintain PV surface temperatures closer to optimal operating conditions, supporting higher electrical efficiency.

Experimental Insights: Hybrid PV/T System with TiO_2-Graphene Nanofluid

A laboratory-scale hybrid PV/T system using a TiO_2-graphene hybrid nanofluid (0.05 vol%) in water was tested under simulated solar irradiation.

Results

- Electrical efficiency improved from 13.2% (water cooling) to 15.4%.
- Thermal efficiency increased from 57% (water) to 63%.
- Overall exergy efficiency rose by 14%.

4.9.4.4 Performance Metrics from Field Installations

In a pilot-scale installation in Spain, a PV/T system employing a Cu-graphene hybrid nanofluid demonstrated:

- Total energy efficiency (electrical + thermal) of 78%, compared to 62% for a similar water-based system.
- Peak PV temperature reductions up to 20°C on sunny summer afternoons.
- Improved daily hot water yield for residential use.

4.9.5 Experimental and Simulation Studies

4.9.5.1 Experimental Studies: Laboratory and Field-Scale

Case 1: Cu-Al_2O_3 Hybrid Nanofluid in PV/T

A study conducted at the Indian Institute of Technology examined a PV/T system cooled by a Cu-Al_2O_3 hybrid nanofluid in water at 0.1 vol%.

Observations

- 16% increase in electrical output
- 20% improvement in thermal energy collection
- Reduction in thermal resistance between PV cells and coolant

Operational Stability

- Stable nanoparticle dispersion for up to 120 days
- Low fouling observed on PV module surfaces, owing to potential self-cleaning effects

---

Case 2: MWCNT-SiO_2 Hybrid Nanofluid in an Air-Cooled PV/T System

In a semi-arid region in northern Africa, a hybrid nanofluid composed of multi-walled carbon nanotubes (MWCNTs) and SiO_2 nanoparticles was utilized in an indirect air-based PV/T system.

Findings

- PV module surface temperature decreased by 18–22°C.
- Average electrical efficiency improved from 10.5% to 13.5%.
- Air outlet temperatures reached up to 60°C, suitable for space heating or agricultural drying.

4.9.5.2 Simulation Studies: Insights into Hybrid Nanofluid Behavior

Numerical simulations using computational fluid dynamics (CFD) and finite element analysis (FEA) have provided detailed insights into hybrid nanofluid dynamics in PV/T systems.

Simulation Parameters Commonly Evaluated

- Flow distribution and mixing in collector channels
- Local and overall heat transfer coefficients
- Temperature distributions across PV cells and heat exchanger surfaces
- Pressure drop and pumping power requirements

Case Study: CFD Analysis of Graphene-Al_2O_3 Hybrid Nanofluid

A 3D CFD model simulated a flat-plate PV/T system using a graphene-Al_2O_3 hybrid nanofluid at various volume fractions.

Results

- Uniform flow distribution improved, reducing hotspots on PV surfaces.
- Heat transfer coefficients increased by 20–35% compared to pure water.
- Predicted electrical efficiency gains of 12–15% under standard irradiance (1000 W/m^2).

4.9.6 Hybrid PV/T System Simulation: Integrated Economic Analysis

A techno-economic simulation in a Mediterranean climate evaluated the payback period and life cycle cost of a PV/T system with a Cu-TiO_2 hybrid nanofluid.

Findings

- Payback period reduced by two years compared to a conventional PV/T system.
- Twenty-year net energy savings increased by 18%.
- CO_2 emission reductions exceeded 25% over the system lifespan.

Synergistic Benefits Identified from Simulations

- **Thermoelectric coupling opportunities**: Hybrid nanofluids can potentially integrate with thermoelectric modules, enhancing hybrid energy harvesting.
- **Dynamic flow optimization**: Smart flow control strategies can dynamically adjust nanofluid velocity to maximize real-time energy capture.
- **Improved seasonal adaptability**: Hybrid nanofluid formulations can be optimized seasonally to balance between heating demands and cooling needs.

4.9.7 Future Prospects and Research Directions

4.9.7.1 Optimizing Hybrid Nanofluid Formulations

Ongoing research aims to:

- Identify optimal nanoparticle combinations for specific climatic and operational conditions.
- Develop environmentally benign, cost-effective nanoparticles for large-scale use.
- Tailor nanoparticle shapes and surface modifications to maximize cooling and optical performance.

4.9.7.2 Hybrid PV/T System Integration with Storage

Combining PV/T systems with hybrid nanofluid-based thermal storage can enable:

- Round-the-clock energy supply.
- Load-shifting capabilities for both electrical and thermal loads.
- Enhanced system flexibility and reliability.

4.9.7.3 Scalability and System Retrofitting

Future work should address:

- Integration of hybrid nanofluids into existing PV and PV/T infrastructures without major redesign.
- Long-term field data to validate laboratory and simulation predictions.
- Development of automated monitoring and control systems for nanofluid health and performance.

4.9.7.4 Environmental and Life cycle Considerations

While hybrid nanofluids offer substantial performance advantages, large-scale deployment requires careful attention to:

- Nanoparticle recovery and recycling strategies.
- Avoiding environmental contamination during operation and disposal.
- Comprehensive life cycle analysis to quantify environmental benefits versus impacts.

The application of hybrid nanofluids in photovoltaic and hybrid PV/T systems represents a significant leap forward in solar energy utilization. By offering superior thermal conductivity, enhanced convective heat transfer, and tailored optical properties, hybrid nanofluids enable effective cooling of PV panels, boosting electrical efficiency and energy yield.

Moreover, their integration into PV/T systems unlocks higher overall energy efficiencies, making dual energy harvesting of electricity and heat more practical and economically viable.

All these advantages are reaffirmed through experimental and simulation work, recording huge improvements in performance indices, lower operational temperatures, and higher daily and seasonal energy productions. Hybrid nanofluid integration has a bright potential to jumpstart the distributed renewable energy systems, especially where the cooling requirement is high, and the solar insolation is high.

Among the current topics of research and improving scalability, cost, and environmental expectations, hybrid nanofluids can become a core technology of the future generation of solar power production systems that contribute to the low-carbon, resilient, and energy-secure future of the planet.

4.10 FUTURE PROSPECTS AND CHALLENGES

4.10.1 Scalability and Economic Considerations

While laboratory and pilot-scale results are promising, large-scale implementation requires addressing:

- **Nanoparticle synthesis costs**: Developing cost-effective, scalable synthesis routes for hybrid nanoparticles is critical.
- **Dispersion and stability management**: Long-term dispersion stability must be ensured under real operating conditions (e.g., freeze-thaw cycles, prolonged stagnation).
- **Compatibility with existing systems**: Retrofitting hybrid nanofluids into current collector designs needs careful material compatibility studies.

4.10.2 Environmental and Health Aspects

The environmental impact of nanoparticle leakage and disposal needs careful consideration. Regulatory frameworks for safe handling and end-of-life management will play an important role in widespread adoption.

4.10.3 Hybridization with Thermal Storage

Future systems are expected to combine hybrid nanofluids with phase change materials (PCMs) and thermochemical storage to create fully integrated solar thermal energy units capable of 24/7 operation. Hybrid nanofluids can facilitate faster heat exchange during storage charging and discharging cycles.

4.10.4 Dynamic and Adaptive Nanofluid Systems

Emerging concepts like magnetically or electrically responsive hybrid nanofluids open new possibilities for adaptive control of flow and heat transfer characteristics. Such smart fluids could dynamically adjust properties in response to solar flux or load variations, further enhancing system flexibility and efficiency.

Hybrid nanofluids present a transformative opportunity to advance solar thermal energy systems, addressing the fundamental limitations of traditional working fluids through superior thermal and optical properties. By enhancing thermal conductivity, convective heat transfer, absorptivity, and spectral selectivity, they enable solar collectors and concentrators to achieve higher efficiency, reduced sizes, and broader operating ranges.

Case studies worldwide demonstrate that hybrid nanofluids consistently outperform conventional fluids across various configurations, including flat plate, evacuated tube, parabolic trough, and direct absorption collectors. The improvements are evident in higher collector efficiencies, elevated outlet temperatures, and improved system reliability.

Future advances will follow the path toward reduced cost per kilowatt-hour of power produced, large-scale stability, energy storage integration, and dynamic control systems that align with the global effort to decarbonize and build sustainable energy infrastructure. Hybrid nanofluids present valuable opportunities to harness the full potential of solar thermal energy in the future and have potential applications in the next several decades of cleaner, more efficient, and more flexible energy systems.

4.11 CONCLUSION

Hybrid nanofluids have progressed from conceptual laboratory formulations to promising candidates for large-scale deployment in renewable energy systems. With continuous advances in emerging materials, surface modifications, and hybrid formulations, their performance is set to reach new heights.

Integration with smart and adaptive energy systems offers dynamic control, higher efficiencies, and operational flexibility, essential for future renewable-dominated grids. Furthermore, a clear commercialization roadmap anchored in robust policy frameworks, global cooperation, and public engagement—is fundamental to transitioning from pilot demonstrations to mainstream applications.

Addressing the remaining challenges in cost, scalability, stability, environmental safety, and long-term life cycle performance will require sustained interdisciplinary collaboration across academia, industry, and government.

Ultimately, the successful integration of hybrid nanofluids into renewable energy and energy storage systems represents more than just an engineering milestone. It symbolizes a crucial step toward a cleaner, smarter, and more resilient global energy future, aligning scientific innovation with societal aspirations for sustainability and equity.

REFERENCES

1. Dorling, D. (2021). World population prospects at the UN: Our numbers are not our problem?. In *The struggle for social sustainability* (pp. 129–154). Policy Press.
2. Yolcan, O. O. (2023). World energy outlook and state of renewable energy: 10-Year Change, I. P. O. C. (2007). Climate change 2007: The physical science basis. *Agenda, 6*(7), 333.
3. Chen, B., & Kan, H. (2008). Air pollution and population health: A global challenge. *Environmental Health and Preventive Medicine, 13*(2), 94–101.
4. Rena, I. (2023). *Renewable power generation costs in 2022*. Abu Dhabi: International Renewable Energy Agency.
5. Dincer, I., & Rosen, M. A. (2011). *Thermal energy storage: Systems and applications*. John Wiley & Sons.
6. Luo, X., Wang, J., Dooner, M., & Clarke, J. (2015). Overview of current development in electrical energy storage technologies and the application potential in power system operation. *Applied Energy, 137*, 511–536.
7. Skoplaki, E., & Palyvos, J. A. (2009). On the temperature dependence of photovoltaic module electrical performance: A review of efficiency/power correlations. *Solar Energy, 83*(5), 614–624.
8. Olatomiwa, L., Mekhilef, S., Ismail, M. S., & Moghavvemi, M. (2016). Energy management strategies in hybrid renewable energy systems: A review. *Renewable and Sustainable Energy Reviews, 62*, 821–835.
9. Sundar, L. S., Singh, M. K., & Sousa, A. C. (2014). Enhanced heat transfer and friction factor of MWCNT–Fe3O4/water hybrid nanofluids. *International Communications in Heat and Mass Transfer, 52*, 73–83.
10. Choi, S. U., & Eastman, J. A. (1995). *Enhancing thermal conductivity of fluids with nanoparticles* (No. ANL/MSD/CP-84938; CONF-951135-29). Argonne National Lab. (ANL), Argonne, IL (United States).
11. Saidur, R., Leong, K. Y., & Mohammed, H. A. (2011). A review on applications and challenges of nanofluids. *Renewable and Sustainable Energy Reviews, 15*(3), 1646–1668.
12. Sarkar, J., Ghosh, P., & Adil, A. (2015). A review on hybrid nanofluids: Recent research, development and applications. *Renewable and Sustainable Energy Reviews, 43*, 164–177.
13. Sundar, L. S., Ramana, E. V., Graça, M. P. F., Singh, M. K., & Sousa, A. C. (2016). Nanodiamond-Fe3O4 nanofluids: Preparation and measurement of viscosity, electrical and thermal conductivities. *International Communications in Heat and Mass Transfer, 73*, 62–74.
14. Mahian, O., Kianifar, A., Kalogirou, S. A., Pop, I., & Wongwises, S. (2013). A review of the applications of nanofluids in solar energy. *International Journal of Heat and Mass Transfer, 57*(2), 582–594.
15. Tohver, V., Chan, A., Sakurada, O., & Lewis, J. A. (2001). Nanoparticle engineering of complex fluid behavior. *Langmuir, 17*(26), 8414–8421.
16. Manimaran, M., Norizan, M. N., Kassim, M. H. M., Adam, M. R., Abdullah, N., & Norrrahim, M. N. F. (2025). Critical review on the stability and thermal conductivity of water-based hybrid nanofluids for heat transfer applications. *RSC Advances, 15*(18), 14088–14125.
17. Said, Z., & Sohail, M. A. (2022). Introduction to hybrid nanofluids. In *Hybrid nanofluids* (pp. 1–32). Elsevier.
18. Sarkar, J., Ghosh, P., & Adil, A. (2015). A review on hybrid nanofluids: Recent research, development and applications. *Renewable and Sustainable Energy Reviews, 43*, 164–177.

19. Septiadi, W. N., Trisnadewi, I. A. N. T., Putra, N., & Setyawan, I. (2018). Synthesis of hybrid nanofluid with two-step method. In *E3S web of conferences* (Vol. 67, p. 03057). EDP Sciences.
20. Manimaran, M., Norizan, M. N., Kassim, M. H. M., Adam, M. R., Abdullah, N., & Norrahim, M. N. F. (2025). Critical review on the stability and thermal conductivity of water-based hybrid nanofluids for heat transfer applications. *RSC Advances*, *15*(18), 14088–14125.
21. Hemmat Esfe, M., Saedodin, S., Yan, W. M., Afrand, M., & Sina, N. (2016). Study on thermal conductivity of water-based nanofluids with hybrid suspensions of CNTs/Al2O3 nanoparticles. *Journal of Thermal Analysis and Calorimetry*, *124*(1), 455–460.
22. Kadhim, S. A., Ibrahim, O. A. A. M., Al-Ghezi, M. K., & Ashour, A. M. (2025). Critical review of the use of TiO_2 nanofluid as a heat transfer fluid in the double-pipe heat exchanger. *Journal of Thermal Analysis and Calorimetry*, 1–21.
23. Soltani, O., & Akbari, M. (2016). Effects of temperature and particles concentration on the dynamic viscosity of MgO-MWCNT/ethylene glycol hybrid nanofluid: Experimental study. *Physica E: Low-Dimensional Systems and Nanostructures*, *84*, 564–570.
24. Suresh, S., Venkitaraj, K. P., Selvakumar, P., & Chandrasekar, M. (2012). Effect of Al2O3–Cu/water hybrid nanofluid in heat transfer. *Experimental Thermal and Fluid Science*, *38*, 54–60.
25. Kadhim, S. A., Abdul-Ghafoor, Q. J., Hammoodi, K. A., Abdul Wahhab, H. A., Ibrahim, O. A. A. M., & Al-Ghezi, M. K. (2025). Enhancement of the heat transfer performance of a double-pipe heat exchangers in various configurations using Al_2O_3 nanofluid: A focused review. *Journal of Thermal Analysis and Calorimetry*, 1–25.
26. dos Santos, F. S., Amjad, M., Wen, D., & Bandarra Filho, E. P. (2025). Experimental study of thermohydraulic performance of nanofluid flowing in horizontal tubes with sudden contraction. *Powder Technology*, *452*, 120596. https://doi.org/10.1016/j.powtec.2024.120596
27. Minea, A. A. (2020). Pumping power and heat transfer efficiency evaluation on Al2O3, TiO2 and SiO2 single and hybrid water-based nanofluids for energy application. *Journal of Thermal Analysis and Calorimetry*, *139*(2), 1171–1181.
28. Xian, H. W., Sidik, N. A. C., & Saidur, R. (2020). Impact of different surfactants and ultrasonication time on the stability and thermophysical properties of hybrid nanofluids. *International Communications in Heat and Mass Transfer*, *110*, 104389.
29. Kadhim, S. A., Ibrahim, O. A. A. M., Al-Ghezi, M. K., & Ashour, A. M. (2025). Critical review of the use of TiO_2 nanofluid as a heat transfer fluid in the double-pipe heat exchanger. *Journal of Thermal Analysis and Calorimetry*, 1–21.
30. Manimaran, M., Norizan, M. N., Kassim, M. H. M., Adam, M. R., Abdullah, N., & Norrahim, M. N. F. (2025). Critical review on the stability and thermal conductivity of water-based hybrid nanofluids for heat transfer applications. *RSC Advances*, *15*(18), 14088–14125.
31. Giwa, S. O., Sharifpur, M., Ahmadi, M. H., Sohel Murshed, S. M., & Meyer, J. P. (2021). Experimental investigation on stability, viscosity, and electrical conductivity of water-based hybrid nanofluid of MWCNT-Fe2O3. *Nanomaterials*, *11*(1), 136.
32. Bethi, B., Sonawane, S. H., Bhanvase, B. A., & Sonawane, S. S. (2021). Textile industry wastewater treatment by cavitation combined with Fenton and ceramic nanofiltration membrane. *Chemical Engineering and Processing-Process Intensification*, *168*, 108540.
33. Malika, M., & Sonawane, S. S. (2020). Effect of nanoparticle mixed ratio on stability and thermo-physical properties of CuO-ZnO/water-based hybrid nanofluid. *Journal of the Indian Chemical Society*, *97*(3), 414–9.

34. Minea, A. A. (2020). Pumping power and heat transfer efficiency evaluation on Al2O3, TiO2 and SiO2 single and hybrid water-based nanofluids for energy application. *Journal of Thermal Analysis and Calorimetry*, *139*(2), 1171–1181.
35. Sundar, L. S., Singh, M. K., & Sousa, A. C. (2014). Enhanced heat transfer and friction factor of MWCNT–Fe3O4/water hybrid nanofluids. *International Communications in Heat and Mass Transfer*, *52*, 73–83.
36. Malika, M. M., & Sonawane, S. S. (2019). Review on application of nanofluid/nano particle as water disinfectant. *Journal of Indian Association for Environmental Management (JIAEM)*, *39*(1–4), 21–24.
37. Malika, M., Bhad, R., & Sonawane, S. S. (2021). ANSYS simulation study of a low volume fraction CuO–ZnO/water hybrid nanofluid in a shell and tube heat exchanger. *Journal of the Indian Chemical Society*, *98*(11), 100200.
38. Sahoo, R. R. (2020). Experimental study on the viscosity of hybrid nanofluid and development of a new correlation. *Heat and Mass Transfer*, *56*(11), 3023–3033.
39. Sreekumar, S., Shah, N., Mondol, J. D., Hewitt, N., & Chakrabarti, S. (2022). Broadband absorbing mono, blended and hybrid nanofluids for direct absorption solar collector: A comprehensive review. *Nano Futures*, *6*(2), 022002.
40. Malika, M., & Sonawane, S. S. (2021). The sono-photocatalytic performance of a novel water based Ti+ 4 coated Al (OH) 3-MWCNT's hybrid nanofluid for dye fragmentation. *International Journal of Chemical Reactor Engineering*, *19*(9), 901–912.
41. Parate, V. R., Kawadkar, D. K., & Sonawane, S. S. (2011). Study of whey protein concentrate fortification in cookies variety biscuits. *International Journal of Food Engineering*, *7*(2), 1–12.
42. Jin, X., Lin, G., Jin, H., Fu, Z., & Sun, H. (2021). Experimental research on the selective absorption of solar energy by hybrid nanofluids. *Energies*, *14*(23), 8186.
43. Sreekumar, S., Shah, N., Mondol, J. D., Hewitt, N., & Chakrabarti, S. (2022). Broadband absorbing mono, blended and hybrid nanofluids for direct absorption solar collector: A comprehensive review. *Nano Futures*, *6*(2), 022002.
44. Thakur, P., Sonawane, S. S., Sonawane, S. H., & Bhanvase, B. A. (2020). Nanofluids-based delivery system, encapsulation of nanoparticles for stability to make stable nanofluids. In *Encapsulation of active molecules and their delivery system* (pp. 141–152). Elsevier.
45. Sonawane, S., Thakur, P., & Paul, R. (2020). Study on thermal property enhancement of MWCNT based polypropylene (PP) nanocomposites. *Materials Today: Proceedings*, *27*, 550–555.
46. Struchalin, P. G., Kuzmenkov, D. M., Yunin, V. S., Wang, X., He, Y., & Balakin, B. V. (2022). Hybrid nanofluid in a direct absorption solar collector: Magnetite vs. carbon nanotubes compete for thermal performance. *Energies*, *15*(5), 1604.
47. Sattar, A., Farooq, M., Amjad, M., Saeed, M. A., Nawaz, S., Mujtaba, M. A., . . . & Pettinau, A. (2020). Performance evaluation of a direct absorption collector for solar thermal energy conversion. *Energies*, *13*(18), 4956.
48. Thakur, P., Sonawane, S., Potoroko, I., & Sonawane, S. H. (2021). Recent advances in ultrasound-assisted synthesis of nano-emulsions and their industrial applications. *Current Pharmaceutical Biotechnology*, *22*(13), 1748–1758.
49. Sonawane, A. V., Rikame, S., Sonawane, S. H., Gaikwad, M., Bhanvase, B., Sonawane, S. S., . . . & Gaikwad, R. (2024). A review of microbial fuel cell and its diversification in the development of green energy technology. *Chemosphere*, *350*, 141127.
50. Syam Sundar, L., Mesfin, S., Tefera Sintie, Y., Punnaiah, V., Chamkha, A. J., & Sousa, A. (2021). A review on the use of hybrid nanofluid in a solar flat plate and parabolic trough collectors and its enhanced collector thermal efficiency. *Journal of Nanofluids*, *10*(2), 147–171.

5 Mathematical Modeling of Hybrid Nanofluids and Applications for Energy Conservation

5.1 INTRODUCTION TO MATHEMATICAL MODELING IN ENERGY SYSTEMS

There has been a steep transfer of renewable and sustainable energy solutions, which has given rise to the need to develop more cutting-edge thermal management and energy conversion systems [1]. The main revolution required around this is the use of highly efficient heat transfer fluids, of which hybrid nanofluids have emerged as potential alternatives [2]. The intricate nature of fluids makes them very difficult to comprehend, foresee, and even optimize their efficiency in real systems [3]. The challenges can be addressed and overcome by means of mathematical modeling, which gives a comprehensive, methodical, and quantitative description of the underlying physical phenomena [4].

In the following chapter, an integrated understanding of mathematical modeling within energy systems is given concerning the hybrid application of nanofluids [5]. It speaks of the vitality of modeling in the study of the behavior of nanofluids, its critical consideration in the design of a system, and the optimization of system performance, and particularly challenges linked to modeling of hybrid nanofluids [6].

5.1.1 Complexity of Nanofluid Thermophysical Behavior

Nanofluids Engineered colloidal suspensions of nanoparticles dispersed in base fluids exhibit an increased thermal conductivity, higher thermal convective heat transfer coefficients, and possibly greater stability than those of conventional fluids [2]. The complexity is multiplied when transferring to the hybrid nanofluids, when two or more kinds of nanoparticles are used in conjunction [7]. Complex interactions between nanoparticles of different types and base fluid emerge.

Thermophysical properties of hybrid nanofluids are affected by a variety of factors:

- **Particle material and shape**: Metallic, metal oxide, or carbon-based particles can dramatically influence thermal properties and flow behavior.
- **Particle size and distribution**: Nanoscale effects such as Brownian motion and thermophoresis impact heat transfer and stability.

 DOI: 10.1201/9781003628071-5

- **Concentration and mixture ratios**: The volume fraction and relative composition of different nanoparticles affect overall thermal conductivity and viscosity.
- **Interfacial phenomena**: Interactions at the particle–fluid interface play critical roles in determining effective properties.

Mathematical modeling allows researchers and engineers to capture these interdependent factors and understand their contributions to overall system behavior [4].

5.1.2 Insight into Fundamental Transport Mechanisms

- Mathematical models allow one to perform dissection and analysis of fundamental processes in transport, for example:
- **Conduction:** The increased efficacious thermal conduction because of highly conductive nanoparticles [8].
- **Convection:** Variations of the convective heat transfer by altered fluid characteristics and fluid flow pattern [3].
- **Radiation**: The potential role of nanoparticles in absorbing or scattering radiative heat, especially in solar applications [5].

Models provide quantitative predictions that are essential for validating theoretical assumptions and for guiding experimental studies [6]. For example, flow between two parallel plates is modeled as shown in Figure 5.1.

5.1.3 Reduction of Experimental Burden

Experimental investigations of nanofluids, particularly hybrid systems, are time-consuming and resource-intensive [1]. They require precise control of synthesis,

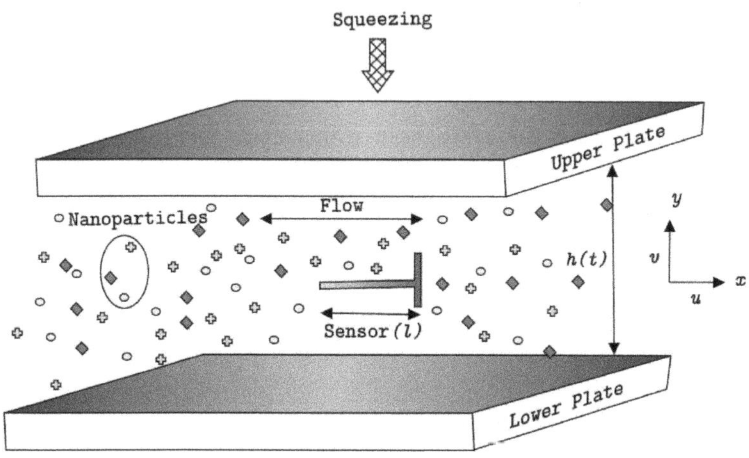

FIGURE 5.1 Coordinate system and conceptual model for flow problem [6].

extensive characterization, and multiple performance tests under varying conditions [7]. Modeling reduces this burden by:

- Allowing virtual screening of different nanoparticle combinations.
- Predicting thermophysical properties and flow behavior prior to synthesis.
- Identifying optimal operational ranges and conditions.

This approach not only saves cost but also accelerates the innovation cycle for new hybrid nanofluid formulations [4].

5.1.4 Support for Scale-Up and System Integration

Nanofluid studies often start at the laboratory scale, but the eventual goal is deployment in large-scale energy systems (e.g., solar thermal plants, geothermal loops) [5]. Mathematical models facilitate the translation of laboratory findings to industrial scales by:

- Predicting scale-dependent phenomena such as flow regime transitions and thermal boundary layer development [8].
- Simulating full-system behavior under realistic operational constraints [6].
- Assessing the impact of nanofluid properties on overall energy system performance and reliability [3].

5.2 ROLE IN DESIGN, OPTIMIZATION, AND PREDICTIVE PERFORMANCE

5.2.1 System Design and Component Optimization

Energy systems incorporating hybrid nanofluids must be carefully designed to harness their potential benefits. Mathematical models are critical tools for this process [9].

5.2.1.1 Heat Exchanger Design

Hybrid nanofluids can significantly increase heat exchanger performance, but they also alter flow dynamics and pressure drop characteristics. Modeling helps:

- Determine optimal heat exchanger geometries and configurations [10].
- Predict trade-offs between heat transfer enhancement and pumping power requirements [11].
- Identify fouling or sedimentation risks and design mitigation strategies [12].

5.2.1.2 Solar Thermal Collectors

For solar collectors, hybrid nanofluids offer improved solar absorption and thermal conversion efficiency [13]. Mathematical models can:

- Analyze the optical properties and absorptivity enhancements offered by different nanoparticle combinations.
- Simulate transient thermal responses to fluctuating solar irradiation [14].
- Optimize flow rates, channel designs, and receiver geometries for maximum efficiency.

5.2.1.3 Energy Storage Systems

In thermal energy storage systems—whether sensible, latent, or thermochemical—hybrid nanofluids improve heat exchange rates and storage capacity. Models enable:

- Prediction of charging and discharging dynamics [15].
- Analysis of phase change kinetics when used with PCMs.
- Evaluation of cyclic stability and degradation effects over extended operation [16].

5.2.2 System-Level Optimization and Control

Beyond component-level design, entire energy systems can be optimized using mathematical models incorporating hybrid nanofluid dynamics [9].

5.2.2.1 Multi-Objective Optimization

Energy systems often involve trade-offs among performance metrics such as thermal efficiency, pressure drop, capital cost, and operational safety. Mathematical modeling supports multi-objective optimization, allowing designers to identify Pareto-optimal solutions that balance these competing goals [11].

5.2.2.2 Dynamic and Transient Analyses

Hybrid nanofluids may exhibit nonlinear thermal responses and time-dependent behavior under transient conditions (e.g., startup, load changes, intermittent solar inputs) [14]. Models enable:

- Simulation of system responses to transient disturbances.
- Design of robust control strategies to maintain stability and performance [9].
- Assessment of thermal lag and storage delays [15].

5.2.2.3 Predictive Maintenance and Fault Detection

By integrating mathematical models into control systems, predictive maintenance schemes can be implemented [12]. Models can:

- Forecast degradation of nanofluid properties (e.g., due to agglomeration or chemical reactions) [16].
- Detect anomalies such as unexpected temperature gradients or flow maldistribution [12].
- Provide early warnings to prevent performance drops and avoid costly shutdowns.

5.2.3 Performance Prediction and Techno-Economic Analysis

Before investing in full-scale deployment, stakeholders require reliable predictions of system performance and economic feasibility [9]. Mathematical models provide the foundation for such predictions.

5.2.3.1 Levelized Cost of Energy (LCOE) Estimation

Models help estimate how hybrid nanofluids impact LCOE by quantifying improvements in thermal efficiency, reductions in auxiliary energy consumption, and potential savings from smaller system footprints [11].

5.2.3.2 Sensitivity and Uncertainty Analyses

By systematically varying input parameters, models can identify:

- Most influential design and operational parameters [14].
- Robust operational windows that maintain high performance under uncertainties [13].
- Economic thresholds for nanoparticle cost, replacement frequency, and maintenance.

5.2.3.3 Life cycle Assessments (LCA)

Mathematical models also integrate environmental considerations, helping to:

- Quantify total energy and emissions savings across the life cycle [16].
- Compare environmental footprints of different hybrid nanofluid formulations.
- Support regulatory compliance and sustainability reporting.

5.3 SPECIFIC CHALLENGES IN HYBRID NANOFLUIDS

While mathematical modeling offers powerful benefits, hybrid nanofluids introduce several unique challenges that complicate accurate prediction and optimization [17].

5.3.1 Non-Homogeneous and Dynamic Dispersion

Unlike simple single-phase fluids, hybrid nanofluids involve two or more types of nanoparticles, which may have different densities, sizes, and surface charges [18]. This introduces complexities such as:

- Non-uniform dispersion and possible stratification.
- Dynamic aggregation and de-aggregation processes during operation.
- Differential settling rates, affecting local thermophysical properties [19].

Modeling these phenomena requires advanced multiphase flow frameworks and sophisticated particle interaction models, often coupled with real-time experimental data for validation [20].

5.3.2 Rheological Complexities

Hybrid nanofluids often exhibit non-Newtonian behavior, especially at higher nanoparticle concentrations or under strong shear conditions [21]. Such behaviors include:

- Shear thinning or thickening, which affects pumping power and flow distribution.
- Viscoelastic effects that influence transient thermal responses [22].

Accurately modeling these rheological properties necessitates advanced constitutive equations and empirical validation.

5.3.3 Thermophysical Property Prediction

Predicting the effective thermal conductivity, viscosity, specific heat capacity, and density of hybrid nanofluids is challenging because:

- Classical models (e.g., Maxwell, Brinkman) were developed for dilute, single-type nanoparticle systems and do not directly account for synergistic effects in hybrid mixtures [23].
- Inter-particle interactions, nanoparticle shape, and clustering dynamics strongly influence effective properties.
- Temperature dependence introduces additional nonlinearity.

To address this, hybrid models often combine theoretical, empirical, and data-driven approaches, each with inherent limitations and assumptions.

5.3.4 Agglomeration and Sedimentation Modeling

Agglomeration leads to changes in effective particle size and concentration, reducing thermal performance and potentially causing operational failures [24]. Modeling these dynamics involves:

- Population balance equations (PBE) to describe particle size distribution evolution.
- Coupled hydrodynamic and thermodynamic models to simulate sedimentation effects.
- Consideration of interparticle forces, including van der Waals attraction, electrostatic repulsion, and steric stabilization.

These models require high computational resources and extensive experimental calibration.

5.3.5 Interfacial and Surface Phenomena

In hybrid nanofluids, interfacial phenomena are critical, especially when particles interact with system surfaces or each other [25]. Challenges include:

- Modeling thermal boundary resistance (Kapitza resistance) at particle–fluid and particle–solid interfaces.
- Capturing Marangoni effects in systems with strong temperature gradients.
- Accounting for chemical reactions or fouling layers on nanoparticle surfaces during extended operation.

These phenomena introduce additional coupled terms in the governing equations, complicating numerical solution strategies.

5.3.6 Numerical Challenges and Computational Cost

High-fidelity models for hybrid nanofluids involve:

- Multi-phase, multi-component transport equations.
- Temperature- and shear-dependent property variations.
- Fine spatial and temporal resolution to capture localized effects [26].

Consequently, simulations become computationally intensive, requiring advanced numerical methods (e.g., adaptive meshing, parallel computing) and high-performance computing resources.

5.3.7 Lack of Standardized Validation Protocols

Unlike traditional fluids, there are no universally accepted standards for validating hybrid nanofluid models [27]. This leads to:

- Discrepancies between different experimental and numerical studies.
- Difficulty in comparing results across research groups.
- Challenges in establishing confidence for industrial adoption.

Developing harmonized experimental protocols and standardized benchmark problems is essential to advance modeling reliability.

5.3.8 Environmental and Life cycle Uncertainties

Hybrid nanofluids may degrade over time through chemical reactions, nanoparticle oxidation, or loss of dispersion stability [28]. Modeling these long-term effects requires integrating:

- Chemical kinetics models for nanoparticle reactions.
- Long-term performance degradation curves.
- Life cycle cost and environmental impact models.

This multi-scale, multi-disciplinary modeling effort is still in its infancy and remains a major research frontier.

5.4 GOVERNING EQUATIONS FOR HYBRID NANOFLUID FLOW AND HEAT TRANSFER

5.4.1 Conservation of Mass (Continuity Equation)

The conservation of mass is expressed through the continuity equation, which ensures that the mass is neither created nor destroyed within a control volume.

For an incompressible fluid, which is a reasonable assumption for most nanofluid applications, the continuity equation is given by:

$$\nabla \cdot \mathrm{v} = 0$$

where v is the velocity vector field [25].

For compressible flows or if density variations are significant (as can occur in high-temperature applications or if large temperature gradients exist), the general form is used:

$$\frac{\partial \rho}{\partial t} + \nabla \cdot (\rho v) = 0$$

Here, ρ represents the density of the hybrid nanofluid, which is a function of both the base fluid and the dispersed nanoparticles.

5.4.2 Conservation of Momentum (Navier–Stokes Equations)

The momentum equation is derived from Newton's second law and describes the motion of fluid particles under various forces. For an incompressible Newtonian hybrid nanofluid, the equation takes the form:

$$\rho \left(\frac{\partial v}{\partial t} + v \cdot \nabla v \right) = -\nabla p + \mu \nabla^2 v + \rho g + F_{extra}$$

where:

- p is the pressure.
- μ is the dynamic viscosity of the hybrid nanofluid.
- g is the gravitational acceleration vector.
- F_{extra} represents any additional body forces, such as those arising from particle interactions or electromagnetic fields [26].

For hybrid nanofluids, the effective viscosity μ\mu and density ρ\rho are functions of nanoparticle concentrations and compositions, requiring careful evaluation.

5.4.3 Conservation of Energy

The energy equation expresses the balance of thermal energy within a control volume. In hybrid nanofluids, the energy conservation equation can be formulated as:

$$\rho C_p\left(\frac{\partial T}{\partial t}+v\cdot\nabla T\right)=\nabla\cdot\left(k\nabla T\right)+Q$$

where:

- C_p is the specific heat capacity at constant pressure.
- T is the temperature field.
- k is the effective thermal conductivity.
- Q represents internal heat generation (e.g., from radiation absorption in solar collectors).

The presence of nanoparticles significantly enhances kk and modifies C_p, thus influencing heat conduction and storage capabilities [27].

5.5 MODIFICATIONS TO ACCOUNT FOR NANOPARTICLE INTERACTIONS

5.5.1 Effective Thermophysical Property Modeling

A primary challenge in hybrid nanofluid modeling is accurately representing their effective thermophysical properties, which differ substantially from those of base fluids and single-component nanofluids.

5.5.1.1 Effective Density

For hybrid nanofluids, effective density is generally modeled as:

$$\rho_{hnf}=\left(1-\phi\right)_{\rho_f}+\phi_1\rho_{p_1}+\phi_2\rho_{p2}$$

where:

- $\phi=\phi_1+\phi_2$ is the total solid volume fraction.
- ϕ_1, ϕ_2 are the volume fractions of the two types of nanoparticles.
- ρ_f is the base fluid density.
- ρ_{p1}, ρ_{p2} are the densities of nanoparticle types 1 and 2 [28].

5.5.1.2 Effective Specific Heat Capacity

The hybrid nanofluid's effective specific heat capacity at constant pressure is given by:

$$\left(\rho c_p\right)_{hnf}=\left(1-\phi\right)\left(\rho cp\right)_f+\phi_1\left(\rho cp\right)p_1+\phi_2\left(\rho cp\right)p_2$$

This relationship indicates the combined heat storage ability of the mixture [28].

5.5.1.3 Effective Viscosity

Viscosity is critically important because it directly affects pumping power and flow characteristics. Hybrid nanofluids tend to have higher viscosities than their base fluids. Empirical or semi-empirical correlations are typically used, such as:

$$\mu_{hnf} = \mu_f\left(1 + 2.5\phi + \beta\phi_2\right)$$

where β is a parameter that accounts for inter-particle interactions and clustering, which are more significant in hybrid nanofluids [29].

5.5.1.4 Effective Thermal Conductivity

Many models extend classical theories (e.g., Maxwell model) to hybrid systems. A simplified form is:

$$k_{hnf} = k_f\left[\frac{k_p + 2k_f - 2\phi\left(k_f - k_p\right)}{k_p + 2k_f + \phi\left(k_f - k_p\right)}\right]$$

For hybrid nanofluids, this expression becomes more complex as it involves contributions from each nanoparticle type and their synergistic effects [30].

5.5.2 Brownian Motion and Thermophoresis

Hybrid nanofluids exhibit enhanced thermal transport partly due to nanoparticle Brownian motion and thermophoresis (movement under temperature gradients).

5.5.2.1 Brownian Motion Term

A diffusive flux JB due to Brownian motion can be added [31]:

$$J_B = -D_B \nabla\phi$$

where D_B is the Brownian diffusion coefficient, often expressed as:

$$D_B = \frac{K_B T}{3\pi\mu d_p}$$

Here, kB is the Boltzmann constant and dp is the particle diameter.

5.5.2.2 Thermophoretic Term

The thermophoretic flux JT can be modeled as:

$$J_T = -D_T \nabla T$$

where D_T is the thermophoretic coefficient. Inclusion of these terms modifies the nanoparticle concentration distribution and, therefore, the local thermophysical properties [32].

5.5.3 Particle–Particle and Particle–Fluid Interactions

Hybrid nanofluids experience complex particle–particle and particle–fluid interactions, which can lead to aggregation, clustering, or migration effects. Modeling these requires:

- Incorporating interaction potentials (e.g., DLVO theory).
- Population balance equations (PBE) to track size distribution dynamics.
- Additional source or sink terms in mass and momentum conservation equations to account for aggregation or sedimentation.

5.5.4 Incorporation of Buoyancy and Thermophysical Property Variations

5.5.4.1 Buoyancy Effects (Boussinesq Approximation)

In many energy systems, hybrid nanofluids operate under conditions with substantial temperature gradients, leading to natural convection driven by buoyancy forces. The Boussinesq approximation simplifies the treatment of buoyancy effects by assuming that density variations are small and only significant in the buoyancy term of the momentum equation.

The modified momentum equation with buoyancy becomes:

$$\rho\left(\frac{\partial v}{\partial t}+v\cdot\nabla v\right)=-\nabla p+\mu\nabla^2 v+\rho g\left[1-\beta\left(T-T_0\right)\right]$$

where

- β is the thermal expansion coefficient of the hybrid nanofluid.
- T0 is the reference temperature.

The coefficient β\beta can be approximated as a weighted combination of base fluid and nanoparticle contributions [33].

5.5.4.2 Variable Thermophysical Properties

In hybrid nanofluids, properties such as thermal conductivity, viscosity, and specific heat are highly sensitive to local temperature, particle concentration, and shear rate [34]. Including variable properties improves prediction accuracy but adds complexity.

5.5.4.2.1 Temperature-Dependent Viscosity

Viscosity may vary significantly with temperature, modeled as:

$$\mu\left(T\right)=\mu_0 exp\left(\frac{B}{T-Ts}\right)$$

where μ_0 and B are empirical constants, and T_s is a characteristic temperature [35].

5.5.4.2.2 Temperature- and Concentration-Dependent Thermal Conductivity

Hybrid nanofluids can exhibit thermal conductivity enhancements with temperature increases due to intensified Brownian motion [36]. A general form is:

$$k(T,\phi)=k_f(T)\left(1+a_1\phi+a_2\phi^2+a_T(T-T_0)\right)$$

where a_1, a_2, a_T are empirical constants determined experimentally.

5.5.4.2.3 Non-Newtonian Behavior and Shear Rate Dependence

At higher concentrations or specific hybrid formulations, hybrid nanofluids may behave as non-Newtonian fluids. Their apparent viscosity μ_a can be modeled using a power-law:

$$\mu_a=k'\gamma^{n-1}$$

where:

- k′ is the consistency index.
- n is the flow behavior index ($n < 1$ for shear-thinning, $n > 1$ for shear-thickening).
- γ is the shear rate.

Incorporating such rheological models into the governing equations requires numerically solving highly nonlinear coupled partial differential equations [37]. Figure 5.2 shows variation in power law for shear thinning, shear thickening, and Newtonian fluid conditions.

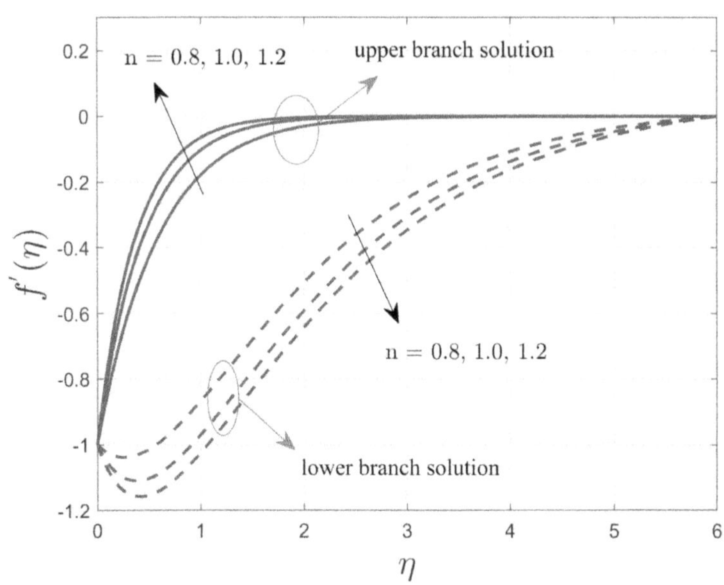

FIGURE 5.2 Variation of power law at different fluid conditions [2].

5.5.4.3 Modified Boundary Conditions

Due to property variations and buoyancy, boundary conditions must be carefully defined [38]:

- **Velocity boundary conditions**: No-slip condition at solid walls, possibly including slip terms if micro/nanoscale effects are relevant.
- **Thermal boundary conditions**: Constant heat flux or constant wall temperature, accounting for nanoparticle deposition or surface fouling.
- **Nanoparticle concentration boundary conditions**: Zero flux at solid boundaries if particle deposition is neglected, or specified flux to model deposition rates.

5.5.5 Grashof and Rayleigh Numbers for Hybrid Nanofluids

Buoyancy-driven flow characterization in hybrid nanofluids uses modified Grashof and Rayleigh numbers [39]:

$$Gr = \frac{g\beta\Delta TL^3}{\nu^2}$$

where ν is the kinematic viscosity, and Pr is the Prandtl number, $\Pr = \nu/\alpha$, with α being thermal diffusivity. Hybrid nanofluids often exhibit lower ν and higher k, resulting in lower Prandtl numbers and modified convective behavior. Rayleigh's number is often used to understand the combined effect of Prandtl number and Grashof number. Rayleigh's number is product of Grashof number and Prandtl number. Figure 5.3 shows the variation of Nusselt number with respect to Rayleigh's number (Gr × Pr).

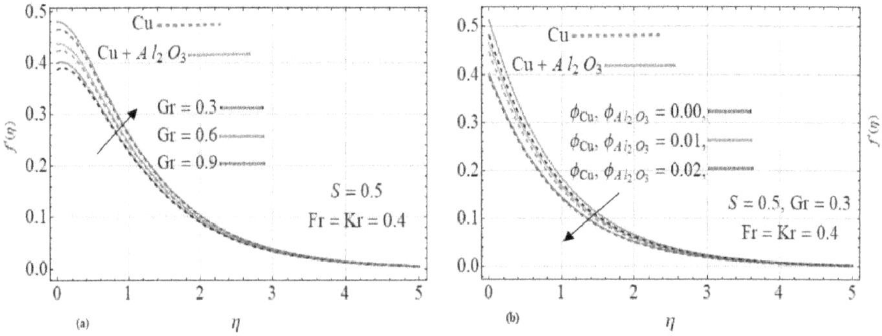

FIGURE 5.3 Variation of Nusselt number with respect to Rayleigh's number (a) for laminar flow (b) for turbulent flow [39].

5.6 NUMERICAL IMPLEMENTATION AND SOLUTION CONSIDERATIONS

5.6.1 Discretization Strategies

The complex coupled equations governing hybrid nanofluids require robust numerical schemes [40]:

- **Finite volume method (FVM)**: Widely used for its conservation properties.
- **Finite element method (FEM)**: Effective for complex geometries and non-linear property variations.
- **Spectral methods**: Suitable for simplified, highly regular geometries with smooth property variations.

5.6.2 Convergence and Stability Issues

The strong non-linearity introduced by temperature-dependent and concentration-dependent properties, coupled with buoyancy effects, can cause convergence challenges [41]. Strategies to address this include:

- Under-relaxation of coupled variables.
- Adaptive time stepping in transient simulations.
- Use of preconditioners in linear solvers.

5.6.3 Validation against Experimental Data

Models must be validated using experimental data on hybrid nanofluid flow and heat transfer [42]. Key performance metrics for validation include:

- Nusselt number (heat transfer enhancement).
- Pressure drop and flow resistance.
- Nanoparticle concentration distribution (via sampling or optical techniques).

5.7 THERMOPHYSICAL PROPERTY MODELS FOR HYBRID NANOFLUIDS

5.7.1 Maxwell Model

The Maxwell model is the classical starting point for estimating the effective thermal conductivity of dilute suspensions of spherical particles. For a single-component nanofluid, the Maxwell equation is [30]:

$$k_{hnf} = k_f \left[\frac{k_p + 2k_f - 2\phi\left(k_f - k_p\right)}{k_p + 2k_f + \phi\left(k_f - k_p\right)} \right]$$

For hybrid nanofluids, the Maxwell model can be extended by introducing an equivalent effective thermal conductivity of the combined solid phase. A simple weighted approach is:

5.7.2 Hamilton–Crosser Model

The Hamilton–Crosser model modifies the Maxwell model to account for nonspherical particle shapes using a shape factor n [43]:

$$knf = \left[\frac{kp + (n-1)kf - (n-1)\phi(kf - kp)}{kp + (n-1)kf + \phi(kf - kp)}\right]kf$$

Here, n is given by: $n = 3/\psi$

where ψ is the particle sphericity (the ratio of surface area of a sphere with the same volume to the actual particle surface area). Lower ψ values correspond to more elongated or irregular shapes, which influence heat transfer pathways.

In hybrid nanofluids, each nanoparticle type may have a different shape, requiring careful averaging or separate treatment in the overall model.

5.7.3 Nanolayer Theories

Nanolayer models introduce the concept of an interfacial nanolayer surrounding each nanoparticle, which possesses different thermal conductivity compared to both the base fluid and the solid nanoparticle.

The nanolayer effect is particularly relevant at high surface-to-volume ratios typical of nanoparticles and is significant in hybrid systems where two nanoparticle types may interact differently with the base fluid.

A representative expression for effective thermal conductivity considering the nanolayer is [44]:

$$k_{eff} = k_f\left[\frac{kp + 2kf + 2\beta(kl - kf)}{kp + 2kf - \beta(kl - kf)}\right]$$

where:

- k_l is the thermal conductivity of the nanolayer.
- β is a parameter representing the nanolayer thickness relative to the particle diameter.

This model can be extended to hybrid nanofluids by introducing different nanolayer parameters for each nanoparticle type and considering interaction effects between layers.

5.7.4 Empirical and Semi-Empirical Models

Due to the complex interactions in hybrid nanofluids, purely theoretical models often fail to capture observed enhancements accurately. Researchers have developed empirical correlations based on experimental data, typically of the form:

$$k_{hnf} = k_f \left(1 + b_1\phi_1 + b_2\phi_2 + b_3\phi_1\phi_2\right)$$

where b_1, b_2, b_3 are empirical constants determined experimentally for specific nanoparticle combinations.

These models are widely used in engineering design, provided the empirical constants are obtained for the particular hybrid formulation under investigation [45].

5.7.5 Specific Heat Capacity, Density, and Viscosity Models

Besides thermal conductivity, accurate modeling of specific heat capacity, density, and viscosity is crucial to describe hybrid nanofluid behavior fully.

5.7.5.1 Effective Specific Heat Capacity

The specific heat capacity at constant pressure governs the fluid's ability to store thermal energy. For hybrid nanofluids, the effective specific heat capacity is generally computed as a weighted sum:

$$cp_{hnf} = \frac{(1-\phi)\rho_f c_{p,f} + \phi_1\rho_{p1}c_{p,p1} + \phi_2\rho_{p2}c_{p,p2}}{\rho_{hnf}}$$

where:

- ρ_f, ρ_{p1}, ρ_{p2} are densities of the base fluid and nanoparticles.
- $c_{p,f}$, $c_{p,p1}$, $C_{p,p2}$ are specific heat capacities of each component.

The denominator ρ_{hnf} is the effective density of the hybrid nanofluid (described next), ensuring the property is expressed per unit mass of the mixture [46].

5.7.5.2 Effective Density

Density affects buoyancy-driven flow, mass transport, and inertia effects in forced convection. For hybrid nanofluids, the effective density is modeled as:

$$\rho_{hnf} = (1-\phi)\rho_f + \phi_1\rho_{p1} + \phi_2\rho_{p2}$$

This straightforward volumetric mixing rule works well for dilute and moderately concentrated suspensions. However, at higher concentrations, interaction effects or nanoparticle clustering may necessitate correction factors [47].

5.7.5.3 Effective Viscosity

Viscosity strongly influences flow characteristics, pressure drop, and required pumping power. Hybrid nanofluids generally exhibit increased viscosity compared to base fluids, potentially offsetting gains in thermal conductivity.

Several models have been proposed to estimate effective viscosity:

5.7.5.3.1 Einstein Model (For Dilute Suspensions)

$$\mu_{nf} = \mu_f \left(1 + 2.5\phi\right)$$

This model applies to low-volume fractions (< 1%) and spherical particles but underestimates viscosity for higher concentrations typical in hybrid nanofluids [48].

5.7.5.3.2 Batchelor Model

$$\mu_{nf} = \mu_f \left(1 + 2.5\phi + 6.2\phi^2\right)$$

This quadratic extension accounts for particle–particle interactions at higher volume fractions [49–51].

5.7.5.3.3 Brinkman Model

$$\mu_{nf} = \mu_f \left(1 - \phi\right)^{-2.5}$$

While originally developed for single-component systems, this model is often generalized for hybrid nanofluids by defining $\phi = \phi_1 + \phi_2$ [52].

5.7.5.3.4 Empirical Correlations for Hybrid Nanofluids

Researchers have proposed empirical relations specific to hybrid systems, often taking the form:

$$\mu_{hnf} = \mu_f \left(1 + a_1\phi_1 + a_2\phi_2 + a_3\phi_1\phi_2 + a_4\phi_2\right)$$

where a_1, a_2, a_3, and a_4 are experimentally determined constants reflecting synergistic effects. Hybrid nanofluids often display non-Newtonian behavior, especially at higher concentrations, requiring even more complex models, such as power-law or Herschel–Bulkley formulations [53–55].

5.7.6 Influence of Hybrid Composition and Nanoparticle Shape

Hybrid nanofluids' thermophysical properties are not merely additive; they depend strongly on the composition and shape of nanoparticles, as well as their interactions [56].

5.7.6.1 Hybrid Composition Effects

5.7.6.1.1 Synergistic Thermal Conductivity Enhancement

One of the most significant advantages of hybrid nanofluids is their ability to achieve higher thermal conductivity than single-component nanofluids at the same total nanoparticle concentration [57–59]. This enhancement results from:

- **Synergistic heat transfer paths**: Combining high-conductivity metal nanoparticles (e.g., Ag, Cu) with metal oxide or carbon-based nanoparticles creates interconnected networks facilitating heat transport.
- **Suppression of particle aggregation**: Certain nanoparticle combinations reduce clustering, maintaining higher effective surface areas [60].

Empirical studies show that hybrid nanofluids can achieve thermal conductivity increases up to 30–40% higher than corresponding single-component systems.

5.7.6.1.2 Adjustable Viscosity

By tuning the relative volume fractions of different nanoparticles, engineers can balance viscosity increases against thermal conductivity improvements. For example, adding low-viscosity-promoting particles (e.g., carbon nanotubes) to high-viscosity metal oxide suspensions can mitigate viscosity rises.

5.7.6.1.3 Enhanced Specific Heat Capacity

Some hybrid combinations increase specific heat capacity, improving thermal storage capacity. This is advantageous for applications like solar thermal storage or phase change heat exchangers [61–63].

5.7.6.2 Influence of Nanoparticle Shape

5.7.6.2.1 Aspect Ratio Effects

Nanoparticles with high aspect ratios, such as nanotubes, nanorods, and nanosheets, enhance thermal conductivity more effectively than spherical particles due to:

- Larger surface contact area with the base fluid.
- Formation of percolation networks, providing continuous heat pathways.

However, they also tend to increase viscosity disproportionately and may pose dispersion challenges.

5.7.6.2.2 Surface Area and Interaction Potential

Particles with larger surface areas per unit volume enhance interfacial heat transfer but can also promote agglomeration. Proper surface functionalization is often required to maintain stability.

5.7.6.2.3 Mixed Shape Hybrid Systems

Hybrid nanofluids combining different shapes (e.g., spherical alumina with tubular CNTs) leverage both enhanced thermal conductivity and manageable viscosity. The combination promotes synergistic thermal networks while mitigating flow resistance.

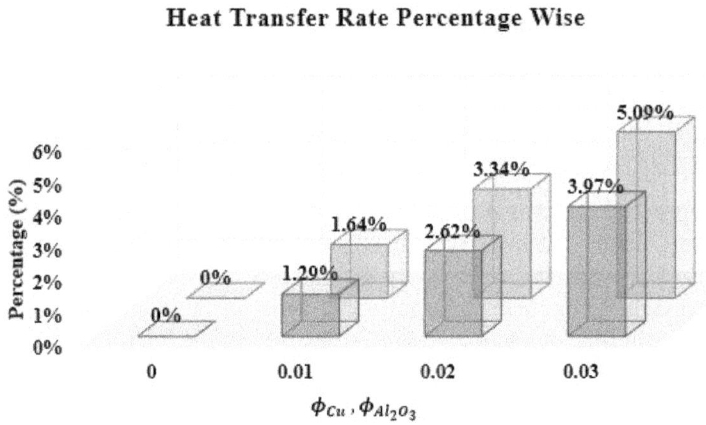

FIGURE 5.4 Heat transfer rate change percentage with respect to volume fraction of nanoparticles [3].

5.7.6.3 Nanoparticle Surface Properties

Surface modifications (e.g., surfactant coatings, functional groups) influence:

- Dispersion stability, affecting long-term thermal performance.
- Interfacial thermal resistance (Kapitza resistance), impacting effective conductivity.
- Rheological behavior, potentially shifting a nanofluid from Newtonian to non-Newtonian.

Hybrid nanofluid models must include these surface effects to achieve accurate predictions. Figure 5.4 shows change in heat transfer rate with respect to volume fractions of nanoparticles.

5.8 EXPERIMENTAL VALIDATION AND MODEL LIMITATIONS

While theoretical and semi-empirical models provide valuable predictions, they must be validated against experimental data due to inherent complexities [64].

5.8.1 Measurement Challenges

Accurate measurements of thermophysical properties of hybrid nanofluids face several difficulties, such as:

- **Sedimentation during measurements**, altering apparent properties.
- **Temperature-dependent behavior**, necessitating measurements over operational temperature ranges.
- **Non-Newtonian effects**, requiring careful rheological characterization.

5.8.2 Uncertainties and Discrepancies

Significant discrepancies often exist between model predictions and experimental results. These arise from:

- Uncertainties in nanoparticle shape, size distribution, and actual dispersion state.
- Simplified assumptions in theoretical models (e.g., ignoring inter-particle forces or Brownian effects).
- Differences in measurement techniques (e.g., transient hot wire vs. steady-state methods for thermal conductivity).

5.8.3 Need for Hybrid-Specific Models

Current models largely originate from single-component nanofluid studies. Hybrid nanofluids introduce additional complexities, including:

- Multiple particle interaction mechanisms.
- Variable clustering behaviors.
- Dynamic changes in local composition.

Thus, the development of hybrid-specific theoretical and empirical models remains a key research priority.

5.9 FUTURE DIRECTIONS IN PROPERTY MODELING

5.9.1 Machine Learning Approaches

Data-driven methods, including machine learning (ML) and artificial intelligence (AI), are increasingly used to model hybrid nanofluid properties. These approaches:

- Utilize large experimental datasets to identify hidden patterns.
- Provide predictive models without explicit functional forms.
- Enable rapid screening of nanoparticle combinations and compositions.

5.9.2 Molecular Dynamics (MD) Simulations

MD simulations can reveal atomistic-level mechanisms of heat transfer enhancement, especially for interfacial effects and nanolayer phenomena. Hybrid nanofluid studies using MD can guide empirical model refinement and new theoretical development.

5.9.3 Integrated Multiscale Modeling

Combining molecular-scale insights with continuum-level models (multiscale modeling) promises:

- More accurate property predictions across scales.
- Better understanding of emergent phenomena (e.g., percolation networks, cooperative diffusion).
- Improved design of hybrid nanofluids tailored for specific applications.

Accurate modeling of thermophysical properties is essential for effectively utilizing hybrid nanofluids in energy systems. Classical models like Maxwell and Hamilton–Crosser provide a starting point, but hybrid systems require advanced modifications to capture the synergistic effects of multiple nanoparticle types [65].

5.10 RHEOLOGICAL MODELING AND NON-NEWTONIAN EFFECTS

5.10.1 Newtonian versus Non-Newtonian Behavior in Hybrid Nanofluids

5.10.1.1 Definition of Newtonian Behavior

A Newtonian fluid is defined by a linear relationship between shear stress τ and shear rate γ, expressed as:

$$\tau = \mu\gamma$$

where μ is the dynamic viscosity, a constant independent of shear rate, and γ is the shear rate [66, 67]. Common fluids like water, air, and most dilute aqueous solutions exhibit Newtonian behavior across a wide range of shear rates and temperatures [68].

5.10.1.2 Non-Newtonian Fluids

Non-Newtonian fluids do not follow this linear relationship. Their apparent viscosity μ_a depends on the shear rate, and their flow behavior can vary widely:

- **Shear-thinning (pseudoplastic) fluids:** Viscosity decreases with increasing shear rate.
- **Shear-thickening (dilatant) fluids:** Viscosity increases with increasing shear rate.
- **Bingham plastics:** Require a yield stress to initiate flow.
- **Herschel–Bulkley fluids:** Exhibit both yield stress and non-linear shear rate dependence [69].

5.10.2 Hybrid Nanofluids and Their Deviations

Hybrid nanofluids often deviate from Newtonian behavior due to several reasons:

1. **Particle–particle interactions:** As concentration increases, particles interact, form clusters, or networks, affecting flow resistance.
2. **Particle shape and aspect ratio:** Non-spherical particles, such as nanotubes and nanosheets, create anisotropic stress fields in the fluid.

3. **Base fluid characteristics:** Using base fluids like ethylene glycol, oils, or polymeric solutions further promotes non-Newtonian behavior.
4. **Surface modifications and surfactants:** These additives can create additional structures or micelle-like assemblies, affecting rheology.

Empirical evidence shows that even at relatively low nanoparticle concentrations (below 1–2 vol%), hybrid nanofluids can already exhibit non-Newtonian signatures [70].

5.10.3 Shear-Thinning/Thickening Characteristics

5.10.3.1 Shear-Thinning Behavior

Shear-thinning (pseudoplastic) behavior is the most common non-Newtonian characteristic observed in hybrid nanofluids. The apparent viscosity decreases as shear rate increases, leading to easier pumping and lower flow resistance at higher velocities [71]. This is beneficial for dynamic cooling systems, microchannel flows, and forced convection applications.

5.10.3.1.1 Mechanism behind Shear-Thinning

Several mechanisms contribute to shear-thinning in hybrid nanofluids:

- **Alignment of anisotropic particles:** Under shear, elongated or plate-like nanoparticles tend to align with the flow, reducing viscous resistance.
- **Breakdown of particle aggregates:** At higher shear rates, nanoparticle clusters disintegrate, effectively lowering viscosity.
- **Thixotropic effects:** Time-dependent structural breakdown of weak networks formed by nanoparticles.

5.10.3.1.2 Modeling Shear-Thinning

The power-law model, also known as the Ostwald–de Waele model, is widely used to describe shear-thinning fluids [67]:

$$\tau = k\gamma^{n}$$

where:

- k is the consistency index.
- n is the flow behavior index ($n<1$ indicates shear-thinning).

The apparent viscosity is then:

$$\mu_a = k\gamma^{n-1}$$

Hybrid nanofluids often exhibit n values between 0.5 and 0.9, depending on concentration, composition, and temperature [70].

5.10.4 Shear-Thickening Behavior

Although less common, some hybrid nanofluids show shear-thickening (dilatant) behavior, particularly at high particle concentrations or when specific interactions dominate (e.g., strong attractive forces between particles) [71].

5.10.4.1 Mechanism behind Shear-Thickening

- **Hydro-cluster formation:** At high shear rates, transient clusters or aggregates form, increasing local resistance.
- **Jamming transition:** Dense suspensions approach a critical volume fraction where particles hinder each other's movement, sharply increasing viscosity.

5.10.4.2 Modeling Shear-Thickening

The power-law model still applies but with $n > 1$ indicating shear-thickening.

5.10.5 Yield Stress and Viscoelastic Behavior

In some hybrid nanofluids, especially at high solid loadings or with strong interparticle networks, a finite yield stress τ_0 must be overcome for flow to start. This behavior is described using models like the following:

5.10.5.1 Bingham Plastic Model

$$\tau = \tau_0 + \mu_p \gamma$$

where μ_p is the plastic viscosity.

5.10.5.2 Herschel–Bulkley Model

A generalized model combining yield stress and non-Newtonian power-law behavior:

$$\tau = \tau_0 + k\gamma^n$$

5.10.5.3 Casson Model

Another viscoelastic model commonly used in food and blood flow studies but sometimes extended to nanofluids:

$$\tau^{0.5} = \tau_0^{0.5} + (k\gamma)^{0.5}$$

These models better capture the complex flow onset and dynamic viscosity variations in hybrid nanofluids [60].

5.11 PRACTICAL IMPLICATIONS AND SYSTEM DESIGN

The rheological behavior of hybrid nanofluids has critical practical implications:

5.11.1 Pumping Power and Pressure Drop

Non-Newtonian hybrid nanofluids with higher apparent viscosities increase pressure drops, demanding higher pumping power. However, shear-thinning behavior can mitigate this in high-flow regions, optimizing energy efficiency.

5.11.2 Flow Regime Transitions

Shear-dependent viscosity influences laminar-to-turbulent transitions, affecting convective heat transfer coefficients and system stability. The Reynolds number for non-Newtonian fluids must be modified to [72]:

$$Re = \frac{\rho v^{2-n} D^n}{k'}$$

where:

- v is the mean velocity.
- D is the characteristic length.
- n and k' are the power-law parameters.

5.11.3 Heat Exchanger Design

Non-Newtonian behavior alters the convective heat transfer correlations, requiring modified Nusselt number relations, for example [73]:

$$Nu = C\, Re^m pr^n \left(\frac{\mu_a}{\mu_w} \right)^r$$

where μ_w is the viscosity at the wall temperature, and r is an empirically determined exponent [74].

5.11.4 Microchannel and Mini-Channel Cooling

Hybrid nanofluids are promising for electronics and photovoltaic cooling, where microchannel flows dominate. Here, non-Newtonian effects are even more pronounced due to high shear rates and confined geometries.

5.12 CHALLENGES AND FUTURE RESEARCH DIRECTIONS

Despite advances, several challenges remain in fully characterizing and modeling hybrid nanofluid rheology:

5.12.1 Particle Dispersion and Stability

Non-uniform dispersion or particle aggregation alters local rheology, making model predictions unreliable. Real-time monitoring techniques and advanced dispersion

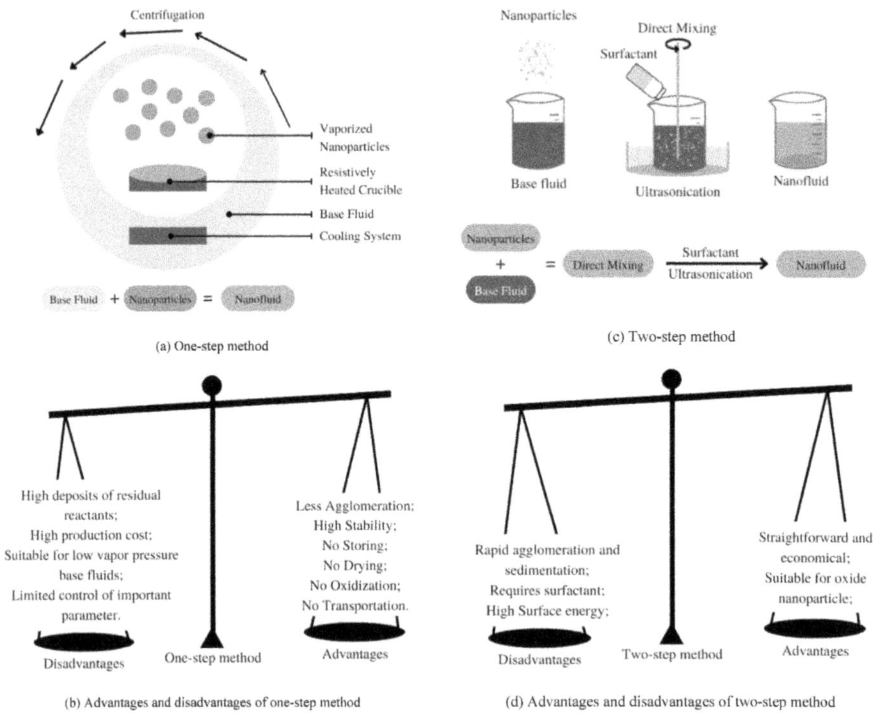

FIGURE 5.5 Advantages and disadvantages of various synthesis methods [65].

strategies are essential. Figure 5.5 gives the details of the advantages and disadvantages of synthesis method.

5.12.2 Time-Dependent Behavior

Many hybrid nanofluids exhibit thixotropic or rheopectic behavior (time-dependent viscosity changes). Most models assume steady-state, neglecting these dynamic effects.

5.12.3 Surface Interactions and Functionalization

Surface chemistry affects interparticle forces, aggregation tendencies, and rheological properties. Incorporating these effects into models remains a major research frontier.

5.12.4 Multi-Scale Modeling

Linking molecular-scale interactions (e.g., from molecular dynamics simulations) to macroscopic rheological models can provide more accurate, predictive tools for engineering design.

5.12.5 Hybrid Nanofluid-Specific Empirical Data

Currently, there is a scarcity of comprehensive experimental datasets for different hybrid nanofluid combinations, shapes, and concentrations, limiting the generalizability of proposed correlations.

Rheological modeling of hybrid nanofluids is essential for accurately predicting their flow and heat transfer performance in practical applications. While some hybrid nanofluids behave approximately Newtonian, many exhibit complex non-Newtonian behavior, including shear-thinning, shear-thickening, and yield stress effects.

A wide range of empirical and semi-empirical models, from classical Einstein-type equations to advanced power-law and viscoelastic models, have been developed to describe these behaviors. However, due to the complexity introduced by hybrid nanoparticle compositions, shape variations, and surface interactions, these models often require case-specific calibration.

Future research integrating experimental data, multi-scale simulations, and machine learning approaches holds great potential to refine rheological models, enabling optimized design and broader adoption of hybrid nanofluids in energy and thermal management systems.

5.13 FUTURE RESEARCH DIRECTIONS

5.13.1 Hybrid Nanofluid-Specific Turbulence Models

Developing new turbulence models incorporating nanoparticle-turbulence interactions explicitly, rather than through property averaging, is an ongoing area of research.

5.13.2 Machine Learning and Data-Driven Models

ML techniques can help correlate complex turbulent flow and heat transfer behavior to fluid parameters and operating conditions, bypassing some limitations of empirical correlations.

5.13.3 High-Fidelity Simulation Benchmarks

LES and DNS studies, though computationally intensive, can provide benchmark datasets for validating lower-order models and exploring new flow enhancement strategies.

5.13.4 Multiphase Modeling Integration

Incorporating multiphase flow effects, such as slip velocities and particle sedimentation, is essential for more realistic predictions, particularly in large-scale industrial applications.

5.14 CONCLUSION

Turbulent flow modeling of hybrid nanofluids is a critical and complex aspect of advancing energy-efficient thermal systems. While laminar models can provide insights for low Reynolds number flows, turbulent regimes are often preferred for their superior heat transfer performance.

The use of advanced turbulence models, including k–ε, k–ω, SST, and LES, is fundamental in capturing the intricate behavior of hybrid nanofluids under turbulent conditions. Modified Nusselt number correlations and empirical adjustments are necessary to account for the enhanced thermal and flow characteristics induced by hybrid nanoparticle suspensions.

REFERENCES

1. Saidur, R., Leong, K. Y., & Mohammad, H. A. (2011). A review on applications and challenges of nanofluids. *Renewable and Sustainable Energy Reviews*, *15*(3), 1646–1668.
2. Khan, U., Zaib, A., Ishak, A., Al-Mubaddel, F. S., Bakar, S. A., Alotaibi, H., & Aljohani, H. A. (2021). Computational modeling of hybrid Sisko nanofluid flow over a porous radially heated shrinking/stretching disc. *Coatings*, 11(10), 1242.
3. Mallawi, F. O., & Ullah, M. Z. (2021). Multiple slip impact on the Darcy–Forchheimer hybrid nano fluid flow due to quadratic convection past an inclined plane. *Mathematics*, *9*(22), 2934.
4. Abu-Nada, E., & Oztop, H. F. (2009). Effects of inclination angle on natural convection in enclosures filled with Cu–water nanofluid. *International Journal of Heat and Fluid Flow*, *30*(4), 669–678.
5. Sarkar, J., Ghosh, P., & Adil, A. (2015). A review on hybrid nanofluids: Recent research, development and applications. *Renewable and Sustainable Energy Reviews*, *43*, 164–177.
6. Rammoorthi, R., & Mohanavel, D. (2024). Entropy generation and sensitivity analysis for a squeezed flow of a Williamson hybrid nanofluid under the supervision of micro-cantilever sensor. *Results in Engineering*, *24*, 103464.
7. Sidik, N. A. C., Adamu, I. M., Jamil, M. M., Kefayati, G. H. R., Mamat, R., & Najafi, G. (2016). Recent progress on hybrid nanofluids in heat transfer applications: A comprehensive review. *International Communications in Heat and Mass Transfer*, *78*, 68–79.
8. Choi, S. U., & Eastman, J. A. (1995). *Enhancing thermal conductivity of fluids with nanoparticles* (No. ANL/MSD/CP-84938; CONF-951135-29). Argonne National Lab. (ANL), Argonne, IL (United States).
9. Mahian, O., Kolsi, L., Amani, M., Estellé, P., Ahmadi, G., Kleinstreuer, C., . . . & Pop, I. (2019). Recent advances in modeling and simulation of nanofluid flows-Part I: Fundamentals and theory. *Physics Reports*, *790*, 1–48.
10. Selvakumar, P., & Suresh, S. (2012). Use of Al_2O_3–Cu/water hybrid nanofluid in an electronic heat sink. *IEEE Transactions on Components, Packaging and Manufacturing Technology*, *2*(10), 1600–1607.
11. Ahmadi, M. H., Mirlohi, A., Nazari, M. A., & Ghasempour, R. (2018). A review of thermal conductivity of various nanofluids. *Journal of Molecular Liquids*, *265*, 181–188.
12. Devendiran, D. K., & Amirtham, V. A. (2016). A review on preparation, characterization, properties and applications of nanofluids. *Renewable and Sustainable Energy Reviews*, *60*, 21–40.

13. Tyagi, H., Phelan, P., & Prasher, R. (2009). Predicted efficiency of a low-temperature nanofluid-based direct absorption solar collector. *Journal of Solar Energy Engineering, 131*(4).
14. Hosseinzadeh, M., Salari, A., Sardarabadi, M., & Passandideh-Fard, M. (2018). Optimization and parametric analysis of a nanofluid based photovoltaic thermal system: 3D numerical model with experimental validation. *Energy Conversion and Management, 160*, 93–108.
15. Aydin, D., Utlu, Z., & Kincay, O. (2015). Thermal performance analysis of a solar energy sourced latent heat storage. *Renewable and Sustainable Energy Reviews, 50*, 1213–1225.
16. Shah, T. R., & Ali, H. M. (2019). Applications of hybrid nanofluids in solar energy, practical limitations and challenges: A critical review. *Solar Energy, 183*, 173–203.
17. Sundar, L. S., Singh, M. K., & Sousa, A. C. (2014). Enhanced heat transfer and friction factor of MWCNT–Fe3O4/water hybrid nanofluids. *International Communications in Heat and Mass Transfer, 52*, 73–83.
18. Sajid, M. U., & Ali, H. M. (2018). Thermal conductivity of hybrid nanofluids: A critical review. *International Journal of Heat and Mass Transfer, 126*, 211–234.
19. Anoop, K. B., Sundararajan, T., & Das, S. K. (2009). Effect of particle size on the convective heat transfer in nanofluid in the developing region. *International Journal of Heat and Mass Transfer, 52*(9–10), 2189–2195.
20. Yıldız, Ç., Arıcı, M., & Karabay, H. (2019). Comparison of a theoretical and experimental thermal conductivity model on the heat transfer performance of Al2O3-SiO2/water hybrid-nanofluid. *International Journal of Heat and Mass Transfer, 140*, 598–605.
21. Sharma, A. K., Singh, R. K., Tiwari, A. K., Dixit, A. R., & Katiyar, J. K. (2020). Rheological behaviour of hybrid nanofluids: A review. *Tribology in Materials and Applications*, 77–94.
22. Mehrali, M., Sadeghinezhad, E., Tahan Latibari, S., Mehrali, M., Togun, H., Zubir, M. N. M., . . . & Metselaar, H. S. C. (2014). Preparation, characterization, viscosity, and thermal conductivity of nitrogen-doped graphene aqueous nanofluids. *Journal of Materials Science, 49*(20), 7156–7171.
23. Murshed, S. S., & Estellé, P. (2017). A state of the art review on viscosity of nanofluids. *Renewable and Sustainable Energy Reviews, 76*, 1134–1152.
24. Pastoriza-Gallego, M. J., Casanova, C., Páramo, R., Barbés, B., Legido, J. L., & Piñeiro, M. M. (2009). A study on stability and thermophysical properties (density and viscosity) of Al2O3 in water nanofluid. *Journal of Applied Physics, 106*(6).
25. Cohen, I. M., & Kundu, P. K. (2004). *Fluid mechanics*. Elsevier.
26. Batchelor, G. K. (2000). *An introduction to fluid dynamics*. Cambridge University Press.
27. Bejan, A. (2013). *Convection heat transfer*. John Wiley & Sons.
28. Pak, B. C., & Cho, Y. I. (1998). Hydrodynamic and heat transfer study of dispersed fluids with submicron metallic oxide particles. *Experimental Heat Transfer an International Journal, 11*(2), 151–170.
29. Brinkman, H. C. (1952). The viscosity of concentrated suspensions and solutions. *Journal of Chemical Physics, 20*(4), 571–571.
30. Maxwell, J. C. (1873). *A treatise on electricity and magnetism* (Vol. 1). Clarendon Press.
31. Buongiorno, J. (2006). Convective transport in nanofluids. *Journal of heat transfer, 128*, 240–250. http://dx.doi.org/10.1115/1.2150834
32. Haddad, Z., Abu-Nada, E., Oztop, H. F., & Mataoui, A. (2012). Natural convection in nanofluids: Are the thermophoresis and Brownian motion effects significant in nanofluid heat transfer enhancement? *International Journal of Thermal Sciences, 57*, 152–162.

33. Incropera, F. P., DeWitt, D. P., Bergman, T. L., & Lavine, A. S. (1996). *Fundamentals of heat and mass transfer* (Vol. 6, p. 116). New York: Wiley.
34. Koo, J., & Kleinstreuer, C. (2004). A new thermal conductivity model for nanofluids. *Journal of Nanoparticle Research*, *6*(6), 577–588.
35. Andrade, E. D. C. (1930). The viscosity of liquids. *Nature*, *125*(3148), 309–310.
36. Chon, C. H., Kihm, K. D., Lee, S. P., & Choi, S. U. (2005). Empirical correlation finding the role of temperature and particle size for nanofluid (Al2O3) thermal conductivity enhancement. *Applied Physics Letters*, *87*(15).
37. Namburu, P. K., Kulkarni, D. P., Misra, D., & Das, D. K. (2007). Viscosity of copper oxide nanoparticles dispersed in ethylene glycol and water mixture. *Experimental Thermal and Fluid Science*, *32*(2), 397–402.
38. Hu, W., & Buongiorno, J. (2005, May). Nanofluid coolants for advanced nuclear power plants. *In Proceedings of ICAPP*, *5*(5705), 15–19.
39. Chandrakar, V., Bhattad, A., Samal, P., Senapati, J. R., & Kashyap, A. K. (2025). A study on different methods to change the Rayleigh number in the analysis of heat transfer. *Scientific Reports*, *15*(1), 25773.
40. Versteeg, H. K. (2007). *An introduction to computational fluid dynamics the finite volume method, 2/E*. Pearson Education India.
41. Ferziger, J. H., Perić, M., & Street, R. L. (2019). *Computational methods for fluid dynamics*. Springer.
42. Sundar, L. S., Singh, M. K., & Sousa, A. C. (2016). Enhanced thermal properties of nanodiamond nanofluids. *Chemical Physics Letters*, *644*, 99–110.
43. Hamilton, R. L., & Crosser, O. K. (1962). Thermal conductivity of heterogeneous two-component systems. *Industrial & Engineering Chemistry Fundamentals*, *1*(3), 187–191.
44. Yu, W., & Choi, A. S. (2003). The role of interfacial layers in the enhanced thermal conductivity of nanofluids: A renovated Maxwell model. *Journal of Nanoparticle Research*, *5*(1), 167–171.
45. Leong, K. Y., Ahmad, K. K., Ong, H. C., Ghazali, M. J., & Baharum, A. (2017). Synthesis and thermal conductivity characteristic of hybrid nanofluids—a review. *Renewable and Sustainable Energy Reviews*, *75*, 868–878.
46. Oztop, H. F., & Abu-Nada, E. (2008). Numerical study of natural convection in partially heated rectangular enclosures filled with nanofluids. *International Journal of Heat and Fluid Flow*, *29*(5), 1326–1336.
47. Xuan, Y., & Li, Q. (2000). Heat transfer enhancement of nanofluids. *International Journal of Heat and Fluid Flow*, *21*(1), 58–64.
48. Einstein, A. (1905). *Eine neue bestimmung der moleküldimensionen* (Doctoral dissertation, ETH Zurich).
49. Malika, M., & Sonawane, S. S. (2021). The sono-photocatalytic performance of a novel water based Ti+ 4 coated Al (OH) 3-MWCNT's hybrid nanofluid for dye fragmentation. *International Journal of Chemical Reactor Engineering*, *19*(9), 901–912.
50. Parate, V. R., Kawadkar, D. K., & Sonawane, S. S. (2011). Study of whey protein concentrate fortification in cookies variety biscuits. *International Journal of Food Engineering*, *7*(2), 1–12.
51. Batchelor, G. K. (1977). The effect of Brownian motion on the bulk stress in a suspension of spherical particles. *Journal of Fluid Mechanics*, *83*(1), 97–117.
52. Brinkman, H. C. (1952). The viscosity of concentrated suspensions and solutions. *Journal of Chemical Physics*, *20*(4), 571–571.

53. Thakur, P., Sonawane, S., Potoroko, I., & Sonawane, S. H. (2021). Recent advances in ultrasound-assisted synthesis of nano-emulsions and their industrial applications. *Current Pharmaceutical Biotechnology*, *22*(13), 1748–1758.
54. Sonawane, A. V., Rikame, S., Sonawane, S. H., Gaikwad, M., Bhanvase, B., Sonawane, S. S., . . . & Gaikwad, R. (2024). A review of microbial fuel cell and its diversification in the development of green energy technology. *Chemosphere*, *350*, 141127.
55. Suresh, S., Venkitaraj, K. P., Selvakumar, P., & Chandrasekar, M. (2011). Synthesis of Al2O3–Cu/water hybrid nanofluids using two step method and its thermo physical properties. *Colloids and Surfaces A: Physicochemical and Engineering Aspects*, *388*(1–3), 41–48.
56. Sundar, L. S., Ramana, E. V., Graça, M. P. F., Singh, M. K., & Sousa, A. C. (2016). Nanodiamond-Fe3O4 nanofluids: Preparation and measurement of viscosity, electrical and thermal conductivities. *International Communications in Heat and Mass Transfer*, *73*, 62–74.
57. Malika, M. M., & Sonawane, S. S. (2019). Review on application of nanofluid/nano particle as water disinfectant. *Journal of Indian Association for Environmental Management (JIAEM)*, *39*(1–4), 21–24.
58. Malika, M., Bhad, R., & Sonawane, S. S. (2021). ANSYS simulation study of a low volume fraction CuO–ZnO/water hybrid nanofluid in a shell and tube heat exchanger. *Journal of the Indian Chemical Society*, *98*(11), 100200.
59. Alfellag, M. A., Kamar, H. M., Sidik, N. A. C., Muhsan, A. S., Kazi, S. N., Alawi, O. A., & Abidin, U. (2023). Rheological and thermophysical properties of hybrid nanofluids and their application in flat-plate solar collectors: A comprehensive review. *Journal of Thermal Analysis and Calorimetry*, *148*(14), 6645–6686.
60. Suresh, S., Venkitaraj, K. P., Selvakumar, P., & Chandrasekar, M. (2011). Synthesis of Al2O3–Cu/water hybrid nanofluids using two step method and its thermo physical properties. *Colloids and Surfaces A: Physicochemical and Engineering Aspects*, *388*(1–3), 41–48.
61. Malika, M., & Sonawane, S. S. (2020). Effect of nanoparticle mixed ratio on stability and thermo-physical properties of CuO-ZnO/water-based hybrid nanofluid. *Journal of the Indian Chemical Society*, *97*(3), 414–419.
62. Sonawane, S., Thakur, P., & Paul, R. (2020). Study on thermal property enhancement of MWCNT based polypropylene (PP) nanocomposites. *Materials Today: Proceedings*, *27*, 550–555.
63. Hemmat Esfe, M., Goodarzi, M., & Esfandeh, S. (2021). Experimental investigation of thermo-physical properties of MgO-MWCNT (75–25%)/10W40 as a new nano-lubricant. *The European Physical Journal Plus*, *136*(5), 1–15.
64. Yu, W., France, D. M., Routbort, J. L., & Choi, S. U. (2008). Review and comparison of nanofluid thermal conductivity and heat transfer enhancements. *Heat Transfer Engineering*, *29*(5), 432–460.
65. Thakur, P., Sonawane, S. S., Sonawane, S. H., & Bhanvase, B. A. (2020). Nanofluids-based delivery system, encapsulation of nanoparticles for stability to make stable nanofluids. In *Encapsulation of active molecules and their delivery system* (pp. 141–152). Elsevier.
66. Bethi, B., Sonawane, S. H., Bhanvase, B. A., & Sonawane, S. S. (2021). Textile industry wastewater treatment by cavitation combined with Fenton and ceramic nanofiltration membrane. *Chemical Engineering and Processing-Process Intensification*, *168*, 108540.
67. Choi, S. U., & Eastman, J. A. (1995). *Enhancing thermal conductivity of fluids with nanoparticles* (No. ANL/MSD/CP-84938; CONF-951135-29). Argonne National Lab. (ANL), Argonne, IL (United States).

68. Bird, R. B. (2002). Transport phenomena. *Applied Mechanics Reviews*, *55*(1), R1–R4.
69. Barnes, H. A., Hutton, J. F., & Walters, K. (1989). *An introduction to rheology* (Vol. 3). Elsevier.
70. Sundar, L. S., Singh, M. K., & Sousa, A. C. (2013). Investigation of thermal conductivity and viscosity of Fe3O4 nanofluid for heat transfer applications. *International Communications in Heat and Mass Transfer*, *44*, 7–14.
71. Parashar, N., & Yahya, S. M. (2022). Thermophysical and rheological properties of hybrid nanofluids: A review on recent studies. *Journal of Thermal Analysis and Calorimetry*, *147*(7), 4411–4449.
72. Chhabra, R. P., & Patel, S. A. (2025). *Non-Newtonian flow and applied rheology: Engineering applications*. Elsevier.
73. Afrand, M., Toghraie, D., & Ruhani, B. (2016). Effects of temperature and nanoparticles concentration on rheological behavior of Fe_3O_4–Ag/EG hybrid nanofluid: An experimental study. *Experimental Thermal and Fluid Science*, *77*, 38–44.
74. Shah, R. K., & Sekulić, D. P. (2003). *Fundamentals of heat exchanger design*. John Wiley & Sons.

6 Mathematical Modeling of Ionanofluids and Applications for Energy Conservation

6.1 INTRODUCTION

6.1.1 Definition and Concept of Ionanofluids (IONFs)

Ionanofluids (IONFs) are a novel class of heat transfer and functional working fluids that combine the distinct advantages of ionic liquids (ILs) and nanofluids [1, 2]. At their core, IONFs consist of nanoparticles dispersed in ionic liquid bases rather than in conventional molecular solvents like water or ethylene glycol [3].

The term "ionanofluid" represents the synergy between nanotechnology and ionic liquid chemistry. By integrating solid nanoparticles (metallic, metal oxide, carbon-based, or hybrid) into a highly tunable ionic liquid medium, these fluids achieve superior thermal, electrical, and chemical performance. Unlike traditional nanofluids, which typically involve a base fluid with low electrical conductivity and limited thermal stability, IONFs provide a conductive, chemically stable, and thermally robust environment that enhances nanoparticle dispersion and overall fluid properties [4].

6.1.2 Evolution from Traditional Nanofluids and Ionic Liquids

The development of ionanofluids is rooted in two major advances: nanofluids and ionic liquids.

Nanofluids, first systematically studied in the 1990s, consist of nanoscale solid particles suspended in base fluids to improve thermal conductivity and heat transfer performance [5]. Their development was driven by the need to overcome the limitations of conventional coolants in applications like electronics cooling, automotive radiators, and solar collectors. While they significantly enhanced thermal properties, issues such as particle agglomeration, sedimentation, and limited high-temperature stability restricted their broader adoption.

Ionic liquids, on the other hand, are salts in liquid form at or near room temperature. They have gained attention due to their unique features: negligible vapor pressure, wide liquid range, high thermal stability, and remarkable solvating power [6]. However, ionic liquids alone typically have lower thermal conductivity than metallic or oxide nanoparticles, limiting their standalone heat transfer capabilities.

DOI: 10.1201/9781003628071-6

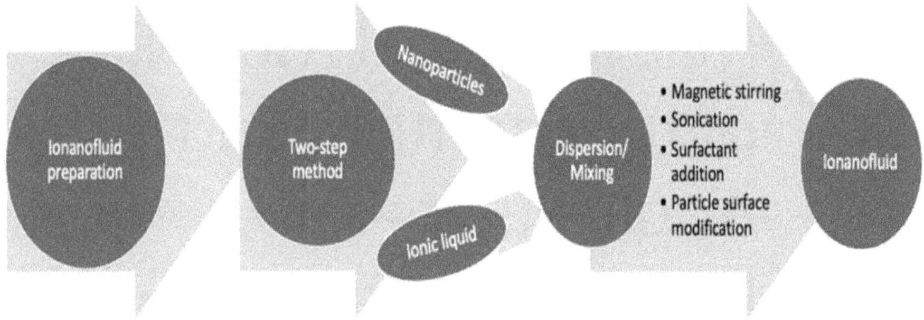

FIGURE 6.1 Ionic liquid-based nanofluid synthesis [6].

The concept of ionanofluids emerged to merge these complementary benefits. By dispersing nanoparticles into ionic liquids, researchers aimed to create a new hybrid fluid that not only improved heat transfer but also maintained excellent stability under extreme thermal and chemical conditions. This synergy marks a significant evolution beyond the individual limitations of nanofluids and ionic liquids. Figure 6.1 shows the synthesis mechanism of ionic liquid-based nanofluids.

6.1.3 Unique Properties (Thermal, Electrochemical, Tunability)

Ionanofluids exhibit a range of exceptional properties, setting them apart from both their nanofluid and ionic liquid precursors.

6.1.3.1 Enhanced Thermal Conductivity

The incorporation of thermally conductive nanoparticles (e.g., Al_2O_3, CuO, graphene) leads to a substantial improvement in overall thermal conductivity, enabling better heat dissipation and higher efficiency in thermal systems.

6.1.3.2 Electrochemical Stability and Conductivity

Ionic liquids are naturally ionic in nature and can support high ionic conductivity and electrochemical stability. When combined with nanoparticles, these properties facilitate enhanced charge transport, beneficial for applications like electrochemical storage devices (batteries and supercapacitors) and electro-thermal systems.

6.1.3.3 Tunable Viscosity and Specific Heat

The rheological behavior of ionanofluids can be precisely tailored by adjusting nanoparticle concentration, shape, surface functionalization, and ionic liquid composition. This tunability allows system designers to optimize flow characteristics and thermal storage capacity according to specific application needs.

6.1.3.4 Wide Thermal Stability Range

IONFs maintain stability over a broader temperature range compared to water- or glycol-based nanofluids. Many ionic liquids resist thermal decomposition up to

300–400°C, allowing ionanofluids to operate in high-temperature regimes without significant performance degradation.

6.1.3.5 Low Volatility and Non-Flammability

Unlike many conventional heat transfer fluids, ionic liquids (and hence ionanofluids) have negligible vapor pressure and are often non-flammable. These features enhance operational safety, especially in high-temperature or confined environments.

6.1.4 Motivation for Energy Conservation Applications

The global push for sustainable and energy-efficient technologies has intensified the search for advanced heat transfer and thermal storage fluids [7]. In sectors ranging from renewable energy (solar thermal, geothermal) to high-performance electronics and industrial waste heat recovery, efficient thermal management is critical for improving energy utilization and reducing emissions.

Ionanofluids offer a compelling solution in this context due to their ability to:

- **Enhance thermal transport efficiency:** Improving heat transfer coefficients leads to reduced energy losses and lower operational costs.
- **Support compact and lightweight designs:** Higher thermal conductivity enables smaller heat exchangers and thermal systems, reducing material usage and footprint.
- **Operate under harsh conditions:** Their stability at extreme temperatures and pressures allows for application in concentrated solar power plants, high-temperature reactors, and advanced heat storage systems.
- **Enable multifunctional roles:** Beyond heat transfer, ionanofluids' electrochemical capabilities make them suitable for integrated systems combining heat and electrical energy storage, further supporting hybrid energy solutions.

By addressing limitations of conventional fluids, IONFs align with global energy conservation goals and pave the way for more compact, efficient, and safe thermal systems.

6.1.5 Current Research Trends and Gaps

Research on ionanofluids is rapidly growing but still in its early stages compared to nanofluids and traditional heat transfer fluids. Key trends include the following:

6.1.5.1 Advanced Synthesis and Dispersion Techniques

Researchers are exploring innovative methods to ensure uniform nanoparticle dispersion and prevent agglomeration, such as surface functionalization of nanoparticles and in situ nanoparticle generation within ionic liquids.

6.1.5.2 Thermophysical and Rheological Characterization

Systematic studies are being conducted to measure and model thermal conductivity, specific heat, viscosity, and density over wide temperature and shear rate ranges. These studies are essential for accurate mathematical modeling and practical design.

6.1.5.3 Electrochemical Applications

The dual functionality of ionanofluids (thermal and ionic conduction) is being investigated for battery cooling and electro-thermal energy storage systems. Such hybrid applications could offer significant advances in energy density and operational flexibility.

6.1.5.4 Long-Term Stability and Recyclability

Assessing and improving the long-term chemical and physical stability of ionanofluids remains a critical challenge. Understanding degradation mechanisms, sedimentation behavior, and reusability will be key for large-scale and continuous applications.

6.1.5.5 Environmental and Economic Considerations

The environmental impacts of ionic liquids and their combination with nanoparticles are under study, with a focus on life cycle analysis, potential toxicity, and end-of-life management. Cost reduction through scalable synthesis routes and recyclable designs is also an active research area.

6.1.6 Gaps and Future Needs

Despite promising results, significant gaps persist, including:

- Lack of standardized thermophysical property databases for various IONFs.
- Limited large-scale demonstration projects to validate laboratory findings.
- Insufficient coupling of experimental data with advanced multi-physics simulation models.
- Need for robust regulatory frameworks addressing safety and environmental impact.

6.2 CORRELATIONS AND MODELING NEEDS

While general trends are similar to traditional nanofluids, IONFs require unique viscosity and specific heat models that incorporate electrostatic interactions, ionic layer formation, and temperature-dependent dissociation of ionic liquids. Existing correlations often fail to fully capture these effects, highlighting a critical area for further research and experimental validation [8]. Figure 6.2 shows the relations of prediction and estimation in mathematical modeling.

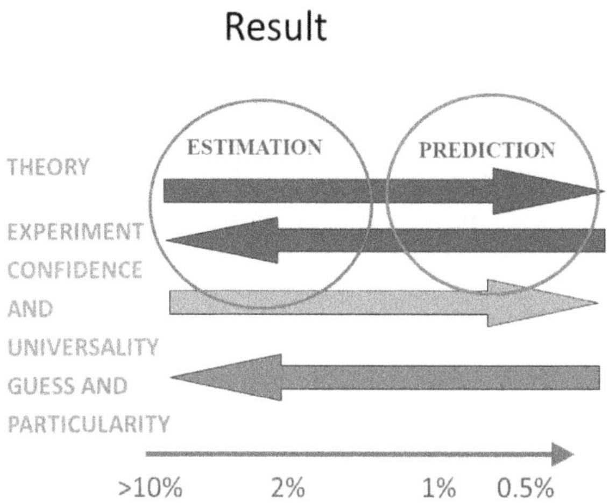

FIGURE 6.2 A relation between prediction and estimation [2].

6.2.1 Ionic Conductivity and Electrochemical Behavior

6.2.1.1 Fundamentals of Ionic Conductivity

Ionic liquids inherently possess ionic conductivity due to the presence of free-moving cations and anions. In ionanofluids, nanoparticle addition can influence ionic conductivity via:

- Providing conductive pathways.
- Modifying ion distribution near nanoparticle surfaces.
- Inducing localized electric field gradients [9].

6.2.1.2 Impact of Nanoparticles

Metallic or conductive nanoparticles (e.g., silver, graphene) can enhance ionic conductivity by forming percolation networks. In contrast, oxide nanoparticles (e.g., Al_2O_3, SiO_2) may reduce conductivity slightly due to their insulating nature and potential to trap ions at their surfaces [10].

6.2.1.3 Electrochemical Stability

Ionanofluids exhibit high electrochemical stability windows, allowing their use in applications requiring exposure to high voltages or harsh chemical environments, such as flow batteries or electro-thermal energy storage systems. However, interactions between nanoparticles and ionic liquids can alter redox potentials and stability windows, necessitating careful compatibility assessments [11].

6.2.1.4 Applications Leveraging Ionic Behavior

IONFs can be tailored for multifunctional roles, such as simultaneous thermal and electrochemical energy storage. For example, in hybrid cooling-electrochemical systems, high ionic conductivity supports both efficient heat removal and enhanced charge transport, improving overall system efficiency [12].

6.2.2 Influence of Nanoparticle Type, Size, and Surface Modifications

6.2.2.1 Nanoparticle Type

Different nanoparticles impart distinct thermophysical and electrochemical characteristics:

- **Metallic particles (e.g., Cu, Ag):** High thermal and electrical conductivity, strong enhancement in heat transfer, potential to improve ionic conductivity.
- **Oxides (e.g., Al_2O_3, TiO_2):** Chemically stable, enhance thermal conductivity moderately, usually reduce ionic conductivity.
- **Carbon-based materials (e.g., graphene, CNTs):** Excellent thermal and electrical properties, can form percolation networks, improve both heat transfer and charge transport.

6.2.2.2 Nanoparticle Size

Smaller nanoparticles generally exhibit larger surface areas, promoting stronger interactions with the ionic liquid and more significant property changes. However, extremely small particles are prone to agglomeration due to higher surface energy, potentially compromising stability.

6.2.2.3 Surface Functionalization

Surface modification strategies aim to:

- Improve dispersion and prevent agglomeration.
- Enhance compatibility with ionic liquids.
- Modify surface charge, affecting both rheology and electrochemical behavior.

For instance, functionalizing graphene with ionic liquid-compatible groups can significantly improve stability and maintain high thermal conductivity.

6.2.2.4 Hybrid Nanoparticle Systems

Combining different types of nanoparticles (e.g., metal and oxide, or carbon and metal) can exploit synergistic effects, achieving tailored thermal and electrical properties that single-component systems cannot provide. Such hybrid strategies are increasingly explored for advanced ionanofluid formulations.

6.2.3 Temperature Effects

Thermophysical properties of ionanofluids are strongly temperature-dependent:

- **Thermal conductivity:** Generally increases with temperature due to enhanced phonon transport and reduced viscosity, facilitating better nanoparticle movement and clustering effects.

- **Viscosity:** Typically decreases with temperature, improving flowability and reducing pumping power requirements. However, very high temperatures can also promote agglomeration if not properly stabilized.
- **Specific heat capacity:** May vary non-linearly with temperature, especially if nanoparticle phase transitions occur.
- **Ionic conductivity:** Often increases with temperature due to increased ion mobility [13].

6.3 THERMAL AND MASS TRANSPORT PHENOMENA

6.3.1 Thermal and Mass Transport Phenomena in Ionanofluids

In this section, we examine the thermal and mass transport phenomena governing IONFs, emphasizing enhanced heat transfer mechanisms, the influence of ionic motion and nanoparticle migration, convective and diffusive contributions, electro-osmotic and electrophoretic effects, and practical case studies [14].

6.3.2 Enhanced Heat Transfer Mechanisms

6.3.2.1 Intrinsic Thermal Conductivity Enhancement

A defining advantage of IONFs is their superior thermal conductivity compared to conventional fluids and base ionic liquids. The inclusion of nanoparticles, metals, metal oxides, or carbon-based nanostructures disrupts the phonon scattering pathways and facilitates energy transfer via solid-phase conduction mechanisms.

The effective thermal conductivity of an IONF is influenced by several factors:

- **Nanoparticle type and intrinsic conductivity:** Materials like silver, copper, graphene, and carbon nanotubes offer exceptionally high conductivities.
- **Particle volume fraction and distribution:** Higher loadings enhance conductivity but can also raise viscosity.
- **Particle shape and aspect ratio:** High-aspect-ratio particles (e.g., nanotubes or nanosheets) provide efficient thermal bridges.
- **Ionic liquid–nanoparticle interface:** The interfacial layer (or nanolayer) around nanoparticles can contribute significantly to overall conductivity [15].

Figure 6.3 shows the thermal conductivity values of various Ionic liquid-based nanofluids.

6.3.2.2 Micro-Convection Effects

The Brownian motion of nanoparticles induces microscale convection within the fluid medium. While these random motions are thermally driven, they create local perturbations that can enhance mixing and heat transport.

Moreover, in forced convection applications, these micro convective effects supplement bulk convective heat transfer, boosting overall system performance [16].

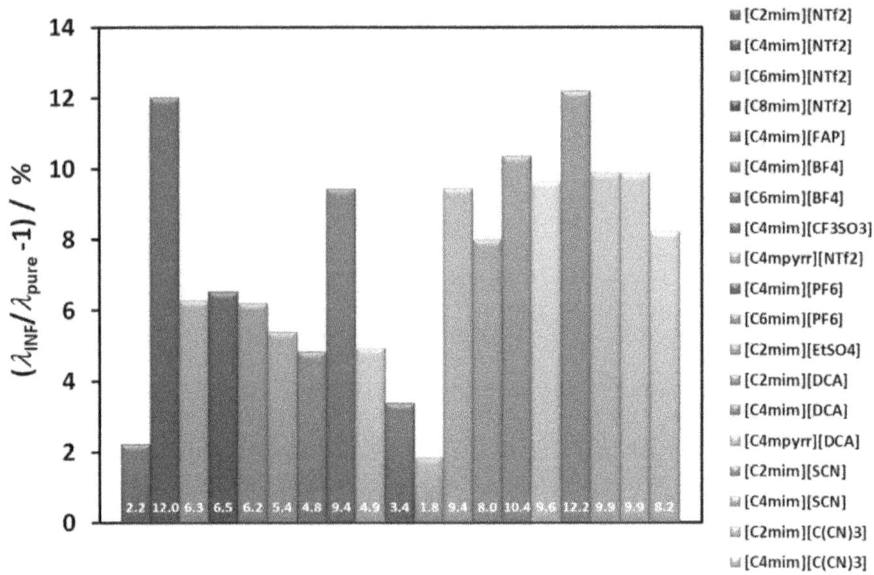

FIGURE 6.3 Enhancement of thermal conductivity in various ionanofluids [2].

6.3.3 Electro-Thermal Coupling

In IONFs, the ionic component introduces electro-thermal coupling. When exposed to electric fields, ions move, carrying both charge and thermal energy. This migration can create localized heating (Joule heating) and contribute to temperature homogenization in microscale applications.

Electro-thermal flows can be engineered to focus heat removal from hotspots, providing targeted thermal management solutions [17]. Figure 6.4 shows the models developed by researchers to predict the thermal conductivity enhancement in.

6.3.4 Mass Transport Influenced by Ionic Motion and Nanoparticle Migration

6.3.4.1 Ion-Driven Mass Transport

Ionic liquids naturally support efficient mass transport through ion migration under electric fields. In IONFs, the presence of nanoparticles modifies this behavior by interacting with the ionic environment.

Key aspects include:

- **Drift phenomena:** Ions move toward oppositely charged electrodes, facilitating mass redistribution and mixing.
- **Ionic wind effect:** Ionic migration can induce bulk fluid motion, known as ionic wind, enhancing mixing and mass transport even in the absence of external mechanical pumping [18].

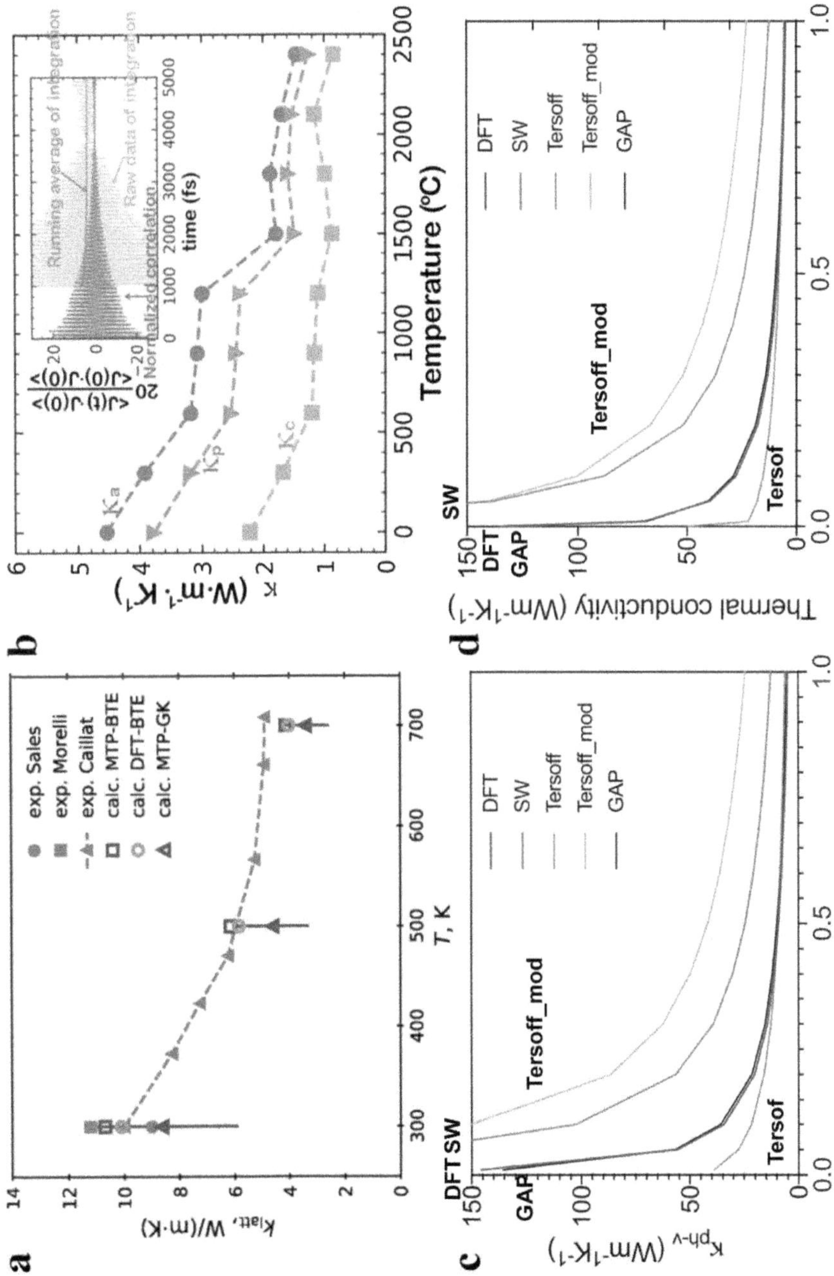

FIGURE 6.4 Thermal conductivity enhancement predictions by machine learning approach [13].

6.3.4.2 Nanoparticle Migration

Nanoparticles in IONFs do not remain static; their migration is affected by:

- **Thermophoresis:** Movement along temperature gradients, often pushing particles from hot to cold regions.
- **Diffusiophoresis:** Migration driven by concentration gradients of ions or other solutes.
- **Electrophoresis:** Migration under applied electric fields, dependent on surface charge and zeta potential.

The coupled effects of these migration phenomena lead to a dynamic and adaptive mass transport profile, which can be strategically leveraged to enhance process intensification and control localized concentration or temperature fields [19].

6.3.5 Diffusive and Convective Transport Contributions

6.3.5.1 Diffusive Transport

Diffusion remains a foundational mass transport mechanism in IONFs. The effective diffusivity of nanoparticles and ions is determined by:

$$D_{eff} = D_0 + D_B + D_T$$

where:

- D_0: Molecular (Brownian) diffusivity.
- D_B: Brownian diffusion contribution, significant at nanoscales.
- D_T: Thermophoretic diffusivity.

For ions, Nernst–Planck equations describe diffusion under combined concentration, electric, and convective forces [20].

6.3.5.2 Convective Transport

Convective transport in IONFs is greatly enhanced compared to pure ionic liquids or water-based nanofluids due to:

- Improved thermal conductivity, facilitating stronger buoyancy forces.
- Lower effective viscosity under shear-thinning behavior (in some formulations), leading to higher Reynolds numbers and enhanced convective mixing.
- Electrohydrodynamic (EHD) flows resulting from ion migration and field-induced body forces.

In heat exchangers and microchannels, forced convection mechanisms allow IONFs to remove or deliver thermal energy with higher efficiency.

6.3.5.3 Interplay between Diffusion and Convection

In practical systems, diffusive and convective mechanisms operate synergistically. High Peclet numbers indicate convection-dominated transport, but nanoscale diffusion remains critical in regions with low velocities (e.g., near walls or corners), preventing particle accumulation and ensuring uniform property distribution.

6.3.5.4 Electroosmotic and Electrophoretic Effects

6.3.5.4.1 Electroosmosis in IONFs

Electroosmosis refers to the movement of a liquid relative to a stationary charged surface under an electric field. In IONFs, the presence of an electric double layer (EDL) around solid surfaces and nanoparticles makes electroosmotic flows particularly relevant.

Key features:

- Enhanced fluid pumping at microscale without mechanical moving parts
- Precise control of flow rate via electric field strength and surface charge manipulation
- Potential for localized heat removal, especially in micro-electronic cooling applications

6.3.5.4.2 Electrophoresis of Nanoparticles

Electrophoresis involves direct migration of charged nanoparticles under an applied electric field. Unlike purely ionic solutions, IONFs exhibit more complex behavior:

- The dense ionic environment may screen particle charges, affecting mobility.
- Nanoparticles functionalized with specific ionic groups can exhibit tailored electrophoretic responses, enabling particle positioning or separation within a system.

6.3.5.4.3 Applications of Electrokinetic Effects

Electroosmotic and electrophoretic effects are harnessed in:

- **Microfluidic devices:** Enabling pumping and mixing without mechanical actuators.
- **Directed assembly:** Aligning nanoparticles to form conductive or thermal pathways.
- **On-demand heat spreading:** Modulating localized flow fields to adjust thermal transport in response to transient heating conditions.

6.4 ENERGY SYSTEM APPLICATIONS OF IONANOFLUIDS

The emergence of ionanofluids (IONFs), a hybrid class combining the unique properties of ionic liquids with nanoparticle suspensions, marks a major milestone in advanced thermal and energy storage technologies [21]. Their superior thermal, electrical, and rheological properties make them ideally suited for a wide spectrum of energy applications, ranging from microscale electronics to large-scale renewable energy systems.

This section delves into key energy system applications of IONFs, including advanced heat exchangers, solar thermal and hybrid PV/T systems, thermal energy storage, battery and supercapacitor thermal management, and waste heat recovery, including geothermal applications [22].

6.4.1 Advantages of IONFs

IONFs overcome many of these limitations through:

- **High thermal conductivity:** The combination of nanoparticles and ionic liquids offers up to two to three times higher conductivity than base fluids.
- **Stable dispersion:** Ionic liquids naturally stabilize nanoparticles, reducing sedimentation and clogging.
- **Low volatility and high thermal stability:** Essential for high-temperature or vacuum applications [23].

6.4.2 Design Improvements in Heat Exchangers

The enhanced thermal properties enable more compact designs with:

- **Reduced heat transfer surface area:** Lowering capital and maintenance costs.
- **Higher heat flux capabilities:** Supporting higher operational loads.
- **Lower pumping power requirements:** Some IONFs exhibit shear-thinning behavior, reducing viscosity under flow.

6.4.2.1 Examples and Case Studies

Recent laboratory and pilot-scale studies have shown:

- Shell-and-tube heat exchangers using IONFs achieve up to 30% higher heat transfer coefficients.
- Plate heat exchangers employing IONFs show improved temperature uniformity and lower approach temperatures.
- Microchannel heat exchangers with IONFs enable higher heat flux removal in electronic cooling [24].

6.4.3 Solar Thermal and Hybrid PV/T Systems

6.4.3.1 Enhancing Solar Thermal Collectors

Solar thermal systems rely on efficient heat capture and transport. Standard fluids often limit absorber temperature and overall efficiency.

IONFs improve performance through:

- **Higher solar absorptivity:** Nanoparticles can be tailored to enhance spectral selectivity, absorbing a wider range of solar wavelengths.

- **Reduced thermal losses:** The low vapor pressure of ionic liquids minimizes evaporation losses at high temperatures.
- **Higher thermal conductivity:** Enables faster heat removal, maintaining lower surface temperatures and preventing thermal degradation.

6.4.3.2 Hybrid PV/T (Photovoltaic/Thermal) Systems

PV/T systems generate electricity and simultaneously recover thermal energy. However, overheating of photovoltaic cells reduces their efficiency.

6.4.3.2.1 IONF-Based Cooling

- **Efficient cell cooling:** Maintaining optimal PV operating temperatures (around 25–35°C) boosts electrical output.
- **Improved thermal energy recovery:** Higher heat transfer coefficients enable more effective thermal utilization.
- **Electrostatic-driven flow control:** In micro-channel PV/T systems, electroosmotic flows using IONFs can enable uniform cooling without pumps.

6.4.3.2.2 Performance Metrics

Studies have demonstrated:

- Overall system efficiency improvements up to 15–20% with IONFs.
- Increase in electrical efficiency by 2–4% due to effective temperature regulation.
- Higher thermal output, allowing integration with secondary uses such as domestic water heating or absorption cooling [25].

6.4.4 Thermal Energy Storage (Latent and Sensible)

6.4.4.1 Importance of Thermal Energy Storage (TES)

TES systems are critical for balancing supply and demand in renewable energy, improving grid stability, and enabling off-peak utilization.

6.4.4.2 Challenges with Traditional TES Materials

- Low thermal conductivity leading to slow charge/discharge rates
- Thermal stratification and non-uniform temperature distributions
- Limited operational temperature ranges and phase change stability issues

6.4.4.3 IONFs as TES Enhancers

6.4.4.3.1 Sensible Heat Storage

IONFs, with their high heat capacity and conductivity, improve sensible storage media:

- Faster thermal response due to higher heat transfer rates.
- Higher volumetric energy density, reducing storage tank size.
- Wide operational temperature windows due to ionic liquid stability.

6.4.4.3.2 *Latent Heat Storage*

When integrated with phase change materials (PCMs):

- Nanoparticles act as nucleation agents, reducing supercooling and improving solid–liquid interface movement.
- Enhanced thermal conductivity leads to faster melting/freezing cycles.

6.4.4.3.3 *Hybrid Configurations*

Examples include IONFs encapsulated within PCM modules, providing both heat transfer fluid and storage medium functions. Experimental studies report:

- Up to 35% faster charging/discharging cycles.
- Enhanced cyclic stability over 500+ thermal cycles.
- Reduction in temperature gradients within storage units, preventing hot/cold spots [26].

6.4.5 Battery and Supercapacitor Thermal Management

6.4.5.1 Thermal Challenges in Electrochemical Energy Storage

High power densities in batteries and supercapacitors generate significant heat, potentially leading to:

- Capacity fading and reduced lifespan.
- Thermal runaway and safety hazards.
- Reduced charge/discharge efficiency.

6.4.5.2 IONFs as Cooling Solutions

IONFs offer multiple advantages for battery thermal management systems (BTMS):

- **High thermal conductivity:** Facilitates rapid heat extraction from cell surfaces.
- **Low flammability and high thermal stability:** Essential for safety-critical applications.
- **Electrical conductivity tuning:** Ionic component allows for passive or active electrical interactions without compromising insulation requirements.

6.4.5.3 Supercapacitors and Hybrid Energy Storage

In high-power supercapacitors, transient heat spikes during rapid charge/discharge cycles can exceed the thermal limits of conventional coolants. IONFs can be integrated in:

- Cooling jackets around cell modules.
- Direct immersion cooling, where modules are submerged in IONFs for maximum contact and uniformity.

6.4.5.4 Performance Outcomes

Prototype studies and simulations demonstrate:

- 20–40% reductions in maximum cell temperatures under high load cycles.
- Increased charge/discharge efficiency due to more stable thermal profiles.
- Extended operational life, with up to 30% improvement in cycle life under stress tests [27].

6.4.6 Waste Heat Recovery and Geothermal Applications

6.4.6.1 Waste Heat Recovery (WHR)

Industrial processes, power plants, and transportation sectors release vast amounts of low-to-medium grade waste heat. Recovering this energy improves overall system efficiency and reduces greenhouse gas emissions.

6.4.6.1.1 IONF Advantages in WHR

- Capability to operate at higher temperatures where conventional fluids degrade
- Enhanced heat recovery rates due to superior convective heat transfer
- Lower corrosion rates in harsh industrial environments

6.4.6.1.2 Applications

- Heat exchangers recovering exhaust heat from gas turbines
- Recovery from chemical reactors and process heaters
- Organic Rankine cycle (ORC) systems using IONFs as secondary fluids for electricity generation

6.4.6.2 Geothermal Energy Systems

6.4.6.2.1 Current Fluid Limitations

Traditional geothermal fluids (water, brine) face scaling, corrosion, and limited heat extraction capacity, especially from low-enthalpy geothermal sources.

6.4.6.2.2 IONFs for Geothermal Extraction

- **Higher thermal capacity:** Improves heat extraction efficiency from low-temperature reservoirs.
- **Reduced scaling and corrosion:** Ionic liquids provide inherent stability against mineral deposition.
- **Enhanced flow control:** Electrokinetic manipulation enables optimized circulation and targeted heat recovery.

6.4.6.2.3 Case Study Outcomes

Field pilot tests have reported:

- 10–20% improvement in heat extraction rates.
- Better maintenance of reservoir temperature over long operational periods.
- Reduced operational downtime due to clogging and scaling.

6.4.6.3 Hybrid Geothermal–Solar Systems

IONFs also enable hybrid configurations where geothermal energy is supplemented by solar thermal collectors. The high thermal capacity and conductivity ensure continuous, stable heat supply regardless of solar intermittency [28–30].

6.4.7 Environmental and Economic Considerations

6.4.7.1 Sustainability Benefits

By improving thermal efficiencies, IONFs contribute directly to:

- Lower greenhouse gas emissions.
- Reduced fuel and electricity consumption.
- Smaller system footprints, reducing material and land use.

6.4.7.2 Cost Challenges

While performance benefits are clear, widespread adoption faces challenges, such as:

- Higher initial costs of ionic liquids and specialized nanoparticles.
- Processing and recycling complexities.
- Need for robust system designs to prevent leakage and material incompatibility [31].

6.5 MODELING IONANOFLUID–ENERGY DEVICE INTEGRATION

6.5.1 System-Level Modeling Frameworks

Traditional component-level analyses often fail to capture the holistic interactions and synergistic effects present in IONF-integrated energy devices. A system-level modeling approach allows for:

- Predicting overall energy efficiency and heat recovery rates.
- Balancing thermal and electrical performance in hybrid devices.
- Evaluating system dynamics under transient and variable operational scenarios (e.g., variable solar irradiance, fluctuating electrical loads) [32].

6.5.2 Multi-Scale Modeling

IONF behavior must be resolved across multiple scales:

- **Microscale:** Nanoparticle motion, EDL (electric double layer) dynamics, and localized heat generation.
- **Mesoscale:** Flow and thermal fields within heat exchangers, microchannels, and battery modules.
- **Macroscale:** Overall system energy balances, exergy analyses, and life-cycle assessments.

Coupling these scales demands advanced numerical techniques, such as:

- CFD-embedded reduced-order models.
- Lattice Boltzmann methods (LBM) for meso-level particle dynamics.
- Finite element and finite volume approaches for global energy device simulations [33–35].

6.5.3 Dynamic Modeling and Control

Dynamic models incorporate time-dependent boundary conditions, enabling transient simulations during startup, shutdown, or load changes. When integrated with control algorithms, such models support:

- Real-time thermal management in battery systems.
- Adaptive flow modulation in microchannel networks using electrokinetic effects.
- Predictive maintenance scheduling based on system stress predictions [36].

6.5.4 Hybrid Systems Combining IONFs with Phase Change Materials

6.5.4.1 Rationale for Hybrid Thermal Storage

Combining IONFs with PCMs merges the high thermal conductivity and tunable flow characteristics of IONFs with the high energy storage density of PCMs. These hybrid systems address key limitations of standalone PCMs, such as slow heat transfer rates and temperature stratification.

6.5.4.2 Modeling Challenges

The coexistence of solid–liquid phase transitions, particle migration, and ionic effects introduces modeling complexities, such as:

- Tracking the solid–liquid interface in PCMs under non-uniform heating or cooling.
- Modeling nanoparticle-enhanced thermal conductivity within PCMs (nano-enhanced PCMs).
- Capturing dynamic viscosity and flow property changes during phase transitions.

6.5.4.3 Numerical Techniques

Approaches for simulating IONF-PCM systems include:

- Enthalpy-porosity methods to handle latent heat effects and phase interface movements.
- Two-phase mixture models for simulating nanoparticle dispersion in molten PCMs.
- Coupled CFD and molecular dynamics simulations to resolve interfacial phenomena at nanoscale.

6.5.4.4 Performance Predictions

Simulations show that IONF-PCM systems can achieve:

- Up to 40% faster charging and discharging rates compared to pure PCMs.
- Improved cyclic stability and thermal uniformity, reducing hotspots.
- Higher overall storage efficiency, especially in solar thermal and concentrated solar power (CSP) applications [37].

6.5.5 Coupling with Renewable Energy Sources

6.5.5.1 Solar Thermal and PV/T Systems

In solar collectors and PV/T hybrids, system models incorporating IONFs must account for:

- Spectral absorptivity and radiative heat transfer enhancements.
- Dynamic interactions between electrical output (PV) and thermal harvesting (T).
- Control of fluid flow rates using electroosmotic or electromagnetic methods to adapt to solar intermittency.

Models often combine:

- Radiation transport equations (Monte Carlo or discrete ordinates methods).
- Fluid and thermal fields governed by Navier–Stokes and energy equations.
- Electrical circuit analogies to simulate PV cell behavior.

6.5.5.2 Geothermal Systems

IONFs can be modeled within geothermal loops to predict:

- Enhanced heat extraction from low-enthalpy reservoirs.
- Electrokinetically driven flow improvements reducing mechanical pumping energy.
- Long-term reservoir thermal stability.

These models integrate porous media flow equations, heat exchanger dynamics, and energy balance constraints across the entire geothermal plant.

6.5.5.3 Wind Energy Hybridization

While not directly used in wind turbines, IONFs can be part of hybrid energy hubs where wind, solar, and thermal systems interact. Integrated system models consider:

- Shared thermal storage buffers using IONF–PCM combinations.
- Load balancing strategies using energy flows between renewable sources and thermal energy storage.
- Environmental and economic trade-offs in optimizing hybrid layouts [38–40].

6.5.6 Multi-Objective Optimization (Performance, Cost, Environmental Impact)

6.5.6.1 Why Multi-Objective Approaches Are Essential

Energy systems integrating IONFs must meet multiple, often conflicting, objectives, such as:

- Maximizing thermal performance and energy conversion efficiency.
- Minimizing life cycle costs and operational expenses.
- Reducing environmental impacts such as emissions and resource depletion.

6.5.6.2 Optimization Frameworks

Optimization methodologies include:

- **Pareto-based genetic algorithms:** Identify sets of solutions representing trade-offs among competing objectives.
- **Multi-criteria decision analysis (MCDA):** Incorporates stakeholder preferences and weighting factors.
- **Gradient-based and surrogate model techniques:** Accelerate convergence and enable rapid sensitivity analyses.

6.5.6.3 Key Performance Indicators (KPIs)

Examples of KPIs in IONF-integrated energy systems include:

- System thermal efficiency (overall and component-specific).
- Exergy destruction and entropy generation rates.
- Payback period and net present value (NPV) for economic viability.
- Global warming potential (GWP) and embodied energy for environmental assessments.

6.5.6.4 Case Study Examples

In a hybrid PV/T system using IONFs, optimization studies have shown:

- Up to 20% higher overall energy conversion efficiencies with an acceptable cost increase of 8–12%.
- Reduced greenhouse gas emissions by approximately 15–18% compared to baseline water-based PV/T setups.
- Favorable economic returns when integrated into smart grids with dynamic pricing [41].

6.5.7 Real-World Deployment Challenges

6.5.7.1 Material Compatibility and Corrosion

Though ionic liquids offer better corrosion resistance than water or glycol, certain alloys and seals may still degrade over time. Models must predict:

- Ion-metal interactions leading to material loss.
- Deposition and scaling tendencies under fluctuating thermal conditions.

6.5.7.2 Long-Term Stability and Recycling

Nanoparticle agglomeration and ionic decomposition can alter system behavior over long durations. Challenges include:

- Tracking time-dependent changes in viscosity and thermal conductivity.
- Designing reconditioning or recycling loops within system models to maintain performance.

6.5.7.3 Operational Complexity and Training

Advanced energy systems using IONFs often require sophisticated control and monitoring infrastructures. Key issues include:

- Developing predictive maintenance models to minimize downtime.
- Training operators in electrokinetic flow control, safe handling of ionic liquids, and nanoparticle management.
- Establishing safety protocols to mitigate risks of leaks, exposure, or thermal runaway [42–44].

6.6 EXPERIMENTAL VALIDATION AND UNCERTAINTY ANALYSIS

This section examines the role of empirical benchmarking, experimental techniques for flow and thermal property characterization, model sensitivity and robustness analysis, sources of uncertainty and their mitigation, and best practices for integrating models with experimental findings [45].

6.6.1 Importance of Empirical Benchmarking

6.6.1.1 Bridging Theory and Practice

Mathematical models rely on assumptions, simplifications, and empirical correlations. While these models provide insights into fluid dynamics, heat transfer, and system performance, they inherently approximate real-world behavior. Empirical benchmarking ensures that these approximations are valid under practical conditions [46–48].

6.6.1.2 Design Validation and Optimization

Experimental data enable:

- Verification of thermal and flow performance predictions.
- Calibration of empirical constants in correlations (e.g., Nusselt number, friction factor).
- Identification of performance-limiting factors not captured in simulations.

6.6.1.3 Building Confidence in Scalability

Especially for emerging fluids like IONFs, laboratory-scale validation supports scaling predictions for pilot and industrial systems. Accurate scale-up models rely on experimentally validated dimensionless parameters and heat transfer characteristics.

6.6.1.4 Regulatory and Safety Compliance

Validated experimental data provide essential evidence for:

- Meeting performance standards set by regulatory bodies.
- Proving safety and reliability claims to stakeholders and investors.
- Supporting life cycle assessments and environmental certifications.

6.6.2 Experimental Techniques for Flow and Thermal Property Measurement

6.6.2.1 Thermophysical Property Characterization

6.6.2.1.1 Thermal Conductivity

Measurement techniques include:

- **Transient hot-wire method:** Highly accurate for liquids, where a heated wire's temperature change is tracked to infer thermal conductivity.
- **Laser flash analysis:** Used mainly for solidified or gel-like states of IONFs, measuring thermal diffusivity, which can be combined with density and specific heat to determine conductivity.

6.6.2.1.2 Specific Heat Capacity

Measured via:

- **Differential scanning calorimetry (DSC):** Provides precise heat flow data as a function of temperature, widely used to capture temperature-dependent specific heat.
- **Modulated DSC:** Allows separation of reversing (heat capacity-related) and non-reversing (e.g., transitions) heat flows.

6.6.2.1.3 Viscosity

Assessed using:

- **Rotational rheometers:** Capture both Newtonian and non-Newtonian behavior, crucial for shear-dependent IONFs.
- **Capillary viscometers:** Simple but less suitable for non-Newtonian or highly viscous samples.

6.6.2.1.4 Density

Measured using:

- **Oscillating U-tube densimeters:** Accurate over a wide temperature range.
- **Hydrometers or pycnometers:** More basic methods, used for approximate or rapid measurements.

6.6.2.1.5 Ionic Conductivity

Electrochemical impedance spectroscopy (EIS) provides frequency-dependent ionic conductivity and insights into ion mobility, critical for applications involving electrokinetic effects.

6.6.2.2 Flow and Heat Transfer Experiments

6.6.2.2.1 Microchannel and Macro-Channel Setups

For validating convective heat transfer models, experimental rigs with:

- Transparent microchannels for flow visualization.
- Embedded thermocouples and infrared thermography for detailed temperature profiling.
- Pressure drop sensors to assess hydrodynamic performance.

6.6.2.2.2 Heat Exchanger Test Rigs

Laboratory-scale heat exchangers (e.g., shell-and-tube, plate, micro-compact) using IONFs allow measurement of:

- Overall heat transfer coefficients.
- Effectiveness-NTU (number of transfer units) performance.
- Fouling and sedimentation behaviors over operational cycles.

6.6.2.2.3 Battery and Electronics Cooling Modules

Experimental setups simulate real-world operating conditions, including:

- Thermal mock-ups or actual battery modules.
- Controlled heat fluxes to emulate charging/discharging cycles.
- Temperature uniformity and transient thermal response tracking.

6.6.2.2.4 Solar Thermal Collector Prototypes

For solar applications:

- Outdoor or indoor solar simulators with controllable irradiance.
- IONF-based circulation loops to measure thermal gain, outlet fluid temperatures, and system efficiency.

6.6.2.3 Advanced Diagnostic Techniques

6.6.2.3.1 Particle Image Velocimetry (PIV)

Visualizes velocity fields within transparent channels or cells, revealing recirculation zones, boundary layer development, and electroosmotic effects.

6.6.2.3.2 Infrared Thermography

Non-contact measurement of surface and fluid temperatures, valuable for mapping thermal distributions and identifying hotspots.

6.6.2.3.3 High-Speed Imaging

Captures transient flow phenomena, especially useful in electrokinetic-driven flows or dynamic phase change interactions [49].

6.6.3 Model Sensitivity and Robustness Analysis

6.6.3.1 Importance of Sensitivity Analysis

Mathematical models often contain numerous input parameters—thermophysical properties, boundary conditions, and empirical coefficients. Sensitivity analysis identifies which parameters most significantly affect model outputs, guiding prioritization of experimental measurements.

6.6.3.2 Approaches

- **Local sensitivity analysis:** Examines partial derivatives of outputs with respect to individual inputs, assuming small perturbations.
- **Global sensitivity analysis:** Evaluates model behavior across entire parameter ranges, using techniques such as Sobol indices or Morris method.

6.6.3.3 Parameter Prioritization

Through sensitivity analysis, researchers can:

- Focus experimental efforts on the most influential properties (e.g., viscosity at operational shear rates).
- Reduce uncertainty in model predictions by tightly controlling critical input parameters.
- Design experiments to generate data for refining uncertain model inputs.

6.6.3.4 Robustness and Uncertainty Quantification

Robustness studies simulate model outputs under varied operating conditions and input perturbations. They assess:

- Model stability under realistic variability in inputs.
- Predictive reliability for long-duration or extreme operating scenarios.

Quantifying uncertainty (e.g., via Monte Carlo simulations) further supports decision-making on design tolerances and operational safety margins [40].

6.6.4 Uncertainty Sources and Error Mitigation

6.6.4.1 Experimental Uncertainties

Sources include:

- Instrument precision and calibration errors (e.g., thermocouple inaccuracies).
- Environmental influences (e.g., ambient temperature fluctuations).
- Human errors in sample preparation or measurement reading.

6.6.4.2 Modeling Uncertainties

Uncertainties arise from:

- Simplified physical assumptions (e.g., neglecting minor electrohydrodynamic effects).
- Empirical correlation inaccuracies outside validated ranges.
- Numerical discretization and convergence errors in computational models.

6.6.4.3 Mitigation Strategies

6.6.4.3.1 Redundant Measurements

Using multiple measurement techniques for the same property (e.g., both DSC and modulated DSC for heat capacity) increases confidence and enables cross-validation.

6.6.4.3.2 Calibration and Standardization

Regular calibration against certified reference materials and inter-laboratory comparisons reduce systematic errors.

6.6.4.3.3 Refinement Loops

Iterative loops between experiments and models, where initial experimental data refine model parameters, and improved models predict new scenarios for further validation.

6.6.4.3.4 Error Propagation Analysis

Quantitative assessment of how input uncertainties affect final outputs, enabling researchers to set acceptable error bands for design decisions.

6.6.5 Best Practices for Model–Experiment Integration

6.6.5.1 Co-Design of Experiments and Models

Rather than developing models and experiments in isolation, co-design ensures that:

- Experiments generate data that directly inform critical model parameters.
- Models identify measurement requirements and guide experimental setup design.

6.6.5.2 Iterative Validation Cycles

Best practices involve:

1. **Initial baseline experiments:** Obtain preliminary data for parameter estimation.
2. **Model calibration:** Adjust models based on initial experimental findings.
3. **Extended validation:** Test under wider operating ranges, including off-design conditions.
4. **Final model refinement:** Achieve robust, generalizable predictive capabilities.

6.6.5.3 Data Sharing and Reproducibility

Publishing detailed experimental protocols, raw data, and model source codes promotes reproducibility, builds community trust, and accelerates further advancements.

6.6.5.4 Digital Twins and Real-Time Validation

Emerging digital twin frameworks integrate real-time operational data with predictive models. These systems continuously compare predictions against live data, enabling:

- Ongoing model refinement during system operation.
- Early detection of deviations indicating potential faults or performance degradation.
- Adaptive control strategies to maintain optimal performance.

6.6.5.5 Cross-Disciplinary Collaboration

Successful integration of IONF models and experiments requires expertise in:

- Fluid dynamics and heat transfer.
- Electrochemistry and electrokinetics.
- Materials science (for nanoparticle and ionic liquid behavior).
- Systems engineering and control.

Establishing multi-disciplinary teams ensures a holistic understanding and better system-level outcomes.

6.7 FUTURE DIRECTIONS AND RESEARCH OUTLOOK

6.7.1 Emerging Advanced Formulations (Hybrid Ions, Functionalized Nanoparticles)

6.7.1.1 Beyond Classical Ionic Liquids

The first generation of IONFs primarily combined simple nanoparticles (e.g., Al_2O_3, SiO_2, CNTs) with standard ionic liquids like [BMIM][PF_6] or [EMIM][BF_4]. Although these formulations have demonstrated enhanced thermal and electrochemical properties, further progress requires:

- **Hybrid ions:** New ionic liquids incorporating functional groups (e.g., hydroxyl, amine, fluorinated chains) enable precise tuning of polarity, thermal stability, and electrochemical windows. This flexibility improves nanoparticle dispersion and allows targeted interactions with device surfaces.
- **Task-specific ionic liquids (TSILs):** Designed to optimize specific properties such as ultra-high thermal conductivity, minimal viscosity, or selective solubility with certain nanoparticles.

6.7.1.2 Functionalized Nanoparticles

Surface engineering of nanoparticles addresses key issues such as sedimentation, aggregation, and interfacial resistance. Recent developments include:

- **Covalent functionalization:** Surface modification with organic or polymeric chains enhances compatibility with ionic liquids and prevents agglomeration.
- **Core-shell structures:** Nanoparticles with metallic or ceramic cores and functional shells combine high conductivity with tunable surface chemistry.

6.7.1.3 Hybrid Multi-Component Systems

- Improved thermal conductivity beyond linear mixing rules.
- Simultaneous enhancement of thermal and electrical conductivity for electro-thermal energy storage.
- Reduced viscosity despite higher particle loadings due to tailored interparticle interactions.

6.7.2 Advantages of AI and ML Integration

Machine learning (ML) and artificial intelligence (AI) offer powerful tools to:

- **Handle complex, nonlinear relationships:** ML algorithms can identify hidden patterns in high-dimensional experimental and simulation data.
- **Accelerate optimization:** Rapid exploration of large design spaces enables real-time decision-making for performance improvements.
- **Predict dynamic system responses:** Recurrent neural networks (RNNs) and long short-term memory (LSTM) architectures capture transient and time-dependent behaviors.

6.7.3 Applications in IONF Systems

- **Thermophysical property prediction:** ML models trained on experimental and molecular simulation data can predict thermal conductivity, viscosity, and specific heat across wide compositional and temperature ranges.
- **Design of advanced formulations:** Generative models (e.g., generative adversarial networks, variational autoencoders) propose new nanoparticle–ionic liquid combinations optimized for specific applications.
- **System-level performance forecasting:** Integrated ML models provide rapid predictions of overall system efficiency, pressure drops, and energy savings.

6.7.4 Challenges and Research Needs

- Access to large, high-quality datasets for training remains limited, highlighting the importance of collaborative data sharing.
- Interpretability of ML models needs improvement to understand the underlying physical mechanisms.

- Integration frameworks that combine ML with physics-based models (so-called physics-informed machine learning) require further development.

6.7.5 Digital Twin Concepts and Real-Time Control

6.7.5.1 Concept of Digital Twins

A digital twin is a virtual representation of a physical system that updates in real-time using operational data, allowing predictive monitoring, optimization, and control. For IONF-based energy systems, digital twins offer transformative potential by continuously bridging simulations and experimental performance.

6.7.5.2 Components of IONF Digital Twins

- **Physical model:** Incorporates thermophysical properties, flow dynamics, and heat transfer behavior derived from experiments and validated simulations.
- **Data layer:** Real-time measurements from sensors (temperature, flow rate, pressure, conductivity).
- **AI/ML layer:** Advanced algorithms to detect deviations, forecast future performance, and suggest corrective actions.

6.7.5.3 Advantages for IONF Energy Systems

- **Predictive maintenance:** Early identification of fouling, sedimentation, or system degradation before failures occur.
- **Dynamic optimization:** Continuous adjustment of flow rates, pump speeds, or thermal inputs for peak efficiency.

6.7.5.4 Research Opportunities

- Development of robust, high-resolution sensors compatible with IONFs and harsh operating conditions.
- Real-time integration platforms with minimal computational lag.
- Adaptive algorithms that learn and evolve with system operation to improve predictive accuracy over time.

6.7.6 Pilot-Scale Demonstrations

6.7.6.1 Pilot Plants Are Important for the Following Reasons

- Refine techno-economic models.
- Identify unforeseen operational challenges.
- Build stakeholder confidence for commercial investment.

6.7.6.2 Research Needs

- Exploration of bio-derived or recyclable ionic liquids to address sustainability.
- Development of modular system designs to facilitate retrofitting and minimize operational disruptions.

6.7.7 Environmental, Safety, and Regulatory Considerations

6.7.7.1 Environmental Impacts

- **Nanoparticle release:** Accidental leakage can lead to nanoparticle dispersion into ecosystems.
- **Ionic liquid degradation:** Some ionic liquids may form toxic degradation products under thermal or oxidative stress.

Life cycle assessments (LCAs) are essential to quantify net environmental benefits and guide responsible material selection.

6.7.7.2 Regulatory Frameworks

Adoption of IONFs must align with evolving global regulations, including:

- REACH (Registration, Evaluation, Authorization, and Restriction of Chemicals) in Europe.
- TSCA (Toxic Substances Control Act) in the United States.
- Local environmental protection standards governing nanoparticle use and disposal.

6.7.7.3 Mitigation Strategies

- **Material design for safety (DfS):** Developing inherently safer formulations, using biocompatible nanoparticles and benign ionic liquids.
- **Closed-loop systems:** Designing energy systems to fully contain and recycle IONFs, minimizing environmental exposure.
- **Robust monitoring and control:** Implementation of advanced leak detection, containment strategies, and real-time system health monitoring.

6.7.7.4 Community Engagement and Policy Advocacy

Proactive engagement with regulators, industry bodies, and the public is essential to:

- Build trust in the safety and environmental performance of IONF-based systems.
- Influence policy development to support innovative thermal and energy technologies.
- Develop educational initiatives for safe handling and maintenance practices.

REFERENCES

1. De Castro, C. N., Ribeiro, A. P., Vieira, S. I., França, J. M., Lourenço, M. J., Santos, F. V., . . . & Hardacre, C. (2013). Synthesis, properties and physical applications of ionanofluids. In *Ionic liquids-new aspects for the future*. IntechOpen.
2. Paredes, X., Lourenço, M. J., Castro, C. N. D., & Wakeham, W. (2021). Thermal conductivity of ionic liquids and ionanofluids. Can molecular theory help? *Fluids*, *6*(3), 116.
3. Gardas, R. L., & Coutinho, J. A. (2008). A group contribution method for viscosity estimation of ionic liquids. *Fluid Phase Equilibria*, *266*(1–2), 195–201.

4. Choi, S. U., & Eastman, J. A. (1995). *Enhancing thermal conductivity of fluids with nanoparticles* (No. ANL/MSD/CP-84938; CONF-951135-29). Argonne National Lab. (ANL), Argonne, IL (United States).
5. Buongiorno, J. (2006). Convective transport in nanofluids. *Journal of Heat Transfer*, 128, 240–250. http://dx.doi.org/10.1115/1.2150834
6. Minea, A. A., & Sohel Murshed, S. M. (2021). Ionic liquids-based nanocolloids: A review of progress and prospects in convective heat transfer applications. *Nanomaterials*, *11*(4), 1039.
7. Liu, X., & Chen, Y. (2013). Transient thermal performance analysis of micro heat pipes. *Applied Thermal Engineering*, *58*(1–2), 585–593.
8. Minea, A. A., & Sohel Murshed, S. M. (2021). Ionic liquids-based nanocolloids: A review of progress and prospects in convective heat transfer applications. *Nanomaterials*, *11*(4), 1039.
9. Khan, I., Bhat, A. H., Sharma, D. K., Usmani, M. A., & Khan, F. (2019). Overview of nanofluids to ionanofluids: Applications and challenges. *Nanomaterials for Healthcare, Energy and Environment*, 199–227.
10. Huminic, G., & Huminic, A. (2021). Heat transfer capability of ionanofluids for heat transfer applications. *International Journal of Thermophysics*, *42*(1), 12.
11. Chereches, E. I., Bejan, D., Ibanescu, C., Danu, M., & Minea, A. A. (2021). Ionanofluids with [C2mim][CH3SO3] ionic liquid and alumina nanoparticles: An experimental study on viscosity, specific heat and electrical conductivity. *Chemical Engineering Science*, *229*, 116140.
12. Huminic, G., & Huminic, A. (2021). Heat transfer capability of ionanofluids for heat transfer applications. *International Journal of Thermophysics*, *42*(1), 12.
13. Luo, Y., Li, M., Yuan, H., Liu, H., & Fang, Y. (2023). Predicting lattice thermal conductivity via machine learning: A mini review. *NPJ Computational Materials*, *9*(1), 4.
14. Fabre, E., & Murshed, S. S. (2021). A review of the thermophysical properties and potential of ionic liquids for thermal applications. *Journal of Materials Chemistry A*, *9*(29), 15861–15879.
15. França, J. M., Lourenço, M. J. V., Murshed, S. S., Pádua, A. A., & Nieto de Castro, C. A. (2018). Thermal conductivity of ionic liquids and ionanofluids and their feasibility as heat transfer fluids. *Industrial & Engineering Chemistry Research*, *57*(18), 6516–6529.
16. Topal, I., & Servantie, J. (2019). Molecular dynamics study of the thermal conductivity in nanofluids. *Chemical Physics*, *516*, 147–151.
17. Pereira, J., Souza, R., Moita, A., & Moreira, A. (2023). Nanofluids and ionic fluids as liquid electrodes: An overview on their properties and potential applications. *Processes*, *11*(11), 3189.
18. Pereira, J., Souza, R., Moreira, A., & Moita, A. (2024). An overview of the ionic liquids and their hybrids operating in electrochemical cells and capacitors. *Ionics*, *30*(8), 4343–4385.
19. Hothar, M., Sundén, B., & Wu, Z. (2022). Rheological behaviour of IoNanofluids based on [emim][DCA] and [emim][TCM]. *Journal of Molecular Liquids*, *348*, 118064.
20. Das, S. K., Putra, N. S. D., Thiesen, P., & Roetzel, W. (2003). Temperature dependence of thermal conductivity enhancement for nanofluids. *Journal of Heat Transfer*, *125*(4), 567–574.
21. MacFarlane, D. R., Tachikawa, N., Forsyth, M., Pringle, J. M., Howlett, P. C., Elliott, G. D., . . . & Angell, C. A. (2014). Energy applications of ionic liquids. *Energy & Environmental Science*, *7*(1), 232–250.

22. Minea, A. A., & Sohel Murshed, S. M. (2021). Ionic liquids-based nanocolloids: A review of progress and prospects in convective heat transfer applications. *Nanomaterials, 11*(4), 1039.
23. Paul, T. C., Tikadar, A., Mahamud, R., Salman, A. S., Morshed, A. M., & Khan, J. A. (2021). A critical review on the development of ionic liquids-based nanofluids as heat transfer fluids for solar thermal energy. *Processes, 9*(5), 858.
24. Das, L., Rubbi, F., Habib, K., Saidur, R., Islam, N., Saha, B. B., . . . & Irshad, K. (2021). Hydrothermal performance improvement of an inserted double pipe heat exchanger with Ionanofluid. *Case Studies in Thermal Engineering, 28*, 101533.
25. Bo, L., Zhang, X., Luo, Z., Saboori, T., Dehghan, M., Ghasemizadeh, M., . . . & Mahian, O. (2022). An overview of the applications of ionic fluids and deep eutectic solvents enhanced by nanoparticles. *Journal of Thermal Analysis and Calorimetry, 147*(14), 7589–7601.
26. Faraji, S., Shekaari, H., Zafarani-Moattar, M. T., Mokhtarpour, M., & Asghari, E. (2023). Thermal properties of phase change materials ionic liquid/fatty acids for thermal energy storage applications. *Journal of Energy Storage, 67*, 107464.
27. Ray, A., & Saruhan, B. (2021). Application of ionic liquids for batteries and supercapacitors. *Materials, 14*(11), 2942.
28. Thakur, P., Sonawane, S. S., Sonawane, S. H., & Bhanvase, B. A. (2020). Nanofluids-based delivery system, encapsulation of nanoparticles for stability to make stable nanofluids. In *Encapsulation of active molecules and their delivery system* (pp. 141–152). Elsevier.
29. Bethi, B., Sonawane, S. H., Bhanvase, B. A., & Sonawane, S. S. (2021). Textile industry wastewater treatment by cavitation combined with Fenton and ceramic nanofiltration membrane. *Chemical Engineering and Processing-Process Intensification, 168*, 108540.
30. Minea, A. A., & Murshed, S. S. (2018). A review on development of ionic liquid based nanofluids and their heat transfer behavior. *Renewable and Sustainable Energy Reviews, 91*, 584–599.
31. Czaplicka, N., Grzegórska, A., Wajs, J., Sobczak, J., & Rogala, A. (2021). Promising nanoparticle-based heat transfer fluids—environmental and techno-economic analysis compared to conventional fluids. *International Journal of Molecular Sciences, 22*(17), 9201.
32. Joseph, A., Sobczak, J., Żyła, G., & Mathew, S. (2022). Ionic liquid and ionanofluid-based redox flow Batteries: A mini review. *Energies, 15*(13), 4545.
33. Malika, M., & Sonawane, S. S. (2020). Effect of nanoparticle mixed ratio on stability and thermo-physical properties of CuO-ZnO/water-based hybrid nanofluid. *Journal of the Indian Chemical Society, 97*(3), 414–419.
34. Sonawane, S., Thakur, P., & Paul, R. (2020). Study on thermal property enhancement of MWCNT based polypropylene (PP) nanocomposites. *Materials Today: Proceedings, 27*, 550–555.
35. He, Y. L., & Tao, W. Q. (2012). Multiscale simulations of heat transfer and fluid flow problems. *Journal of Heat Transfer, 134*(3).
36. Liao, G., Wang, W., Zhang, F., Chen, J., & Leng, E. (2022). Thermal performance of lithium-ion battery thermal management system based on nanofluid. *Applied Thermal Engineering, 216*, 118997.
37. Kiani, M., Ansari, M., Arshadi, A. A., Houshfar, E., & Ashjaee, M. (2020). Hybrid thermal management of lithium-ion batteries using nanofluid, metal foam, and phase change material: An integrated numerical–experimental approach. *Journal of Thermal Analysis & Calorimetry, 141*(5).

38. Malika, M. M., & Sonawane, S. S. (2019). Review on application of nanofluid/nano particle as water disinfectant. *Journal of Indian Association for Environmental Management (JIAEM)*, *39*(1–4), 21–24.
39. Malika, M., Bhad, R., & Sonawane, S. S. (2021). ANSYS simulation study of a low volume fraction CuO–ZnO/water hybrid nanofluid in a shell and tube heat exchanger. *Journal of the Indian Chemical Society*, *98*(11), 100200.
40. Alarcón, M., Luna-Abad, J. P., Seco-Nicolás, M., Moulefera, I., & Víllora, G. (2024). Study of ionanofluids behavior in PVT solar collectors: Determination of thermal fields and characteristic length by means of HEATT® platform. *Energies (19961073)*, *17*(22).
41. Li, J., & Zhao, H. (2021). Multi-objective optimization and performance assessments of an integrated energy system based on fuel, wind and solar energies. *Entropy*, *23*(4), 431.
42. Thakur, P., Sonawane, S., Potoroko, I., & Sonawane, S. H. (2021). Recent advances in ultrasound-assisted synthesis of nano-emulsions and their industrial applications. *Current Pharmaceutical Biotechnology*, *22*(13), 1748–1758.
43. Sonawane, A. V., Rikame, S., Sonawane, S. H., Gaikwad, M., Bhanvase, B., Sonawane, S. S., . . . & Gaikwad, R. (2024). A review of microbial fuel cell and its diversification in the development of green energy technology. *Chemosphere*, *350*, 141127.
44. Kaur, H., Thakur, A., Thakur, R. C., & Kumar, A. (2025). A review on multifaceted role of ionic liquids in modern energy storage systems: From electrochemical performance to environmental sustainability. *Energy & Fuels*, *39*(8), 3703–3734.
45. Venkateswarlu, B., Chavan, S., Joo, S. W., & Kim, S. C. (2023). A numerical study on heat transfer performance using nanofluids in liquid cooling for cylindrical battery modules. *Journal of Molecular Liquids*, *391*, 123257.
46. Malika, M., & Sonawane, S. S. (2021). The sono-photocatalytic performance of a novel water based Ti+ 4 coated Al (OH) 3-MWCNT's hybrid nanofluid for dye fragmentation. *International Journal of Chemical Reactor Engineering*, *19*(9), 901–912.
47. Parate, V. R., Kawadkar, D. K., & Sonawane, S. S. (2011). Study of whey protein concentrate fortification in cookies variety biscuits. *International Journal of Food Engineering*, *7*(2), 1–12.
48. Asim, M., & Siddiqui, F. R. (2022). Hybrid nanofluids—next-generation fluids for spray-cooling-based thermal management of high-heat-flux devices. *Nanomaterials*, *12*(3), 507.
49. Oster, K., Hardacre, C., Jacquemin, J., Ribeiro, A. P., & Elsinawi, A. (2019). Ionic liquid-based nanofluids (ionanofluids) for thermal applications: An experimental thermophysical characterization. *Pure and Applied Chemistry*, *91*(8), 1309–1340.

7 Molecular Dynamics Study of Nanofluid Applications

7.1 INTRODUCTION

Nanofluids, a colloidal suspension of nanoparticles in base fluids, have captured the attention of researchers worldwide due to their potential to enhance heat and mass transfer in energy, electronics, biomedical, and environmental applications. While experimental studies provide crucial insights into macroscale behavior, they often fail to reveal atomic- and nanoscale interactions that govern fundamental transport mechanisms. Molecular Dynamics (MD) simulation has emerged as an indispensable computational tool to bridge this knowledge gap, enabling researchers to explore phenomena at scales that are otherwise difficult or impossible to capture experimentally.

This chapter begins with the historical development of MD simulations, introduces the basic principles and governing equations, and discusses the relevance of MD for studying nanofluid behavior at the atomic and nanoscale levels [1].

7.1.1 Historical Development of MD Simulations

7.1.1.1 Early Foundations

The conceptual roots of MD can be traced back to the early 20th century, when theoretical physicists sought to understand the statistical behavior of systems comprising large numbers of interacting particles. The foundational work of Ludwig Boltzmann and Josiah Willard Gibbs laid the groundwork for statistical mechanics, providing theoretical tools to connect microscopic motions with macroscopic thermodynamic properties.

However, it wasn't until the advent of digital computers in the mid-20th century that MD simulation became a practical research tool. In 1957, Alder and Wainwright conducted the first computer simulations of hard-sphere systems, demonstrating phase transitions and transport properties arising from particle collisions. This pioneering work marked the transition from purely analytical methods to computational approaches capable of handling many-body interactions [2].

7.1.1.2 Advancements in Algorithms and Computing Power

Throughout the 1960s and 1970s, algorithmic improvements facilitated the simulation of increasingly complex systems. Rahman's landmark simulation of liquid argon in 1964 was one of the first to incorporate continuous interatomic potentials, moving beyond

DOI: 10.1201/9781003628071-7

hard-sphere approximations. Simultaneously, Verlet introduced an efficient integration scheme (the Verlet algorithm), which remains a cornerstone of MD simulations.

The subsequent decades saw rapid developments:

- **1980s–1990s:** Introduction of advanced force fields, including those capable of describing biomolecules, polymers, and surfaces.
- **2000s:** Emergence of parallel computing and specialized hardware (e.g., GPUs), dramatically accelerating MD simulations and expanding the accessible system sizes and timescales.
- **Recent years:** Integration of machine learning (ML) to develop more accurate and transferable potential energy surfaces, enabling simulations with quantum-level fidelity at reduced computational costs [3–4].

7.1.2 MD in Nanofluid Research

Initially focused on simple fluids and crystalline solids, MD's scope expanded to soft matter, including colloids, polymers, and complex fluids. The unique challenges of modeling nanofluids involving heterogeneous interfaces, nanoparticle interactions, and non-continuum effects motivated the application of MD. Over the past two decades, MD has become central to understanding heat transfer enhancements, nanoparticle dispersion stability, and interfacial phenomena critical to nanofluid performance. Figure 7.1 shows the geometry development for molecular dynamics analysis for nanofluidic devices [2]

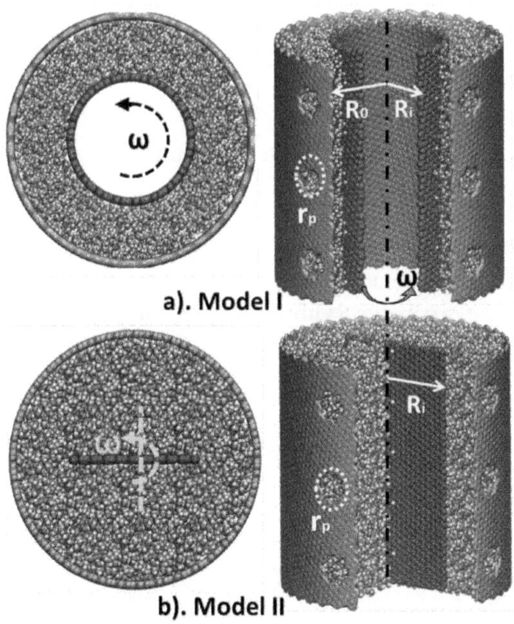

FIGURE 7.1 Geometry development for molecular dynamics analysis for nanofluidic devices [2].

7.1.2.1 Basic Principles and Governing Equations

7.1.2.1.1 Newton's Equations of Motion

At the heart of MD lies Newton's second law, which governs the motion of each particle [5]:

$$m_i \frac{d^2 r_i}{dt^2} = F_i$$

Here, m_i is mass of particle i, r_i is its position vector, and F_i is force acting on it.

7.1.2.1.2 Integration Algorithms

Discretization of continuous time enables numerical solution of Newton's equations. Popular integration schemes include:

- **Verlet algorithm:** Combines simplicity and stability, updating positions using previous positions and current accelerations.
- **Velocity Verlet:** Extends Verlet to directly calculate velocities, often used for thermostated simulations.
- **Leapfrog scheme:** Updates positions and velocities at staggered time steps, providing improved energy conservation in certain cases [6].

7.1.2.1.3 Boundary Conditions

In finite simulations, artificial boundaries can introduce artifacts. To mimic bulk behavior, periodic boundary conditions (PBCs) are often employed: when a particle exits one side of the simulation box, it re-enters from the opposite side, effectively creating an infinite lattice of boxes.

7.1.2.1.4 Temperature and Pressure Control

Real systems typically exist at constant temperature and pressure. To simulate such environments:

- **Thermostats** (e.g., Berendsen, Nosé-Hoover) regulate system temperature by rescaling particle velocities or introducing fictitious degrees of freedom.
- **Barostats** (e.g., Parrinello-Rahman, Berendsen barostat) adjust the simulation box volume to maintain constant pressure.

7.1.2.2 Ensemble Types

Depending on the thermodynamic constraints, different statistical ensembles are simulated:

- **Microcanonical (NVE):** Constant number of particles (N), volume (V), and energy (E).
- **Canonical (NVT):** Constant N, V, and temperature (T).
- **Isothermal–isobaric (NPT):** Constant N, pressure (P), and T.

Selection of an ensemble depends on the target physical scenario.

7.1.2.3 Interatomic Potentials and Force Fields

Forces in MD arise from potential energy functions. Common forms include:

- **Lennard–Jones (LJ):** Models van der Waals interactions and repulsive overlap.

$$U_{LJ}(r) = 4\varepsilon\left[\left(\frac{\sigma}{r}\right)^{12} - \left(\frac{\sigma}{r}\right)^{6}\right]$$

In this equation, r is distance between two particles and ε is depth of the potential well (interaction strength), σ = distance at which the potential is zero, $\left(\frac{\sigma}{r}\right)^{12}$ is strong repulsion at short range, and $\left(\frac{\sigma}{r}\right)^{6}$ is attractive van der Waals term at longer range.

- **Coulombic interactions:** For charged species, often evaluated using Ewald summation or particle mesh methods.
- **Embedded atom models (EAM):** Describe metallic bonding by considering electron density contributions.
- **Bonded potentials:** For polymers or complex molecules, including harmonic bonds and angular potentials [6].

7.1.2.4 Data Collection and Analysis

Simulations generate trajectories of particle positions and velocities over time. Analysis yields macroscopic observables:

- **Thermodynamic properties:** Temperature, pressure, energy.
- **Transport coefficients:** Thermal conductivity, viscosity (e.g., via Green–Kubo relations).
- **Structural data:** Radial distribution functions, coordination numbers.

Relevance of MD to studying nanofluid behavior at atomic and nanoscale levels

7.1.2.5 Limitations of Continuum Models

Classical continuum approaches (e.g., Navier–Stokes equations, Fourier's law) assume local thermodynamic equilibrium and continuum behavior, valid at macroscale but increasingly inaccurate as dimensions approach nanometer scales. In nanofluids:

- **Interfacial effects dominate:** The surface area-to-volume ratio is high, making interfacial thermal resistance (Kapitza resistance) crucial.
- **Discrete molecular interactions:** Continuum models cannot capture detailed nanoparticle–fluid interactions or near-field heat transfer mechanisms.
- **Non-Newtonian behavior emerges:** Effective viscosity and shear behavior change at small scales and high shear rates.

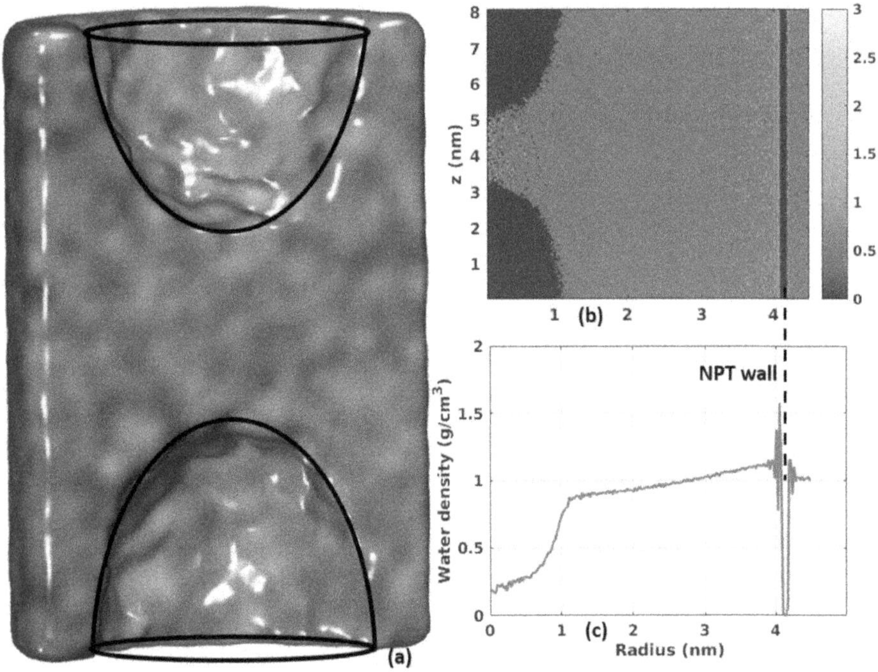

FIGURE 7.2 Vortex development and Taylor instability [2].

MD provides an atomistic perspective, explicitly simulating molecular interactions without relying on continuum assumptions [7]. Figure 7.2 shows the vortex in fluid and Taylor instability.

7.2 INSIGHTS INTO THERMAL CONDUCTIVITY ENHANCEMENT

Experimental evidence shows that nanofluids often display thermal conductivities higher than predicted by classical models (e.g., Maxwell's equation). MD simulations help elucidate mechanisms behind this anomaly [8]:

- **Interfacial thermal transport:** Quantify Kapitza resistance and its dependence on particle size, surface modification, and fluid properties.
- **Nanoparticle Brownian motion:** Evaluate contribution to energy transport, although its actual significance remains debated.
- **Formation of liquid layering:** Structured layers of fluid molecules around nanoparticles can create preferential thermal pathways [9].

7.2.1 Nanoparticle Dispersion and Stability

A critical challenge in nanofluid applications is preventing nanoparticle agglomeration and sedimentation. MD simulations allow researchers to:

- Investigate van der Waals, electrostatic, and steric forces affecting particle–particle and particle–fluid interactions [10].
- Evaluate the effect of surface functionalization on dispersion stability.
- Study dynamic aggregation and breakup processes under flow conditions [11].

7.2.2 Slip Flow and Microchannel Behavior

In confined geometries such as microchannels or nanoporous media, slip effects at solid–liquid interfaces significantly impact flow and heat transfer:

- MD can directly simulate velocity profiles near walls, revealing slip lengths and deviations from no-slip boundary conditions.
- Analysis of nano-confinement effects on viscosity and thermal conductivity supports design of miniaturized cooling devices [12].

7.2.3 Electrokinetic and Magnetic Field Interactions

IONFs and some functionalized nanofluids respond to electric or magnetic fields:

- MD enables modeling of electrophoretic motion, dielectrophoretic effects, and induced charge distributions.
- Insights from MD support design of tunable thermal transport systems, such as field-controlled heat exchangers [13].

7.2.4 Phase Change and Nucleation Phenomena

MD simulations provide atomistic views of:

- Vapor bubble nucleation on heated nanoparticle surfaces.
- Solid–liquid phase transitions influenced by nanoparticle presence.
- Role of interfacial energy barriers and nanoparticle wettability in controlling phase change heat transfer [14].

7.2.5 Validation and Parameter Development

MD-derived data, such as interfacial thermal conductance or effective transport coefficients, can inform larger-scale models (e.g., computational fluid dynamics, finite element models). This multi-scale approach integrates atomistic accuracy into system-level simulations [15].

7.2.6 Emerging Hybrid Simulation Frameworks

Coupling MD with other simulation methods, including:

- **Monte Carlo simulations:** Enhance sampling of equilibrium configurations.
- **Coarse-grained models:** Reduce computational costs while retaining essential molecular details.

- **Continuum–atomistic coupling (e.g., Quasi-Continuum, Adaptive Resolution):** Bridge scales for problems involving large domains but localized nanoscale phenomena.

Molecular dynamics (MD) simulation has revolutionized our understanding of nanofluid behavior by providing atomistic insights into transport and interfacial phenomena that traditional continuum models cannot capture. From its roots in theoretical physics to its modern-day applications enhanced by machine learning and hybrid frameworks, MD serves as an essential pillar of nanofluid research.

By accurately modeling thermal, structural, and dynamic interactions at the molecular scale, MD helps explain observed enhancements in thermal conductivity, clarify mechanisms of nanoparticle stability, and reveal new pathways for optimizing energy and thermal management systems. Its integration with experimental data and continuum models enables comprehensive multi-scale analyses critical for advancing nanofluid applications in energy conservation and high-performance thermal systems.

The following chapters will delve deeper into the specific force fields, simulation techniques, and practical applications of MD in nanofluid systems, highlighting how this versatile tool contributes to both scientific discovery and technological innovation [16].

7.3 INTERATOMIC POTENTIALS AND FORCE FIELDS FOR NANOFLUIDS

Nanofluids, which consist of nanoparticles suspended in base fluids, display remarkable thermal and transport properties that exceed those of traditional heat transfer fluids. To accurately capture the underlying physics governing these enhancements, Molecular Dynamics (MD) simulations rely on carefully chosen interatomic potentials and force fields. These mathematical representations of atomic and molecular interactions dictate the accuracy of predicted thermophysical properties, stability behaviors, and heat transfer mechanisms at the nanoscale.

In this chapter, we explore the most commonly used potential functions, the selection criteria for solid–liquid and nanoparticle–fluid interactions, and the validation and limitations of force fields specific to nanofluid studies. Understanding these components is crucial for constructing reliable MD models that bridge the gap between atomistic phenomena and macroscopic observables.

7.3.1 Overview of Common Potential Functions

7.3.1.1 Lennard–Jones (LJ) Potential

One of the most widely used interatomic potentials in MD simulations is the Lennard–Jones (LJ) potential. It captures the balance between attractive van der Waals forces and short-range Pauli repulsion. The potential energy U(r)U(r) between two atoms separated by a distance rr is given by:

$$U_{LJ}(r) = 4\varepsilon\left[\left(\frac{\sigma}{r}\right)^{12} - \left(\frac{\sigma}{r}\right)^{6}\right]$$

In this equation, r is distance between two particles and ε is depth of the potential well (interaction strength), σ = distance at which the potential is zero, $\left(\frac{\sigma}{r}\right)^{12}$ is strong repulsion at short range, and $\left(\frac{\sigma}{r}\right)^{6}$ is attractive van der Waals term at longer range.

Applications in Nanofluids

- Simulating non-bonded interactions among fluid atoms (e.g., argon) or simple fluid–nanoparticle interactions.
- Studying layering effects and surface adsorption phenomena.

Advantages

- Simple and computationally efficient.
- Requires only two parameters (ϵ, σ), facilitating straightforward parameterization.

Limitations

- Inadequate for systems involving strong directional bonding or chemical reactions.
- Less suitable for polar fluids and complex surface chemistries [17].

7.3.1.2 Embedded Atom Method (EAM)

The EAM was developed to better capture the metallic bonding in systems involving metals and alloys. It accounts for the contribution of surrounding electron densities to the cohesive energy of each atom. The total energy EE is given by:

$$E = \sum_{i} F_i(\rho_i) + \frac{1}{2}\sum_{i \neq j} \varnothing(r_{ij})$$

where $F_i(\rho_i)$ is the embedding energy for atom i (function of the local electron density ρ_i) and $\varnothing(r_{ij})$ is a pair potential between atoms i and j.

Applications in Nanofluids

- Modeling metal nanoparticle cores (e.g., copper, gold) embedded in a liquid medium.
- Understanding the mechanical stability and surface restructuring of nanoparticles.

Advantages

- Accurately captures many-body interactions essential for metals.
- Suitable for surface and defect studies of metallic nanoparticles.

Limitations

- Less accurate for non-metallic or ionic systems.
- Parameterization depends heavily on experimental or quantum mechanical data [18].

7.3.2 Coulombic Potentials

Charged systems, such as ionanofluids or nanofluids involving charged nanoparticles, require accurate treatment of electrostatic interactions. The Coulombic potential between two charges, q_i and q_j, separated by r_{ij} is given by:

$$U_{(r_{ij})} = \frac{1}{4\pi\epsilon 0} \frac{q_i q_j}{r^2_{ij}}$$

In periodic simulations, techniques such as Ewald summation or Particle Mesh Ewald (PME) are used to handle long-range electrostatics efficiently.

Applications in Nanofluids

- Surface charge effects on nanoparticle dispersion.
- Double-layer interactions in aqueous suspensions.

Advantages

- Essential for modeling ionic interactions.
- Enables study of zeta potential and electrostatic stabilization mechanisms.

Limitations

- Computationally intensive without proper approximations.
- Requires careful treatment to avoid artifacts in periodic systems [19].

7.3.3 Bonded Interaction Potentials

For simulations involving molecular chains or functionalized nanoparticles, bonded interactions include:

- **Harmonic bond potential:**

$$U_{(r)} = \frac{1}{2} k_b \left(r - r_0\right)^2$$

where k_b is the bond stiffness constant, and $r - r_0$ is the equilibrium bond length.

- **Angle potential:**

$$U_{(\theta)} = \frac{1}{2} k_b \left(\theta - \theta_0\right)^2$$

- **Dihedral potential:** Often expressed as Fourier series to capture torsional behaviors.

Applications

- Surface-grafted polymers on nanoparticles for stability enhancement.
- Surfactant molecules in nanofluids.

7.3.4 Other Advanced Potentials

- **Tersoff and Stillinger–Weber potentials:** Used for covalent systems like silicon and carbon nanotubes.
- **Morse potential:** Better suited for systems with softer bonding interactions.
- **Reactive force fields (ReaxFF):** Allow bond formation and breaking, enabling simulation of chemical reactions [20].

7.4 SELECTION CRITERIA FOR SOLID–LIQUID, NANOPARTICLE–FLUID INTERACTIONS

Choosing an appropriate potential is critical for accurate nanofluid simulations. The selection criteria depend on the following:

7.4.1 Nature of the Materials

- **Metallic nanoparticles:** EAM potentials are preferred for core descriptions, possibly coupled with LJ or Coulombic potentials for surface and solvent interactions.
- **Ceramic or oxide nanoparticles:** Often modeled using Buckingham or Tersoff potentials due to strong covalent or ionic bonding.
- **Base fluids:** Water typically requires specialized models like SPC/E, TIP3P, or TIP4P; organic solvents may use OPLS-AA or CHARMM force fields.

7.4.2 Surface Functionalization and Coating

Surface-modified nanoparticles demand force fields that account for the surface chemistry:

- Explicit bonded potentials for ligands and surfactants.
- Cross-interaction parameters to represent van der Waals and electrostatic contributions.

7.4.3 Interfacial Phenomena

Key properties, such as wetting, contact angle, and interfacial thermal resistance, are sensitive to solid–liquid interaction potentials. Accurate modeling requires:

- Validation of interaction parameters to reproduce experimentally measured contact angles.
- Adjustments to match observed adsorption behavior and hydration structures.

7.4.4 Thermal and Electrostatic Behavior

For applications involving thermal transport or electric fields:

- Potentials must capture local polarization effects, which can significantly impact heat transfer at interfaces.
- Electrostatic interactions should consider solvent screening effects and ion distributions.

7.4.5 Computational Feasibility

While more detailed potentials provide higher accuracy, they also impose heavier computational costs. Researchers often balance between:

- **All-atom force fields:** High accuracy but limited to smaller systems and shorter timescales.
- **Coarse-grained models:** Reduced resolution but enable larger system simulations.

7.4.6 Parameter Availability and Transferability

Reliable parameter sets are essential for consistent predictions. Parameters should be validated across various temperatures, pressures, and compositions relevant to the nanofluid application.

Force fields determine the reliability of simulation outputs. Validation involves comparing simulated properties with experimental or higher-fidelity computational data. Key benchmarks include:

- **Thermophysical properties:** Density, viscosity, and thermal conductivity.
- **Interfacial characteristics:** Surface tension, contact angles, and adsorption isotherms.
- **Structural properties:** Radial distribution functions (RDFs), coordination numbers.

7.4.7 Approaches to Validation

- **Single-component property matching:** Validate fluid and nanoparticle potentials independently before mixing.
- **Composite system tests:** Compare predicted effective thermal conductivity or viscosity of nanofluids with measured values.
- **Cross-scale validation:** Use ab initio or quantum mechanical results as reference data for parameter tuning.

7.4.8 Limitations and Challenges

7.4.8.1 Transferability

Force fields optimized for a specific system may perform poorly when applied to different temperatures, pressures, or compositions. For example, parameters fitted to room-temperature liquid water might fail under supercooled or superheated conditions.

7.4.8.2 Limited Chemical Reactivity

Most traditional force fields (e.g., Lennard–Jones, EAM) do not account for bond breaking or formation. Reactive force fields like ReaxFF can address this but at substantially higher computational cost and with their own parameterization challenges.

7.4.8.3 Surface Effects and Finite Size Artifacts

Nanoparticle surfaces often exhibit defects, roughness, or dynamic restructuring not captured in idealized models. Additionally, simulation box sizes can introduce finite-size effects that alter computed interfacial properties.

7.4.8.4 Long-Range Interactions

Accurate treatment of long-range van der Waals or electrostatic interactions remains computationally demanding, especially in periodic systems. While Ewald summation and PME methods mitigate this, they increase simulation time.

7.4.8.5 Time-Scale Limitations

MD simulations typically span nanoseconds to microseconds. Processes like long-term aggregation, sedimentation, or large-scale thermal fluctuations extend beyond these time frames, requiring extrapolation or coupling with mesoscale methods (e.g., Dissipative Particle Dynamics or Monte Carlo techniques).

7.4.8.6 Lack of Experimental Data

For many novel nanofluid systems, especially those involving ionic liquids or complex hybrid nanoparticles, experimental thermophysical data are scarce. This complicates parameter validation and confidence in predictive modeling.

Strategies to overcome limitations

- **Hybrid force fields:** Combine strengths of different models (e.g., EAM for cores, LJ for surface layers).
- **Multi-scale approaches:** Bridge atomistic and continuum scales using coarse-grained or mean-field approximations.
- **Machine learning potentials:** Recent advancements allow the development of data-driven potentials that capture complex interactions with quantum-level accuracy but at lower computational cost [21].

7.4.9 Case Studies in Nanofluid Research

7.4.9.1 Thermal Conductivity Enhancement Mechanisms

MD studies using validated force fields have elucidated mechanisms like liquid layering, phonon scattering, and enhanced energy transport near nanoparticles.

7.4.9.2 Nanoparticle Dispersion Stability

By simulating interactions among charged or functionalized nanoparticles, researchers have identified optimal surface chemistries and surfactants to prevent aggregation.

7.4.9.3 Interfacial Thermal Resistance

MD models have quantified Kapitza resistance, showing how surface treatments can drastically reduce thermal barriers and improve overall heat transfer.

7.4.10 Future Directions

Continued refinement of force fields, informed by ab initio simulations and experimental data, will expand the applicability of MD in nanofluid studies. Key directions include:

- Development of universal force fields capable of describing diverse hybrid nanoparticle systems.
- Improved reactive models for studying chemical transformations and thermal decomposition at extreme conditions.
- Integration with AI to automate parameter optimization and enable rapid adaptation to new materials [22].

7.5 THERMOPHYSICAL PROPERTY PREDICTIONS VIA MD

Nanofluids, consisting of nanoparticles dispersed in base fluids, have demonstrated extraordinary potential in enhancing heat and mass transfer in various energy and thermal management systems. To harness their full potential, it is crucial to understand and predict their thermophysical properties, such as thermal conductivity, viscosity, density, heat capacity, and surface tension, with high accuracy. Experimental techniques, while invaluable, often face limitations in resolving nanoscale interactions and revealing the fundamental mechanisms responsible for observed behaviors.

Molecular dynamics (MD) simulations have emerged as an indispensable tool for bridging this knowledge gap. By explicitly modeling atomic and molecular interactions, MD allows researchers to probe the microscopic origins of macroscopic thermophysical properties, offering unique insights into enhancement mechanisms, nanoparticle–fluid interactions, and confinement effects.

This chapter explores MD-based approaches to predicting key thermophysical properties of nanofluids, highlighting thermal conductivity and its enhancement mechanisms, viscosity, and shear behaviors at the nanoscale, and methods for calculating density, heat capacity, and surface tension.

7.5.1 Thermal Conductivity and Mechanisms of Enhancement

7.5.1.1 Importance of Thermal Conductivity in Nanofluids

One of the primary motivations for developing nanofluids is their significantly higher thermal conductivity compared to conventional heat transfer fluids. Experimental studies have reported enhancements ranging from 10% to over 100%, depending on particle concentration, shape, and surface chemistry. Understanding the mechanisms behind this enhancement is crucial for designing next-generation thermal management systems, including advanced heat exchangers, electronic cooling devices, and solar thermal collectors [23].

7.5.1.2 Predicting Thermal Conductivity via MD

7.5.1.2.1 Equilibrium Methods: Green–Kubo Relations

The Green–Kubo formalism links thermal conductivity kk to fluctuations in the heat current in an equilibrium system. In MD, it is expressed as:

$$k = \frac{1}{3Vk_BT^2}\int_0^\infty \langle J(0).J(t)\rangle dt$$

where:

- V is the system volume.
- k_B is Boltzmann's constant.
- T is temperature.
- $\langle J(0).J(t)\rangle$ is the heat flux vector at time t.

By analyzing the time autocorrelation function of the heat flux, thermal conductivity can be derived directly from microscopic dynamics.

7.5.1.2.2 Non-Equilibrium MD (NEMD) Approaches

In NEMD simulations, an explicit temperature gradient is imposed across the system. The resulting steady-state heat flux J is related to thermal conductivity via Fourier's law:

$$k = -\frac{J}{\nabla T}$$

NEMD provides a more intuitive picture of heat transport and is often easier to implement for nanofluid systems with strong interfacial effects.

7.5.1.3 Mechanisms of Thermal Conductivity Enhancement

7.5.1.3.1 Liquid Layering around Nanoparticles

MD simulations have revealed that fluid molecules form highly ordered layers near nanoparticle surfaces, often resembling solid-like structures. This "liquid layering" can create high-conductivity channels, effectively facilitating heat transfer.

7.5.1.3.2 Brownian Motion of Nanoparticles

The random movement of nanoparticles contributes to micro-convection effects, which may slightly enhance energy transport. However, MD studies have shown that this contribution is generally minor at low particle concentrations and moderate temperatures.

7.5.1.3.3 Percolation Networks

At higher concentrations, nanoparticles can form interconnected networks or clusters, providing continuous thermal pathways through the fluid. MD models can capture the formation and breakdown of these networks, offering insights into concentration-dependent conductivity behavior.

7.5.1.3.4 Phonon Transport across Interfaces

The interfacial thermal resistance, or Kapitza resistance, significantly impacts overall heat transfer. MD allows quantification of this resistance and exploration of how surface functionalization, roughness, and wettability modify phonon transmission.

7.5.1.4 Case Studies and Validation

Several MD studies have successfully matched experimental enhancements in thermal conductivity:

- **Al_2O_3-water nanofluids:** Simulations confirmed the role of strong hydrogen bonding at the particle surface in reducing interfacial resistance.
- **Graphene-based nanofluids:** MD predicted exceptional enhancements due to high intrinsic conductivity of graphene and strong layering effects.

These examples underline the predictive power of MD and its ability to disentangle complex heat transfer mechanisms in nanofluids.

7.5.2 Viscosity and Shear Behavior at the Nanoscale

7.5.2.1 Significance of Viscosity in Nanofluid Applications

While enhanced thermal conductivity is desirable, an increase in viscosity can hinder fluid flow and increase pumping power requirements. Understanding and controlling viscosity at the nanoscale are thus critical for practical implementations [24].

7.5.2.2 Predicting Viscosity via MD

7.5.2.2.1 Green–Kubo Method for Viscosity

Analogous to thermal conductivity, viscosity η\eta can be derived from stress autocorrelation functions:

$$\eta = \frac{V}{k_B T} \int_0^\infty \langle P_{\alpha\beta}(0) P_{\alpha\beta}(t) \rangle dt$$

where $P_{\alpha\beta}$ denotes off-diagonal elements of the pressure tensor.

7.5.2.2.2 Shear Flow Simulations

NEMD simulations can also impose shear flow directly, allowing the measurement of steady-state stress responses to determine viscosity:

$$\eta = \frac{\tau}{\gamma}$$

where τ is the shear stress and γ the shear rate.

7.5.2.3 Nanoscale Shear Behaviors

7.5.2.3.1 Shear Thinning and Thickening

MD simulations have shown that nanofluids can exhibit non-Newtonian behavior, including:

- **Shear thinning:** Viscosity decreases with increasing shear rate, often due to alignment of nanoparticle aggregates along the flow direction.
- **Shear thickening:** Viscosity increases at high shear rates, typically caused by transient formation of particle clusters or jamming.

7.5.2.3.2 Confinement Effects

When confined in nanochannels or pores, viscosity can differ significantly from bulk values. MD reveals that:

- Strong wall-fluid interactions may lead to layered fluid structures, increasing apparent viscosity.
- Slip flow at the walls can reduce effective viscosity, depending on surface chemistry and roughness.

7.5.2.3.3 Nanoparticle Concentration and Shape Effects

MD studies confirm that:

- Higher particle concentrations generally lead to increased viscosity due to enhanced hydrodynamic interactions and clustering.
- Non-spherical particles, such as rods or platelets, increase viscosity more dramatically compared to spherical ones due to larger effective volume fractions and alignment effects.

7.5.2.4 Applications and Implications

Accurate prediction of nanoscale viscosity informs the design of nanofluid-based systems, balancing enhanced heat transfer against pumping requirements. MD insights guide the selection of particle shapes, concentrations, and surface treatments to achieve optimal performance.

7.5.3 Density, Heat Capacity, and Surface Tension Calculations

7.5.3.1 Density Predictions

MD simulations determine density by directly calculating mass per unit volume from atomic configurations [25]. For nanofluids:

- Particle inclusion slightly increases overall density compared to base fluids.
- Density variations near nanoparticle surfaces due to layering effects can be analyzed using radial density profiles.

7.5.3.2 Heat Capacity Calculations

7.5.3.2.1 Constant Volume Heat Capacity (CV)

CV can be obtained from energy fluctuations in the canonical ensemble:

$$C_v = \frac{\langle E^2 \rangle - \langle E^2 \rangle}{k_B T^2}$$

where E is the system's total energy.

7.5.3.2.2 Constant Pressure Heat Capacity (CP)

While more challenging in MD, **CV** can be approximated by simulating isobaric conditions or derived from **CV** using thermodynamic relations.

7.5.3.2.3 Impact of Nanoparticles

MD studies show that nanoparticles generally increase the effective heat capacity slightly due to their higher specific heat contributions and interfacial energy storage. The degree of enhancement depends on particle material, shape, and volume fraction.

7.5.3.3 Surface Tension Determination

Surface tension γ\gamma is crucial for understanding boiling, wetting, and interfacial stability. In MD, γ can be calculated from the pressure tensor:

$$\gamma = \frac{L_z}{2}\left[P_N - \frac{P_T}{2}\right]$$

where:

- L_z is the simulation box length normal to the interface.
- P_N and P_T are normal and tangential components of the pressure tensor, respectively.

7.5.3.3.1 Effects of Nanoparticles

Nanoparticles adsorbed at interfaces can significantly reduce surface tension, stabilizing emulsions or promoting bubble nucleation. MD reveals:

- The balance between adsorption energy and entropy governs particle attachment to interfaces.
- Surface-functionalized nanoparticles exhibit tunable surface activity.

7.5.4 Case Studies

7.5.4.1 Water-Based Nanofluids

MD simulations have predicted minor increases in density and heat capacity but significant interfacial effects on surface tension, consistent with experimental trends.

7.5.4.2 Ionic Liquid-Based Nanofluids

Due to strong ion–particle interactions, these systems show unique density layering and surface tension reductions, enabling stable dispersions even at high temperatures [26].

7.5.5 Validation and Accuracy

Comparing MD predictions with experimental data confirms the method's accuracy when appropriate force fields are used. Sensitivity analyses further strengthen confidence in the predictions by quantifying uncertainty from finite-size effects, force field choices, and simulation time scales.

Molecular Dynamics simulations provide unparalleled access to the fundamental thermophysical properties that govern nanofluid behavior. By resolving atomic and molecular details, MD enables accurate predictions and mechanistic insights into:

- **Thermal conductivity:** Disentangling contributions from liquid layering, Brownian motion, percolation, and interfacial resistance.
- **Viscosity:** Capturing complex shear-dependent behaviors and confinement effects, vital for system-level design.
- **Density, heat capacity, and surface tension:** Revealing subtle nanoparticle-induced modifications crucial for phase change, wetting, and stability applications.

The continued development of accurate force fields, combined with hybrid and multi-scale modeling strategies, will further improve the predictive power of MD. As nanofluid research advances, MD-derived thermophysical data will be essential for designing high-performance, energy-efficient systems across thermal management, renewable energy, and advanced manufacturing sectors [27–29].

7.6 NANOPARTICLE DISPERSION AND AGGLOMERATION BEHAVIOR

Nanoparticle dispersion is a fundamental aspect of nanofluid performance and stability. The superior thermal and transport properties of nanofluids rely on the homogeneous distribution of nanoparticles within the base fluid. However, due to the high surface energy and van der Waals attraction between particles, nanoparticles have a strong tendency to agglomerate and form clusters, leading to sedimentation and a decline in the desired thermophysical enhancements.

Molecular Dynamics (MD) simulations provide unique insights into the microscopic mechanisms underlying dispersion and agglomeration phenomena. This chapter explores the molecular mechanisms governing stability, the role of surface functionalization and surfactants, and aggregation dynamics and their impacts on effective properties.

7.6.1 Molecular Mechanisms Governing Stability

7.6.1.1 Fundamental Forces in Nanoparticle Interactions

The stability of nanoparticles in a fluid medium is governed by a delicate balance among several intermolecular forces, including:

- **Van der Waals forces:** These attractive forces arise from induced dipoles and are universal among atoms and molecules. They promote agglomeration when nanoparticles come into close proximity.
- **Electrostatic repulsion:** In charged systems, electric double layers form around nanoparticles, creating repulsive forces that can counteract van der Waals attraction.
- **Steric hindrance:** When nanoparticles are coated with polymers or surfactants, the attached molecules create a physical barrier that prevents close approach and reduces aggregation.
- **Hydration forces:** In aqueous systems, water structuring near hydrophilic surfaces can generate additional repulsion.

MD simulations can explicitly model these interactions, providing insights into the net potential of mean force (PMF) between nanoparticles. The PMF represents the effective interaction potential as a function of interparticle distance and includes all solvent-mediated effects [30].

7.6.1.2 DLVO Theory and Its Limitations

Traditionally, the Derjaguin–Landau–Verwey–Overbeek (DLVO) theory combines van der Waals attraction and electrostatic repulsion to predict colloidal stability. However, for nanoscale systems, DLVO theory faces limitations:

- It oversimplifies by neglecting steric and hydration forces.
- It assumes continuum solvent properties, which fail at nanoscale separations.
- It cannot capture dynamic restructuring or local variations in surface charge.

MD simulations overcome these limitations by including explicit solvent molecules and detailed surface structures [31].

7.6.1.3 Interfacial Structuring and Layering Effects

MD has revealed that fluid molecules tend to form structured layers around nanoparticles, especially near smooth or crystalline surfaces. This structuring influences stability in two ways:

- **Repulsion enhancement:** Well-ordered hydration layers can increase effective repulsion between particles, preventing close approach.
- **Induced attraction:** In some cases, fluid-mediated depletion forces can drive particles together, promoting aggregation [32–34].

7.6.1.4 Temperature and Concentration Effects

At elevated temperatures, increased thermal motion can destabilize ordered solvent layers, reducing repulsion and increasing the likelihood of aggregation. MD studies have shown:

- **Low temperatures:** Strong hydration structures promote stability.
- **High temperatures:** Disruption of structuring can lead to rapid agglomeration.

Similarly, at higher nanoparticle concentrations, the probability of collision and subsequent aggregation increases, often leading to percolated networks or large flocs [30].

7.6.2 Role of Surface Functionalization and Surfactants

7.6.2.1 Purpose and Mechanisms of Functionalization

Surface functionalization involves modifying the nanoparticle surface with chemical groups, polymers, or surfactants to improve dispersion stability. The goals include:

- **Providing steric barriers:** Attached molecules prevent direct contact between particle surfaces.
- **Enhancing electrostatic repulsion:** Functional groups can increase surface charge, augmenting electric double-layer repulsion.
- **Improving chemical compatibility:** Tailoring surface chemistry to match the solvent reduces the enthalpic penalty of dispersion [35].

7.6.2.2 Polymeric Coatings

Polymers grafted or adsorbed onto nanoparticles create steric hindrance. MD simulations reveal the following:

- **Chain length and density:** Longer or denser polymer chains offer greater stabilization but may also increase viscosity.
- **Conformational behavior:** Under shear or confinement, polymers may compress or align, affecting stability dynamically.

For example, polyethylene glycol (PEG)-functionalized nanoparticles show enhanced dispersion in water and biocompatible solvents due to their flexible hydrophilic chains.

7.6.2.3 Surfactant-Assisted Stabilization

Surfactants lower interfacial tension and form micellar structures around nanoparticles. MD studies have elucidated:

- **Adsorption mechanisms:** Surfactant head groups anchor to particle surfaces, while tails extend into the solvent.

- **Critical micelle concentration (CMC):** Above CMC, excess surfactants form micelles, potentially trapping particles and affecting long-term stability.
- **Dynamic exchange:** Surfactant molecules continuously adsorb and desorb, leading to transient changes in surface coverage [36–38].

7.6.2.4 Ligand Exchange and Multi-Functionalization

Advanced approaches involve introducing multiple functional groups or ligands to achieve synergistic effects, such as combining hydrophilic and hydrophobic moieties to tune interfacial behavior.

MD enables detailed mapping of ligand distributions and conformations, revealing how surface heterogeneity influences interaction potentials and local structuring.

7.6.2.5 Electrostatic Stabilization through Surface Charge

By introducing ionizable functional groups, nanoparticles acquire surface charge, enhancing double-layer repulsion. MD captures ion distribution around charged surfaces, providing:

- **Potential profiles:** Visualization of electric double layers and counter-ion condensation.
- **Ion-specific effects:** Different ions can vary in their capacity to stabilize or destabilize dispersions, depending on valency and hydration properties [39].

7.6.2.6 Trade-Offs in Functionalization

While surface modifications improve stability, they may introduce undesirable effects:

- **Increased viscosity:** Thick polymer or surfactant layers hinder flow.
- **Thermal resistance:** Coatings can impede heat transfer, reducing thermal conductivity benefits.
- **Chemical reactivity:** Functional groups may interact with fluid components, affecting long-term stability.

MD studies allow quantitative assessment of these trade-offs, facilitating optimization of surface treatments.

7.6.3 Aggregation Dynamics and Impacts on Effective Properties

7.6.3.1 Agglomeration Kinetics

Aggregation refers to the irreversible assembly of nanoparticles into larger clusters, while agglomeration implies looser, reversible assemblies. MD provides detailed insights into aggregation kinetics, including:

- **Collision frequency:** Determined by Brownian motion and fluid shear, dictating initial contact rates.

- **Attachment efficiency:** Controlled by energy barriers arising from repulsive interactions.

Simulation studies have shown that even small changes in surface charge or steric stabilization dramatically affect aggregation rates.

7.6.3.2 Cluster Growth Mechanisms

Agglomerates form through different pathways:

- **Diffusion-limited cluster aggregation (DLCA):** Particles stick upon first contact, resulting in open, fractal-like structures.
- **Reaction-limited cluster aggregation (RLCA):** Particles require multiple collisions before binding, leading to denser aggregates.

MD enables real-time observation of these growth processes and quantification of fractal dimensions, internal densities, and structural anisotropies [40].

7.6.3.3 Sedimentation and Gravitational Effects

Once aggregates exceed a critical size, gravitational forces can overcome Brownian stabilization, leading to sedimentation. MD studies incorporate gravitational fields to simulate:

- **Settling velocities:** Depend on aggregate size, density, and fluid viscosity.
- **Cluster restructuring:** As aggregates settle, they may compact further, changing their effective properties [41–43].

7.6.4 Impact on Thermophysical Properties

7.6.4.1 Agglomeration Alters Nanofluid Behavior in Several Ways

7.6.4.1.1 Thermal Conductivity

- **Short-term enhancement:** Small, well-dispersed clusters can form thermal percolation pathways, temporarily increasing conductivity.
- **Long-term reduction:** Large aggregates impede uniform heat transfer and increase thermal resistance.

7.6.4.1.2 Viscosity

- Aggregates increase effective particle size, leading to dramatic viscosity rises, sometimes resulting in gelation or flow blockage.

7.6.4.1.3 Specific Heat Capacity and Density

- Cluster formation may slightly modify these properties by altering nanoparticle dispersion homogeneity.

7.6.4.1.4 Surface Tension

- Changes in nanoparticle distribution at interfaces affect surface tension, potentially influencing boiling heat transfer and wetting behavior.

7.6.4.2 Mitigation Strategies Based on MD Insights

By understanding aggregation mechanisms, strategies can be developed to maintain dispersion:

- **Optimizing surface charge and steric barriers:** Fine-tuning surface potential and functional layer thickness.
- **Dynamic control:** Modifying pH, ionic strength, or temperature to maintain favorable interaction regimes.
- **In situ mixing or sonication:** Periodically breaking up clusters mechanically.

MD provides guidance on the energy barriers and timescales required for effective re-dispersion [44]. Figure 7.3 shows the molecular dynamics results for the cylindrical nanofluidic-based membrane system.

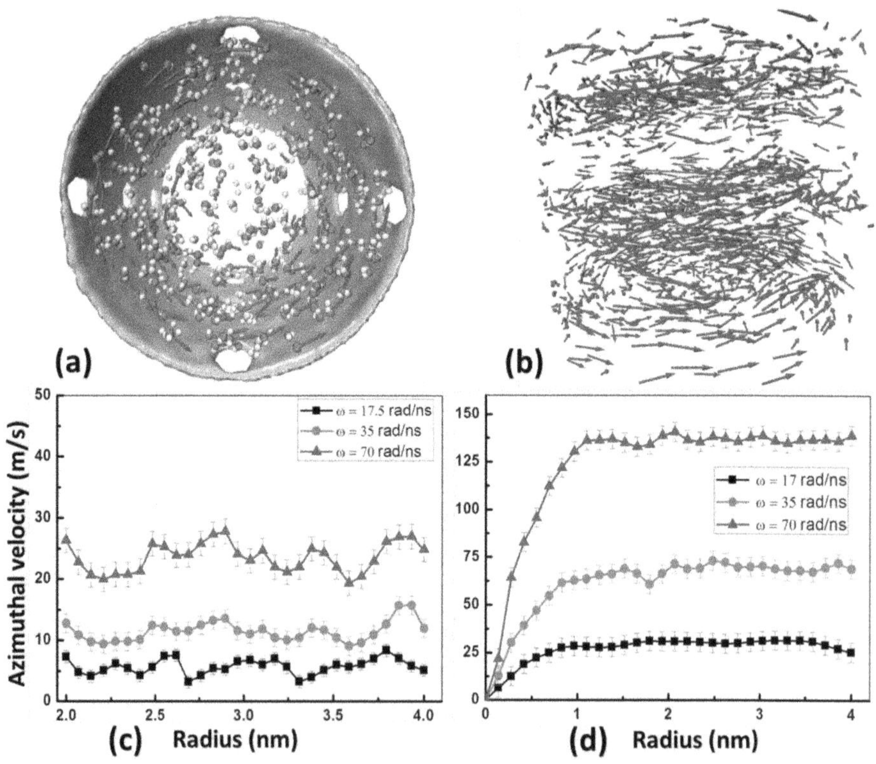

FIGURE 7.3 Molecular dynamics of cylindrical nanofluidic system [20].

7.7 CHALLENGES IN MD SIMULATION OF NANOFLUIDS

Molecular Dynamics (MD) simulations have emerged as a powerful tool for understanding the intricate behaviors of nanofluids at the atomic and molecular levels. By capturing particle interactions, fluid layering, and heat transfer mechanisms with atomistic precision, MD enables researchers to unlock fundamental insights impossible to achieve through purely experimental or continuum approaches.

However, despite their transformative potential, MD simulations of nanofluids are fraught with numerous challenges that limit their widespread application and practical impact. These challenges include the significant computational cost and time-scale limitations inherent to atomistic simulations, uncertainties associated with force fields and parameter accuracy, and the difficulty of bridging nanoscale results to inform macroscale system performance.

This section provides an in-depth exploration of these critical issues, discusses their implications for research and engineering applications, and suggests potential strategies for mitigation.

7.7.1 Computational Cost and Time-Scale Limitations

7.7.1.1 Atomistic Detail and Computational Burden

MD simulations require resolving the trajectories of millions of atoms or molecules over femtosecond-scale timesteps. This level of detail is necessary to capture accurate interatomic forces, local energy exchanges, and dynamic phenomena such as Brownian motion and phonon transport.

However, this fine-grained resolution comes at a high computational cost. Even with advanced parallelization on supercomputers or GPU acceleration, typical MD simulations are restricted to:

- **Spatial domains:** Generally limited to 10–100 nanometers in size.
- **Temporal domains:** Simulations typically cover up to nanoseconds or microseconds.

This makes it challenging to directly simulate realistic engineering-scale systems, which span millimeters to meters and involve operational timescales of seconds to hours.

7.7.1.2 Scaling Challenges with Nanofluids

Nanofluids, by definition, involve dispersed nanoparticles within a base fluid. Simulating these systems demands additional computational resources because of the following:

- **Nanoparticle size and count:** Realistic nanofluids often contain millions of particles dispersed in relatively large volumes. Even simplified models with a few dozen nanoparticles require substantial atomistic detail.

- **Long-range interactions:** Electrostatic and van der Waals forces introduce long-range interactions that must be carefully resolved, further increasing the computational load.
- **Dynamic aggregation and migration:** Simulating processes like agglomeration, sedimentation, or thermophoretic migration requires long simulation times to capture equilibrium states or dynamic behaviors accurately.

7.7.1.3 Strategies to Mitigate Computational Cost

7.7.1.3.1 Coarse-Grained Models

Coarse-grained MD reduces the number of particles by grouping atoms into "superatoms" or larger interaction sites. This simplifies the force calculations and allows longer time and larger spatial scales. However, it sacrifices some atomistic detail, which may obscure critical local phenomena.

7.7.1.3.2 Accelerated Sampling Methods

Techniques such as meta-dynamics, umbrella sampling, and hyperdynamics enable faster exploration of configurational space by biasing the system to overcome energy barriers. These methods can help simulate rare events or slow processes, like nanoparticle aggregation, more efficiently.

7.7.1.3.3 Hybrid Multi-Scale Approaches

Combining MD with continuum methods (e.g., hybrid MD–CFD frameworks) allows focusing atomistic detail only in critical regions, such as near surfaces or in high-gradient zones. While this reduces computational cost, it introduces new challenges in coupling and interface consistency.

7.7.1.3.4 High-Performance Computing (HPC) Resources

Exploiting cutting-edge supercomputers and GPU-accelerated platforms provides a direct route to extend MD capabilities. However, access to such resources is limited, and the high operational cost remains a significant barrier.

Despite advances, fundamental limits in computational capacity mean that fully resolving engineering-scale nanofluid systems using MD alone remains unfeasible. Continued improvements in algorithms, hardware, and hybrid modeling approaches will be critical to address these constraints [45–47].

7.7.2 Accuracy of Force Fields and Parameter Uncertainty

7.7.2.1 Role of Force Fields in MD

MD simulations rely on force fields—mathematical models that describe interatomic and intermolecular interactions. The accuracy of these force fields fundamentally determines the reliability of simulation results.

Commonly used force fields include:

- **Lennard–Jones potential:** Describes van der Waals interactions with a balance of attractive and repulsive terms.
- **Embedded atom method (EAM):** Suitable for metallic systems, incorporating many-body interactions.
- **Tersoff and Stillinger–Weber potentials:** Applied in covalent systems such as silicon- or carbon-based nanoparticles.
- **CHARMM and AMBER:** Widely used for biomolecular and organic systems.

7.7.2.2 Limitations for Nanofluids

While these force fields have been successful in many applications, their use in nanofluid simulations introduces unique challenges:

7.7.2.2.1 Parameter Transferability

Force fields are typically parameterized for specific materials or molecular systems under idealized conditions. Applying them to complex nanofluids involving diverse particle shapes, fluid chemistries, and surface modifications can lead to inaccuracies.

7.7.2.2.2 Surface Functionalization Effects

Many nanofluids utilize surface-modified nanoparticles to improve dispersion stability and tailor interactions. Standard force fields often lack parameters for functionalized surfaces, requiring custom parameterization or empirical adjustments, which introduces uncertainty.

7.7.2.2.3 Interfacial Interactions

Interfacial phenomena, such as wetting, slip, and contact line dynamics, are critically important in nanofluid applications but are challenging to capture accurately with existing force fields. Subtle deviations in interaction strength can lead to significant discrepancies in predicted behavior.

7.7.2.2.4 Electrostatic and Polarization Effects

Nanofluids often exhibit complex electrostatic interactions, especially when ionic surfactants or charged nanoparticles are involved. Fixed-charge models used in many force fields fail to represent polarization effects accurately, limiting fidelity.

7.7.2.3 Parameter Uncertainty and Validation

7.7.2.3.1 Sensitivity to Force Field Parameters

MD predictions of properties such as thermal conductivity, viscosity, and slip length can be highly sensitive to force field parameters. Small variations in interaction coefficients can cause large deviations in macroscopic predictions.

7.7.2.3.2 Lack of Comprehensive Experimental Data

Experimental data to validate force fields for nanofluids are limited, particularly at the nanoscale. This lack of robust benchmarks complicates parameter tuning and error assessment.

7.7.2.3.3 Emerging Force Field Development

Ongoing research is focused on:

- **Reactive force fields (ReaxFF):** Capable of modeling bond formation and breaking, providing better accuracy for dynamic surface chemistry.
- **Machine-learned force fields:** Use data-driven approaches to fit potential energy surfaces from quantum mechanical calculations, offering improved transferability and accuracy.

Improving force field accuracy and reducing parameter uncertainty is a critical priority. A systematic approach integrating quantum calculations, targeted experiments, and machine learning offers a promising pathway to develop more robust, application-specific force fields for nanofluids.

7.7.3 Translating Nanoscale Findings to Macroscale Performance

7.7.3.1 The Scale-Bridging Challenge

While MD provides profound insights into nanoscale processes, translating these findings into engineering-scale designs and performance metrics is non-trivial. Nanofluid applications typically involve flow domains ranging from micrometers to meters and timescales from seconds to hours, far beyond MD's reach.

7.7.3.2 Representative Elementary Volume (REV) Considerations

One approach to connect scales is defining a representative elementary volume, within which nanoscale behaviors are averaged to derive effective macroscopic properties (e.g., thermal conductivity, viscosity). However, defining REVs for nanofluids is challenging because:

- **Inhomogeneity:** Nanoparticle distribution may be highly non-uniform due to aggregation or thermophoretic migration.
- **Time-dependent dynamics:** Local properties can fluctuate dynamically, particularly under transient operating conditions.

7.7.3.3 Effective Property Extraction from MD

7.7.3.3.1 Thermal Conductivity and Viscosity

MD can compute thermal conductivity via equilibrium methods (e.g., Green–Kubo formalism) or non-equilibrium techniques (e.g., imposed temperature gradients). Viscosity is extracted from stress autocorrelation functions or Couette flow setups. While these provide valuable data, discrepancies often arise when comparing to experimental or macroscale conditions.

7.7.3.3.2 Interfacial Slip and Wettability

Slip length data from MD informs modified boundary conditions in continuum models, but translating this accurately requires careful scaling analysis and validation.

Differences in surface roughness, contamination, and real-world operating environments complicate direct transfer.

7.7.3.3.3 Particle Migration and Agglomeration

MD insights into nanoparticle migration under thermal gradients (thermophoresis), flow fields, and electric or magnetic fields offer qualitative guidance but are difficult to quantify at larger scales without complementary continuum or mesoscale models.

7.7.3.4 Coupling to Continuum Models

Hybrid multi-scale models integrate MD-derived properties into continuum solvers (e.g., CFD). However, challenges remain:

- **Nonlinear property dependence:** Effective properties from MD often depend nonlinearly on temperature, shear rate, and concentration. Accurately incorporating these dependencies into continuum models requires extensive interpolation or parametric fits.
- **Interface consistency:** Ensuring conservation of mass, momentum, and energy at MD–continuum interfaces demands meticulous coupling schemes.

7.7.3.5 Validation against Experimental Data

Even with careful multi-scale integration, predicted system-level behaviors must be validated experimentally. Discrepancies can arise due to:

- **Unmodeled phenomena:** Real systems include impurities, complex geometries, and unexpected operational transients.
- **Parameter uncertainties:** Errors in force fields or property extrapolations propagate into macroscale predictions.
- **Simplifying assumptions:** Many continuum models assume single-phase homogeneous fluids, neglecting local particle clustering or interfacial heterogeneities.

Figure 7.4 shows the scale-up mechanism for the molecular dynamic studies for nanofluidic devices.

7.7.4 Toward Digital Twins and Real-Time Models

Emerging concepts such as digital twins, real-time virtual replicas of physical systems aim to integrate multi-scale models, including MD insights. However, challenges in data integration, computational speed, and real-time feedback limit practical deployment.

Effectively translating nanoscale findings into actionable macroscale design insights remains an open frontier. Advances in data assimilation, uncertainty quantification, and robust experimental validation will be essential for bridging this critical gap.

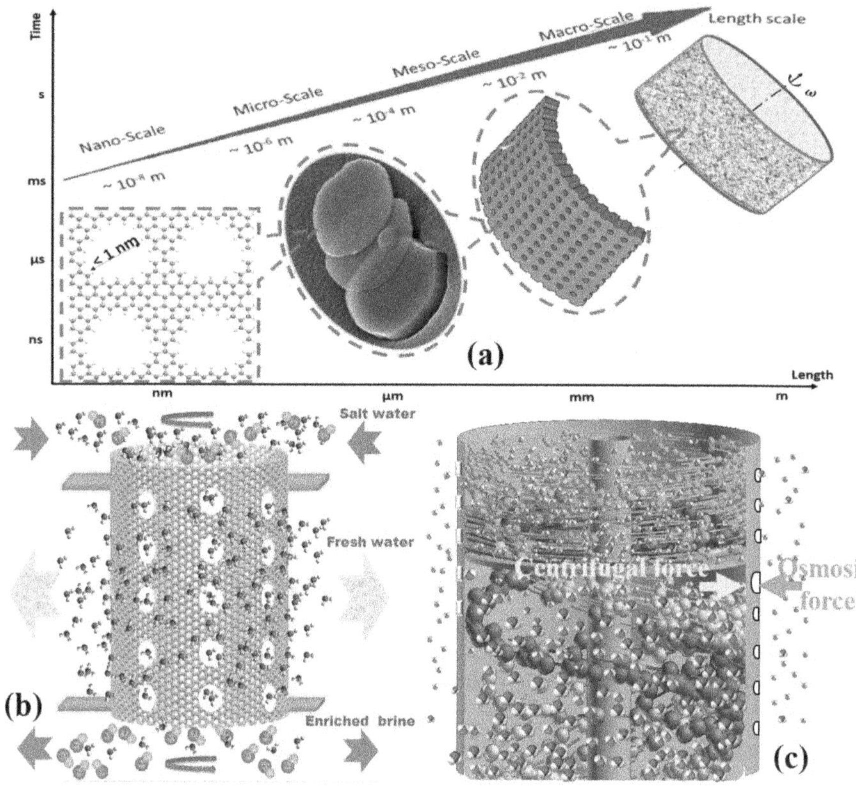

FIGURE 7.4 Scale-up studies for nanofluidic devices [2].

7.8 CONCLUSION

The future of Molecular Dynamics simulation in nanofluid research holds tremendous promise. By integrating machine learning for force field development, MD can overcome traditional limitations in accuracy and transferability, enabling predictive simulations for a wide range of chemistries and operational conditions.

Real-time MD-based predictive control frameworks represent a revolutionary direction, where nanoscale understanding is embedded directly into operational systems, allowing dynamic, on-the-fly optimization of thermal and flow performance.

Beyond energy, the expansion into biomedical applications, electronics cooling, and a multitude of other sectors signals the broad impact MD can have across science and engineering. Whether designing safer and more effective drug delivery vehicles, developing ultra-efficient heat sinks for next-generation electronics, or optimizing advanced manufacturing processes, MD simulations will be central to innovation.

To fully realize this potential, future research must address challenges in computational scalability, data integration, and robust validation against experimental benchmarks. At the same time, cross-disciplinary collaboration will be crucial—merging materials science, computational physics, engineering design, and data science to create holistic solutions.

As we enter an era demanding high-performance, sustainable, and safe materials and systems, Molecular Dynamics stands poised as a cornerstone technology, enabling discoveries and optimizations that push beyond what is possible with traditional approaches alone. The coming years will likely see a convergence of MD, machine learning, advanced experimental methods, and system-level design philosophies, ushering in a new generation of nanofluid-enabled innovations across energy, healthcare, and beyond.

REFERENCES

1. Allen, M. P., & Tildesley, D. J. (2017). *Computer simulation of liquids*. Oxford University Press.
2. Tu, Q., Ibrahimi, W., Ren, S., Wu, J., & Li, S. (2020). A molecular dynamics study on rotational nanofluid and its application to desalination. *Membranes*, *10*(6), 117.
3. Verlet, L. (1967). Computer 'experiments' on classical fluids. I. Thermodynamical properties of Lennard-Jones molecules. *Physical Review*, *159*(1), 98.
4. Darden, T., York, D., & Pedersen, L. (1993). Particle mesh Ewald: An N log (N) method for Ewald sums in large systems. *Journal of Chemical Physics*, *98*, 10089–10089.
5. Srivastava, I., Kotia, A., Ghosh, S. K., & Ali, M. K. A. (2021). Recent advances of molecular dynamics simulations in nanotribology. *Journal of Molecular Liquids*, *335*, 116154.
6. Sarkar, S., & Selvam, R. P. (2007). Molecular dynamics simulation of effective thermal conductivity and study of enhanced thermal transport mechanism in nanofluids. *Journal of Applied Physics*, *102*(7).
7. Keblinski, P., Phillpot, S. R., Choi, S. U. S., & Eastman, J. A. (2002). Mechanisms of heat flow in suspensions of nano-sized particles (nanofluids). *International Journal of Heat and Mass Transfer*, *45*(4), 855–863.
8. Sarkar, S., & Selvam, R. P. (2007). Molecular dynamics simulation of effective thermal conductivity and study of enhanced thermal transport mechanism in nanofluids. *Journal of Applied Physics*, *102*(7).
9. Yu, W., & Xie, H. (2012). A review on nanofluids: Preparation, stability mechanisms, and applications. *Journal of Nanomaterials*, *2012*(1), 435873.
10. Choi, S. U., & Eastman, J. A. (1995). *Enhancing thermal conductivity of fluids with nanoparticles* (No. ANL/MSD/CP-84938; CONF-951135-29). Argonne National Lab. (ANL), Argonne, IL (United States).
11. Sarviya, R. M., & Fuskele, V. (2017). Review on thermal conductivity of nanofluids. *Materials Today: Proceedings*, *4*(2), 4022–4031.
12. Wang, L., Dumont, R. S., & Dickson, J. M. (2012). Nonequilibrium molecular dynamics simulation of water transport through carbon nanotube membranes at low pressure. *The Journal of Chemical Physics*, *137*(4).
13. Vahidhosseini, S. M., Bidi, M. A., & Rashidi, S. (2024). Thermophysical properties of magnetic nanofluids under effects of magnetic field-a review on mechanisms and studies. *Journal of Thermal Analysis and Calorimetry*, 1–36.
14. Pu, J. H., Sun, J., Wang, W., & Wang, H. S. (2020). Generation and evolution of nanobubbles on heated nanoparticles: A molecular dynamics study. *Langmuir*, *36*(9), 2375–2382.

15. Kang, X., Zou, X., Sun, H. M., Ma, X. Y., & Chen, R. P. (2022). Molecular dynamics simulations of microstructure and dynamic shearing behaviors of kaolinite-water-salt system. *Applied Clay Science*, *218*, 106414.
16. Noid, W. G. (2013). Perspective: Coarse-grained models for biomolecular systems. *The Journal of Chemical Physics*, *139*(9).
17. Plimpton, S. (1995). Fast parallel algorithms for short-range molecular dynamics. *Journal of Computational Physics*, *117*(1), 1–19.
18. Daw, M. S., & Baskes, M. I. (1984). Embedded-atom method: Derivation and application to impurities, surfaces, and other defects in metals. *Physical Review B*, *29*(12), 6443.
19. Jorgenson, W. L., & Tirado-Rives, J. (1988). The OPLS potential functions for proteins. Energy minimizations for crystals of cyclic peptides and crambin. *Journal of the American Chemical Society*, *110*(6), 1657–1666.
20. Van Duin, A. C., Dasgupta, S., Lorant, F., & Goddard, W. A. (2001). ReaxFF: A reactive force field for hydrocarbons. *The Journal of Physical Chemistry A*, *105*(41), 9396–9409.
21. Heinz, H., Lin, T. J., Kishore Mishra, R., & Emami, F. S. (2013). Thermodynamically consistent force fields for the assembly of inorganic, organic, and biological nanostructures: The INTERFACE force field. *Langmuir*, *29*(6), 1754–1765.
22. Behler, J. (2016). Perspective: Machine learning potentials for atomistic simulations. *The Journal of Chemical Physics*, *145*(17).
23. Keblinski, P., Eastman, J. A., & Cahill, D. G. (2005). Nanofluids for thermal transport. *Materials Today*, *8*(6), 36–44.
24. Lou, Z., & Yang, M. (2015). Molecular dynamics simulations on the shear viscosity of Al2O3 nanofluids. *Computers & Fluids*, *117*, 17–23.
25. Prasher, R., Evans, W., Meakin, P., Fish, J., Phelan, P., & Keblinski, P. (2006). Effect of aggregation on thermal conduction in colloidal nanofluids. *Applied Physics Letters*, *89*(14).
26. Lu, W. Q., & Fan, Q. M. (2008). Study for the particle's scale effect on some thermophysical properties of nanofluids by a simplified molecular dynamics method. *Engineering Analysis with Boundary Elements*, *32*(4), 282–289.
27. Malika, M., & Sonawane, S. S. (2021). The sono-photocatalytic performance of a novel water based Ti+ 4 coated Al (OH) 3-MWCNT's hybrid nanofluid for dye fragmentation. *International Journal of Chemical Reactor Engineering*, *19*(9), 901–912.
28. Parate, V. R., Kawadkar, D. K., & Sonawane, S. S. (2011). Study of whey protein concentrate fortification in cookies variety biscuits. *International Journal of Food Engineering*, *7*(2), 1–12.
29. Jabbari, F., Rajabpour, A., & Saedodin, S. (2017). Thermal conductivity and viscosity of nanofluids: A review of recent molecular dynamics studies. *Chemical Engineering Science*, *174*, 67–81.
30. Fichthorn, K. A., & Qin, Y. (2006). Molecular-dynamics simulation of colloidal nanoparticle forces. *Industrial & Engineering Chemistry Research*, *45*(16), 5477–5481.
31. Sraelachvili, J. N. (2011). *Intermolecular and surface forces*. Academic Press.
32. Thakur, P., Sonawane, S. S., Sonawane, S. H., & Bhanvase, B. A. (2020). Nanofluids-based delivery system, encapsulation of nanoparticles for stability to make stable nanofluids. In *Encapsulation of active molecules and their delivery system* (pp. 141–152). Elsevier.
33. Bethi, B., Sonawane, S. H., Bhanvase, B. A., & Sonawane, S. S. (2021). Textile industry wastewater treatment by cavitation combined with Fenton and ceramic nanofiltration membrane. *Chemical Engineering and Processing-Process Intensification*, *168*, 108540.

34. Li, Y., Zhai, Y., Ma, M., Xuan, Z., & Wang, H. (2021). Using molecular dynamics simulations to investigate the effect of the interfacial nanolayer structure on enhancing the viscosity and thermal conductivity of nanofluids. *International Communications in Heat and Mass Transfer, 122*, 105181.
35. Mitra, A., Trifkovic, M., & Ponnurangam, S. (2021). Surface functionalization-induced effects on nanoparticle dispersion and associated changes in the thermophysical properties of polymer nanocomposites. *Macromolecules, 54*(9), 3962–3971.
36. Malika, M., & Sonawane, S. S. (2020). Effect of nanoparticle mixed ratio on stability and thermo-physical properties of CuO-ZnO/water-based hybrid nanofluid. *Journal of the Indian Chemical Society, 97*(3), 414–419.
37. Sonawane, S., Thakur, P., & Paul, R. (2020). Study on thermal property enhancement of MWCNT based polypropylene (PP) nanocomposites. *Materials Today: Proceedings, 27*, 550–555.
38. Songolzadeh, R., Behbahani, H. S., & Moghadasi, J. (2025). Molecular dynamics simulation of SDS-driven silica nanoparticle stabilization for wettability alteration in carbonate reservoirs. *Results in engineering, 27*, 106714.
39. Gupta, R., & Rai, B. (2017). Effect of size and surface charge of gold nanoparticles on their skin permeability: A molecular dynamics study. *Scientific Reports, 7*(1), 45292.
40. Zhou, L., Zhu, J., Zhao, Y., & Ma, H. (2022). A molecular dynamics study on thermal conductivity enhancement mechanism of nanofluids: Effect of nanoparticle aggregation. *International Journal of Heat and Mass Transfer, 183*, 122124.
41. Malika, M. M., & Sonawane, S. S. (2019). Review on application of nanofluid/nano particle as water disinfectant. *Journal of Indian Association for Environmental Management (JIAEM), 39*(1–4), 21–24.
42. Malika, M., Bhad, R., & Sonawane, S. S. (2021). ANSYS simulation study of a low volume fraction CuO–ZnO/water hybrid nanofluid in a shell and tube heat exchanger. *Journal of the Indian Chemical Society, 98*(11), 100200.
43. Liao, J., Qing, S., Zhang, X., Huang, X., & Wang, X. (2022). Factors that influence the aggregation structure of nanoparticles in nanofluids: A molecular dynamics simulation. *Journal of Molecular Liquids, 368*, 120716.
44. Zhou, L., Zhu, J., Zhao, Y., & Ma, H. (2022). A molecular dynamics study on thermal conductivity enhancement mechanism of nanofluids: Effect of nanoparticle aggregation. *International Journal of Heat and Mass Transfer, 183*, 122124.
45. Thakur, P., Sonawane, S., Potoroko, I., & Sonawane, S. H. (2021). Recent advances in ultrasound-assisted synthesis of nano-emulsions and their industrial applications. *Current Pharmaceutical Biotechnology, 22*(13), 1748–1758.
46. Sonawane, A. V., Rikame, S., Sonawane, S. H., Gaikwad, M., Bhanvase, B., Sonawane, S. S., . . . & Gaikwad, R. (2024). A review of microbial fuel cell and its diversification in the development of green energy technology. *Chemosphere, 350*, 141127.
47. Plimpton, S. (1995). Fast parallel algorithms for short-range molecular dynamics. *Journal of Computational Physics, 117*(1), 1–19.

8 Life Cycle Assessment of Nanofluid Application in Process Industries

8.1 GLOBAL ENERGY TRENDS AND INDUSTRIAL CHALLENGES

The world today is experiencing an all-time high in energy usage due to increasing population that is accompanied by rapid industrialization, urbanization, and technological development. With the world's countries shifting to an increased level of quality of life and economic growth, the need for energy-consuming goods and services is growing at a faster pace, which heavily strains the energy reserves. The International Energy Agency (IEA) forecasts that global energy demand will increase by around 25% in 2040, in case current trends persist. The industrial sector, especially process industries, is still the center of this consumption of energy.

The process industries constitute several industries of manufacturing, such as chemical production, petrochemicals, pharmaceutical industry, food and beverage industry, textile industries, cement industries, pulp and paper mill industries, and power generation industries. Such industries require a lot of thermal and electrical energy to boost reactions, processes, and efficiency in their operations. It should be noted that heating and cooling functions alone consume more than 50% of the industrial energy. Various unit operations that need thermal energy include evaporation, distillation, drying, crystallization, and heat exchange. Efficient management of thermal energy in the operations is vital to not only optimize the energy expenditure bills but also reduce the carbon footprints via greenhouse gas (GHG) emissions.

Energy efficiency has, in this regard, become an important parameter of the performance of industrial operations. Regulatory organizations and the government are graduating to more stringent energy efficiency standards and emission limits, which are forcing industries to go green. Energy recovery, renewable integration, and heat exchanger technologies are some of the innovations that are undergoing vast use. Thermal management, the ability to control and use thermal energy strategically, turns out to be one of these areas of intervention.

8.1.1 Thermal Management: A Core Pillar of Process Efficiency

Thermal management refers to the techniques and systems employed to regulate temperatures and facilitate effective heat transfer in industrial processes. It is necessary in order to sustain product quality, guarantee reliability of equipment, maximize reaction kinetics, and realize energy preservation. Ineffective thermal control may result in poor production, damage to devices, safety risks, and the waste of excessive energy.

DOI: 10.1201/9781003628071-8

The main elements in which thermal control takes place include heat exchangers, boilers, condensers, cooling towers, and evaporators, which make up the heat transfer systems. The specifications of these systems highly depend on the characteristics of the heat transfer fluid used. The most popular traditional fluids are water, mineral oils, and glycols: they are available, stable, and that is why they can be used in equipment. Their low conductivity in terms of heat, however, has a considerable drawback, such that they may frequently require higher flow rates, or higher differences in temperatures to meet necessary rates of heat transfer.

Although these compensatory measures work, they lead to more energy usage and investments. Taking an example, an increase in the volume of fluid circulating in a system to counter slow thermal conductivity will result in high cost of operation in a system and possible ill effects on the environment. With the world adopting energy-smart and resource-efficient behaviors, there is a necessity to come up with the identification and implementation of advanced thermal fluids that can achieve high thermal performance without an equal value of energy or resource investment.

8.1.2 The Limitations of Conventional Heat Transfer Fluids

As Table 8.1 indicates, no matter how successful the application of conventional heat transfer fluids has been, they are indeed restricted in their capacity due to their thermal characteristics. Water, for example, though relatively high specific heat, has a value of thermal conductivity of around 0.6 W/m K. Another widely used coolant, ethylene glycol, has a thermal conductivity of around 0.25 W/m K. Such values are much lower than metals or high-tech solid ones. The fluids therefore have a lower performance in conducting heat, especially in high-temperature or high-heat-flux applications.

The effect of the low conductivity fluids is that there need to be changes in design to ensure performance of the system. Such changes can comprise:

- Enlarging heat exchange surfaces.
- Increasing fluid velocities.
- Using enhanced surface geometries.
- Employing multiple stages of heat exchange.

TABLE 8.1
Thermal Conductivity of Materials at Room Temperature

Material Category	Examples	Thermal Conductivity Range (W/m·K)
High-Conductivity Metals	Silver, Copper, Gold	317–429
Moderate Metals/Alloys	Aluminum, Iron, Steel	15–237
Non-Metallic Solids	Glass, Plastics, Wood, Cork	0.039–0.81
Liquids	Water, Glycol, Oil, Benzene	0.145–0.6
Gases	Hydrogen, Air	0.018–0.18
Common Nanoparticles	TiO_2, ZnO, CuO, SiO_2, Al_2O_3	8.4–36
High-Performance Nanomaterials	Nanodiamond, CNTs (SWCNT, MWCNT)	1000–6600

While effective, these approaches can escalate system complexity, maintenance requirements, and energy demand. Moreover, the footprints of larger equipment can also conflict with the aims of intensification of processes, where more compact and integrated process units are desired. Alternative heat transfer fluids that hold prospect when it comes to surmounting the drawbacks of conventional coolants have generated increasing interest in this area. Nanotechnology and fluid systems have come up with an innovative solution to the incorporation of nanotechnology in fluid systems, resulting in the development of a new type of material called nanofluids.

8.1.3 Emergence of Nanofluids as Advanced Heat Transfer Media

Nanofluids are constructed suspensions of colloids—nanoparticles typically below 100 nanometers in size—of conventional base fluids. Such nanoparticles can be made of metals (e.g., copper, silver), metal oxides (e.g., alumina, titania, silica), carbon (e.g., carbon nanotubes, graphene), or a composite of some combination of these. A small fraction of these particles added to base fluid can substantially augment the thermal conductivity of the fluid in addition to improving the viscosity and convective heat transfer capability.

The concept of nanofluids was introduced by Choi in the late 1990s at Argonne National Laboratory. His pioneering work demonstrated that even a 1% volume addition of nanoparticles to water could result in a substantial increase in thermal conductivity. Studies of this field have increased extensively since then, as many experimental and theoretical studies have cemented the fact that nanofluids have superior heat transfer capabilities as compared to those of conventional fluids.

The key benefits of nanofluids include:

- Higher thermal conductivity and convective heat transfer coefficient.
- Enhanced thermal response and temperature control.
- Potential for reduced equipment size and weight.
- Improved energy efficiency.
- Customizable properties based on nanoparticle selection.

These attributes make nanofluids highly attractive for process industries that demand precise thermal management and high heat flux removal. Applications range from heat exchangers in chemical plants to cooling systems in power generation, electronics, solar thermal collectors, and HVAC systems [1].

8.1.4 Process Intensification through Nanofluids

Process intensification seeks to lead to smart chemical and industrial treatment that is efficient, sustainable, and miniaturized. The inclusion of nanofluids in heat transfer devices can facilitate this purpose only too well. Nanofluids can enhance the speed of heat transfer in a chemical or a physical process, improving its production yields, quality of the products, and decreasing energy requirements because they can act in or near the optimal thermodynamic level.

Also, nanofluids may help to minimize the size of heat exchangers and other thermal systems, thus minimizing material usage, the physical installation size, and maintenance. It is especially useful in applications such as pharmaceuticals or microelectronics, where space is of the essence and thermal accuracy matters.

The other potential application of nanofluids has been the use in improving the functionality of the renewable energy systems. A good example is in the scenario of solar thermal collectors, where nanofluids can intensify the process of solar radiation absorption and transfer a higher level of solar heat into the system, which in turn will be more effective. On the same token, nanofluids help to utilize low-grade heat sources and are also efficient in geothermal and waste heat recovery.

8.1.5 Barriers to Widespread Industrial Adoption

Despite their demonstrated advantages, the industrial adoption of nanofluids remains limited due to several challenges:

- **Stability and Agglomeration**: Nanoparticles tend to agglomerate over time, leading to sedimentation and loss of thermal performance.
- **Synthesis and Cost**: Producing high-quality nanoparticles with consistent properties can be expensive, especially at a commercial scale.
- **Health and Environmental Risks**: The long-term impact of nanoparticle exposure on human health and ecosystems is not fully understood, necessitating caution and regulation.
- **Compatibility and Corrosion**: The interaction of nanofluids with existing materials and systems must be assessed to prevent corrosion or material degradation.

These barriers highlight the need for comprehensive evaluation tools that go beyond thermal performance and consider the broader implications of nanofluid use in industrial settings. This is where life cycle thinking becomes essential.

8.1.6 The Need for Life Cycle Thinking

To accurately evaluate the sustainability of nanofluids, it is essential to examine their effects throughout their entire life cycle—from raw material extraction and nanoparticle synthesis to operational utilization and end-of-life disposal or recycling. Life Cycle Assessment (LCA) provides a structured methodology for performing these evaluations.

An LCA takes into account:

- Resource inputs (energy, water, raw materials).
- Emissions (air, water, soil).
- Waste generation.
- Environmental impacts (global warming potential, toxicity, eutrophication, etc.).

In the context of nanofluids, LCA can help answer critical questions such as:

- Do the energy savings during operation offset the energy used in nanoparticle production?
- What is the environmental burden of nanoparticle synthesis methods?
- Can nanofluids be safely disposed of or recycled?
- How do nanofluids compare to conventional fluids in overall sustainability metrics?

Integrating LCA in development and implementation of nanofluids will help achieve the goal of shifting innovations based on the real benefits to the environment and ensuring the transfer of burden to different life cycle stages.

The objective of this chapter is to present an in-depth discussion of the life cycle impacts of nanofluids in the process industries. It commences by introducing information about the types, synthesis methods, and applications of the nanofluids. It then gives an excellent LCA framework, specific to nanofluids, detailing life cycle inventory (LCI), impact assessment, and interpretation. Real-life implications are demonstrated by using case studies in different industries. There is also a chapter on the technical-economic, sustainability, and the future prospects of the development of green nanofluid.

The chapter can be useful in the technical, environmental, and economic perspectives with the view of making informed decisions in the application of nanofluids in process industries. It forms a basis on which policymakers, researchers, and even industry players can think of nanofluids as not only a tool of high-performance thermal media but also intelligent elements of a green industrial world [2].

8.2 OVERVIEW OF NANOFLUIDS IN PROCESS INDUSTRIES

8.2.1 Definition and Classification of Nanofluids

Nanofluids are designed as colloidal mixtures of nanomaterials with the traditional base fluids. These nanoparticles have diameters usually below 100 nm and are added only in significant amounts to better the thermophysical properties of the base fluid, which in most significant cases are those of thermal conductivity, viscosity, and heat capacity. The ultimate goal of nanofluids would be to enhance the heat transfer system in such a way that it would not be accompanied by radical changes in the design of equipment or energy consumed.

Classification of nanofluids can be approached in several ways:

a. **Based on nanoparticle type**
   - Metal nanofluids: copper, silver, gold
   - Metal oxide nanofluids: alumina (Al_2O_3), titania (TiO_2), copper oxide (CuO), zinc oxide (ZnO)
   - Carbon-based nanofluids: carbon nanotubes (CNTs), graphene, graphite, fullerenes

    - Magnetic nanofluids: Fe_3O_4, used in controlled or targeted heating systems
    - Hybrid nanofluids: combinations of two or more different nanoparticles for enhanced synergistic effects

b. **Based on base fluid**
    - Water-based nanofluids
    - Oil-based nanofluids
    - Glycol-based nanofluids
    - Ionic liquids and refrigerant-based nanofluids

c. **Based on functional behavior**
    - Heat transfer enhancement fluids
    - Dielectric cooling nanofluids
    - Solar energy harvesting nanofluids

The specific classification depends on the application and targeted performance attributes.

8.2.2 Common Base Fluids and Nanoparticles Used

Base fluids serve as the medium in which nanoparticles are suspended. The most widely used base fluids include:

- **Water**: Most common due to high specific heat, low cost, and availability.
- **Ethylene glycol/Propylene glycol**: Used in automotive and HVAC applications.
- **Oils (mineral/synthetic)**: Employed where high-temperature stability is required.
- **Refrigerants**: Emerging in vapor compression cooling systems.
- **Ionic liquids**: For niche high-performance applications requiring extreme thermal and chemical stability.

The commonly used nanoparticles in nanofluids include:

- **Al_2O_3 (Alumina)**: Economical and stable; offers moderate thermal enhancement.
- **CuO (Copper oxide)**: High thermal conductivity.
- **TiO_2 (Titanium dioxide)**: Environmentally benign and photocatalytic properties.
- **Fe_3O_4 (Magnetite)**: Enables magnetically controllable heat transfer.
- **CNTs and Graphene**: Exceptional thermal properties but expensive and difficult to disperse.

Surface functionalization with surfactants or chemical treatments is often used to enhance the dispersion and stability of nanoparticles in the base fluid, preventing agglomeration and sedimentation.

8.2.3 Key Applications in Process Industries

Nanofluids are increasingly used across various segments of process industries for their superior heat transfer capabilities:

a. **Heat exchangers:** Nanofluids improve the overall heat transfer coefficient in shell-and-tube, plate, and compact heat exchangers. This leads to:
   - Reduction in size and weight of heat exchangers.
   - Lower energy consumption.
   - Improved process efficiency.

b. **Reactors:** In chemical reactors, temperature uniformity and rapid heat dissipation are crucial. Nanofluids help in avoiding hotspots and maintaining optimal conditions for exothermic/endothermic reactions.

c. **Cooling systems:** In industrial-scale cooling systems such as HVAC, engine cooling, and electronic component cooling, nanofluids offer:
   - Faster heat removal.
   - More compact system design.
   - Lower fan/pump power consumption.

d. **Solar thermal collectors:** Nanofluids enhance photothermal conversion efficiency in flat plate and parabolic solar collectors due to better solar energy absorption and heat distribution.

e. **Refrigeration and cryogenics:** Nano-refrigerants improve the coefficient of performance (COP) and reduce compressor work in vapor compression refrigeration systems.

f. **Power generation and waste heat recovery:** Nanofluids increase the efficiency of Rankine and Brayton cycle components and are suitable for low-grade heat recovery.

8.2.4 Benefits for Process Intensification

Process intensification (PI) aims to reduce the size, energy usage, and environmental impact of chemical and process systems. Nanofluids contribute to PI in the following ways:

- **Higher heat transfer rates:** Enhanced thermal conductivity improves convective and conductive heat transfer, leading to faster and more uniform temperature control.
- **Reduced equipment size:** Higher heat fluxes allow downsizing of thermal units like heat exchangers.
- **Energy efficiency:** Lower pumping power and faster heating/cooling cycles reduce overall energy demand.
- **Environmental benefits:** Improved energy efficiency translates to reduced CO_2 and greenhouse gas emissions.
- **Improved safety:** Stable temperature profiles reduce the risk of thermal runaways and equipment failure.
- **Economic competitiveness:** Reduced operational costs, maintenance frequency, and equipment size lead to significant CAPEX and OPEX savings.

Despite these benefits, certain challenges such as nanoparticle cost, stability, and potential toxicity must be addressed through rigorous life cycle assessments and techno-economic analyses.

8.3 LIFE CYCLE ASSESSMENT FRAMEWORK FOR NANOFLUIDS

The Life Cycle Assessments (LCA) is an internationally accepted and standardized technique of quantitatively testing the possible environmental impact of all the life stages of a product: starting with the extraction of raw materials (cradle) and ending with its disposal (grave). This covers both the upstream and downstream processes, like the material sourcing processes and production process, and the subsequent processes like distribution, usage, and waste management. According to ISO 14040 and ISO 14044 standards (ISO, 2006a; ISO, 2006b), LCA gives a comprehensive and quantitative ground for assessing the environmental impact of goods, processes, or systems. It helps the industry, governments, and scientists make better decisions to promote results in sustainability [3].

8.3.1 Key Benefits of LCA

8.3.1.1 Proactive Environmental Management

The strong point of LCA is that it can predict the environmental impacts of a product or a process in advance of implementing it. By looking ahead in terms of the process of design or planning, decision-makers can be able to take precautionary tactics instead of using post-factum solutions, which are rather expensive and ineffective.

8.3.1.2 Consistent and Objective Comparisons

LCA allows comparative evaluation of different systems or technologies on a common and scientifically sound basis. An example is that various formulas of nanofluid or heat exchanger material may be compared in terms of life cycle carbon emissions, energy requirement, and adverse effects of toxicity.

8.3.1.3 Support for Eco-Design and Innovation

LCA helps in the embedding of the principles of sustainability in the designing (eco-design) that promotes the creation of products that are as efficient, limited impact, and are simpler to recycle or reuse.

LCA is commonly carried out in four well-organized and interrelated stages (ISO 14040/44), as in Figure 8.1, which illustrates the Four Steps of the Life Cycle Assessment (LCA) Process.

8.3.2 Goal and Scope Definition

The first and most decisive step in the Life Cycle Assessment (LCA) procedure is the Goal and Scope Definition step. It offers the strategic basis on which the remaining stages, including Life Cycle Inventory (LCI), Life Cycle Impact Assessment (LCIA),

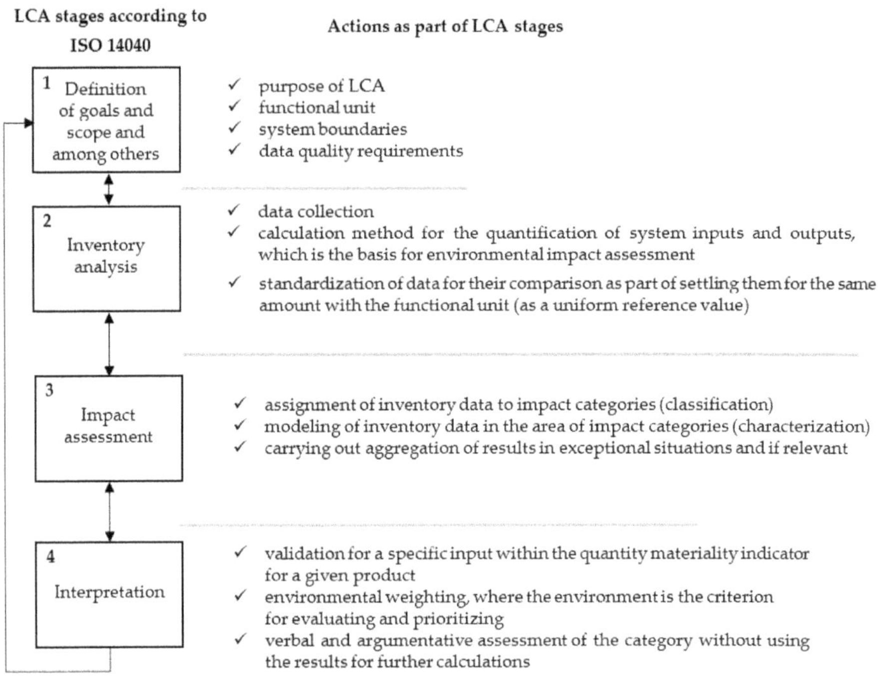

FIGURE 8.1 Four steps of life cycle assessment process [4].

and Interpretation stages, are implemented. This step explains the purpose of an assessment, which system or product it shall examine, and how it shall be tested in respect to functionality, boundaries, as well as impact standards.

The need to have a clear definition of the goal and this characteristic is even more essential when LCA is used in process industries to investigate the nanofluids. The reason is that nanofluids are a complex phenomenon because they are not mere sophisticated heat-carrying materials but ones that imply some environmental and health hazards. Nanofluids include the process of preparing nanoparticles, their formulations, and applications within thermodynamic processes and disposal thereafter: each of these processes possesses specific environmental impacts.

8.3.2.1 Goal Definition

The goal of the LCA answers the fundamental question: "Why is this study being conducted?" It establishes the purpose for which the LCA results will be utilized, the audience, and decision-making environment. In the process industries, the objectives of nanofluids are normally the following:

- **Comparative analysis**: To compare the environmental effects of nanofluid-based thermal systems (e.g., Al_2O_3-water, CuO-water) with the typical heat transfer fluids like water, mineral oil, or ethylene glycol.

- **Sustainability verification**: To confirm the allegations with claims that nanofluids have a better energy and environmental effect in heat transfer processes.
- **Identification of hotspots**: Determination of what environmental areas are hotspots of the life cycle of nanofluids; for example, nanoparticle formulation is energy-intensive, or end-of-life disposal.
- **Assist in decision-making**: Here, it will help the equipment manufacturers, engineers, and decision-makers in policy-making regarding the best environmentally friendly thermal management technologies.
- **Participate in green Innovation**: To develop a reference to eco-design of nanofluids to offer in the future, such aspects as green synthesis techniques, green dispersion procedures, and the circular economy scenarios.

8.3.2.2 Scope Definition

Once the goal is established, the next step is to clearly define the *scope* of the LCA. This step contains establishing boundaries, presuppositions, restrictions, and choosing important methodological aspects. An articulated scope makes the study consistent, transparent, and reproducible.

8.3.2.2.1 Functional Unit

It is used as the point of reference to all impact calculations. Some standard functional units in LCA studies of nanofluids are:

- **"1 MJ of transferred heat"**: This corresponds to the efficiency of the thermal fluid in energy transfer assignments, primarily in heat-changing or reactors or solar reactors.
- **"1 m^2 of heat exchanger area per year of operation"**: This measure is useful in cases when the spatial design, compactness, or lifespan of thermal equipment is under consideration.
- **"1000 operating hours of a thermal system"**: Used where there are time-limited industrial processes that are in continuous production, such as continuous manufacturing.

The selection of a meaningful and accurate functional unit provides an equal comparison of nanofluids with conventional fluids. As an example, provided that nanofluids carry out heat faster or more efficiently, then this heat duty may be attained with a lesser area of surface or shorter time, which has to be captured in the functional unit.

8.3.2.2.2 System Boundaries

The system boundary in Life Cycle Assessment (LCA) has many risky facts to tell or explain about the assessment range of processes covered by the analysis and plays a significant role in the context of the extent and applicability of LCA. In Figure 8.2, four conventional scenarios of the system boundaries *Cradle-to-Gate*, *Gate-to-Gate*, *Gate-to-Grave,* and *Cradle-to-Grave* are shown, which differ in the scope as

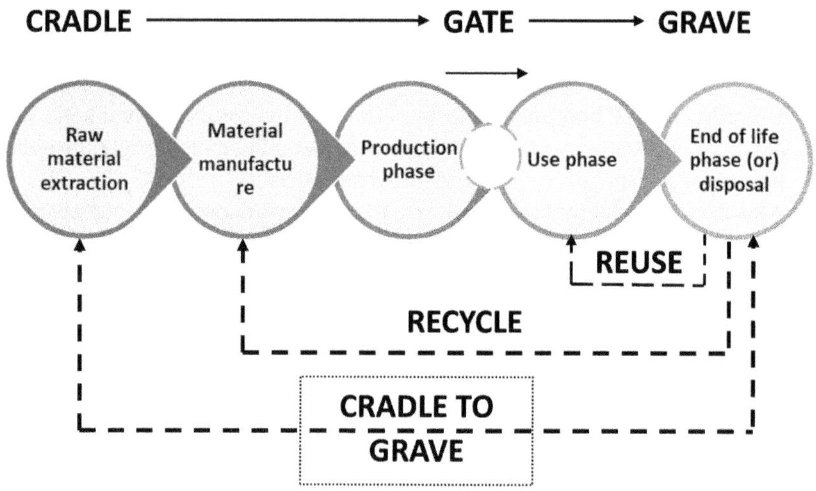

FIGURE 8.2 The system boundary of the LCA model.

well as usage, particularly when considering the environmental effect of nanofluids with regard to process industries.

In *cradle-to-gate* analysis, the system boundary includes the processes obtaining raw materials (i.e., mining of metal oxides to produce nanoparticles) to the production and formulation of nanofluids. It does not however include use phase and end-of-life processes. This boundary can be especially helpful when the emphasis is on the environmental footprint of the production process: in that case, the manufacturers can compare possible alternative synthesis paths or optimize production efficiencies. For example, it may compare the energy and emissions consumed during the synthesis of Al_2O_3 or CuO nanoparticles and one into a base fluid such as water or glycol.

Gate-to-gate boundary further narrows the focus with only one process or operational unit being analyzed; normally, these exclude both upstream and downstream activity. In the nanofluids sense, this could be a case study involving only the analysis of the dispersion and stabilization process in a factory and not thinking of where the raw materials were sourced and the end product consumed or its disposal. This method tends to be applied to in-house process improvement or comparison among various operations of the plant.

Conversely, the *gate-to-grave* limit begins at the stage when the product is usable, therefore, omitting the delivery of raw materials and the production process, and goes to the phases of use and disposal. In the case of nanofluids, this would mean testing the advantage during their use in the heat exchangers (less energy usage, better heat transfer), along with the analysis of potential effects on the environment after their disposal, such as incineration or leaching of nanoparticles during landfill. This boundary assists in knowing how the product will perform in real industrial environment and the impact of the end-of-the-life treatment on environmental sustainability.

Lastly, *cradle-to-grave* boundary presents a fully detailed analysis as all the processes, such as extracting the raw materials and taking them to the final stage of production, utilization, and subsequent recycling or disposal are covered. This method offers a complete life cycle view, and it is favorable in making the comparison between the nanofluids and the conventional heat transfer fluids as a way of evaluating the net environmental benefit. It assists in resolving such multifaceted questions as whether the energy savings implemented throughout the term of its utility outweigh the environmental expenditures spent on nanoparticle production and whether the overall consumption of nanofluids has any positive effects on the objectives of sustainability. In research intended to achieve efficacy in its decision-making and policy formations, cradle-to-grave assessments are crucial, and the use of cradle-to-grave assessments is vital in giving a comprehensive perspective on the use of nanofluid in the process industry.

For example, we can take a cradle-to-grave view with nanofluids in order to take an entire environmental impact. The system normally involves a complete life cycle, where raw materials are extracted and then disposed or recycled. The process begins with the raw material purchasing, whereby the system involves mining and excavation of metal ore in the mining process, like bauxite ore (to produce aluminum oxide Al_2O_3) or copper ore (to produce copper oxide CuO). Such processes tend to produce a lot of energy, and chemical refinement and purification are commonly involved, causing emissions and consumption of resources.

The following one is nanoparticle synthesis, and such techniques as sol-gel processing, hydrothermal techniques, and chemical vapor deposition (CVD) are the most typical methods. They generally involve a large amount of energy input, elevated temperatures, or a vacuum, and use many different solvents, precursors, and surfactants. The synthesis phase also produces byproducts and emissions such as nitrogen oxides (NOx), carbon dioxide (CO_2), and volatile organic compounds (VOC), which all cause environmental effects in terms of global warming and poor air quality.

After synthesis, the formulation of nanofluid consists of the dispersion of these nanoparticles into a base fluid containing any one of the three—water, ethylene glycol, or oils—to generate the product. Another form of energy required at this stage is mixing, typically achieved through ultrasonication or high-shear mixing, and the method necessitates the use of stabilizing agents and surfactants to prevent aggregation and sedimentation.

The manufacturing plant should also keep in check any losses of material and waste created in the process of blending and transfer.

After the formulation, packaging and transportation form the important parts of the system. Nanofluids are usually kept in high-density polyethylene (HDPE) drums or in special containers that allow dispersion stability and avoiding contamination. The transportation stage entails greenhouse gas emissions and energy use, and it mainly depends on how far a product should be transported and the methods of transport.

The use phase is the physical implementation of nanofluids at industrial process machinery, including heat exchangers, condensers, and reactors. In this case, enhanced heat transfer efficiency through the high thermal conductivity of nanofluids leads to operating energy savings, low energy input, compact equipment, and high-efficiency heat exchangers. The benefits reduce the overall energy requirements as well as the emissions, especially where nanofluids substitute regular fluids of lower thermal performance.

Industrial systems also need maintenance and replacement over the years, and this might be affected by the stability and lifespan of nanofluids. Nanoparticle agglomeration, sedimentation, fouling, and corrosion problems may bring about a serious degradation of performance and even the need to replace fluid or clean the system. Such maintenance jobs result in environmental effects by utilizing cleaners, employing extra resources, and producing wastes.

At last, *end-of-life (EoL)* of management involves the disposal or recycling or burning of used nanofluids. Such a stage is particularly important due to the unpredictable environmental behavior of nanoparticles. Poor disposals could cause leaching to the soil or water bodies, causing bioaccumulation and toxicity risks. Many of the base fluids might be reclaimed by separation methods, but metal or oxide nanoparticle products commonly pose dilemmas that require safe disposal or reuse. This is a stage that must be evaluated in order to know how the benefits in terms of use phase compare with the burdens in the end-of-life phase.

Typically, when results are limited by a lack of data or resources, *cradle-to-gate* or *gate-to-gate* (where one step is examined rather than the entire process) is chosen as an assessment method, but it qualifies as a partial method of determining environmental impact because of the lack of data on the outcome of the entire process. Such narrow scopes do not always incorporate downstream effects of use and disposal in most cases, which are important in the determination of the actual sustainability of nanofluids in practice [5].

8.3.2.2.3 Assumptions and Limitations of LCA

Besides system boundaries, all LCA studies should publicly state their assumptions and limitations, especially in application to novel technologies such as nanofluids. The most significant one is access to toxicological data. A number of engineered nanomaterials are comparatively contemporary, and the long-term consequences of engineered nanomaterials on human health and the environment are not fully comprehended. Therefore, sometimes studies are based on qualitative proxies, or they present a narrow estimation of the impact it can have, specifically, in the categories of human toxicity and ecotoxicity. The final notable assumption is the choice of the energy source mix that is applied during the production of nanoparticles. When fossil sources prevail in the production of power, there is a huge environmental impact as opposed to renewable-based systems. In such a way, the structure of the local energy grid should definitely be defined.

Another factor with assumptions that will affect the outcome is the process scaleup. During laboratory-scale synthesis, energy and emission per unit mass of production are usually higher than in scaled-up production. Thus, extrapolations have to be performed with caution, and scale-up principles should be properly recorded. Performance and sustainability are also determined by the stability and life cycle of nanofluids. The useful lifetime of nanofluids is potentially constrained by agglomeration, oxidation, or chemical degradation, necessitating their replacement on a periodic basis. In a case where the stated stability period is not accurate, then the benefits that will be estimated during the used phase may be overestimated. Finally, a crucial role is played by the geographical context. The different regions have different environmental regulations, transportation distances, and local energy sources, and these contextual factors have to be taken into consideration so that the realistic assessment is realistic to the location.

In the case of LCA examination of CuO-water nanofluids, a nanofluid is considered to have stability of dispersion over 12 months of steady operation under intensive heat exchanger conditions. But in practice, because of the sedimentation after four months in service, it will be necessary to replace it three times a year, and then the total energy and environmental effects will be three times higher, which may completely alter the results of the analysis, due to an unfavorable study. This demonstrates that a slight assumption can have a big impact on the meaning and applicability of LCA findings.

8.3.2.2.4 Impact Categories

Impact categories in LCA reflect the environmental consequences of material and energy flows across the product life cycle. In the case of nanofluids, the most appropriate categories are:

- **Global warming potential (GWP)**: This type includes all greenhouse gas emissions, which are usually expressed in *CO2 equivalent*, and assists one in knowing whether the energy saved in operations can counter emissions caused by nanoparticle production or not.
- **Cumulative energy demand (CED):** This is the overall life cycle energy input in the form of both renewable and non-renewable energy. It outlines whether nanofluids carry deck energy gain or loss.
- **Human toxicity potential (HTP):** Estimates possible harm to human health as a result of exposure as part of synthesis, handling, use, or disposal of the substance.
- **Freshwater ecotoxicity:** Estimates the potential toxicity of nanoparticles to aquatic organisms upon leakages or release to the water bodies.
- **Formation of particulate:** An essential factor where the synthesis of nanoparticles is carried out using combustion or high temperatures that produce fine particles.
- **Resource depletion:** Evaluates how non-renewable raw materials resources are used, such as rare metals, which may prove unsustainable in terms of environmental or economic aspects.
- **Water use and eutrophication:** Applicable when the quantity of water employed during a synthesis is in huge volumes or when wastewater effluent contains nutrients that could trigger algal bloom, resulting in the depletion of oxygen in the waterways.

Such internationally accepted methods of study of impact categories include the following:

- CML 2001
- TRACI (EPA)
- ReCiPe
- IMPACT 2002+

The selected categories must align with the *goals of the study* and the *concerns of stakeholders*, such as industry partners, regulators, or policymakers.

8.3.3 Life Cycle Inventory (LCI) Analysis

The second large part of the Life Cycle Assessment (LCA) methodology (after Goal and Scope Definition) is the Life Cycle Inventory (LCI) analysis. During this step, all the input and output data concerning energy, materials, emissions, and waste are quantitatively recorded regarding each stage of the life cycle of the product or process under study. LCI forms the quantitative foundation of LCA, giving an empirical foundation to the next step, Life Cycle Impact Assessment (LCIA) [6].

In the case of nanofluids in the process industries, the LCI should be able to represent the multi-faceted and multifaceted nature of such material, which encompasses extraction of raw material, production of nanoparticles, formulation, application, as well as during discard. Due to the newness and technical complexity of the nanofluid systems, the LCI stage requires not just the inclusion of a sufficient amount of data and the rigor of methodology but also a detailed description of the assumptions and sources, as well as uncertainties.

8.3.3.1 Purpose of Life Cycle Inventory (LCI) in Nanofluid LCA

The Life Cycle Inventory (LCI) phase serves as the quantitative backbone of the Life Cycle Assessment (LCA) framework. Its primary purpose is to systematically identify, compile, and quantify all the flows of materials and energy into and out of a product system. When applied to the evaluation of nanofluids in process industries, the LCI provides a comprehensive dataset that captures every unit of input and output associated with their production, use, and disposal. This detailed mapping enables the accurate assessment of environmental burdens and resource demands, ensuring that sustainability metrics are grounded in real-world data rather than assumptions.

For nanofluids, the LCI begins by accounting for raw material consumption, which forms the initial building block of the product system. This includes all substances used in the synthesis of nanoparticles and the formulation of the nanofluid. Common raw materials include metal salts such as aluminum nitrate ($Al(NO_3)_3$) for producing aluminum oxide (Al_2O_3) nanoparticles, or copper sulfate ($CuSO_4$) for copper oxide (CuO) nanoparticles. Additional inputs such as deionized water, base fluids (e.g., water, ethylene glycol), and surfactants or stabilizers (e.g., SDS, CTAB, oleic acid) are also crucial to ensure nanoparticle dispersion and long-term fluid stability. The quantity, concentration, and purity of these materials are meticulously recorded to understand their contribution to resource depletion, toxicity, and overall life cycle impact.

Next, the LCI addresses the energy requirements of each stage in the nanofluid life cycle. Nanoparticle synthesis often involves energy-intensive steps like thermal treatment, calcination, hydrothermal processing, or chemical vapor deposition (CVD). Each of these processes requires a specific amount of thermal energy (e.g., natural gas, electricity) and electrical energy for operations such as mixing, ultrasonication, high-shear blending, or centrifugation. Similarly, the formulation phase—which involves dispersing nanoparticles into a fluid medium—consumes energy for stirring, sonication, and possibly vacuum degassing to remove air bubbles. All of these energy flows must be quantified, as they contribute to impact categories like Cumulative Energy Demand (CED) and Global Warming Potential (GWP) in the subsequent LCIA phase.

In addition to inputs, the LCI carefully documents all emissions to air, water, and soil that occur throughout the system's life cycle. These include gaseous emissions such as carbon dioxide (CO_2) from energy consumption, nitrogen oxides (NO_x) from high-temperature synthesis, and volatile organic compounds (VOCs) released during solvent evaporation. Liquid effluents may contain unreacted chemicals, surfactants, or nanoparticle residues, posing risks to aquatic ecosystems and requiring proper wastewater treatment. Solid wastes may be generated from filtration processes, discarded synthesis residues, or packaging materials. These emissions and wastes are quantified per functional unit (e.g., per MJ of heat transferred), forming a critical part of the environmental profile of nanofluid use.

The other important point of the LCI is the identification and quantification of the products and co-products that may be generated by the system. The key product is, of course, the formulated nanofluid, that is, a homogeneous suspension of nanoparticles on a carrier fluid. Waste byproducts in other cases might be produced, say, co-products, which might include heat generated during synthesis and is used to recover, or byproducts, which might include secondary nanoparticles. Any solid byproducts, for example, filtration cake or recovered metal oxides, can also be defined as LCI co-products or wastes as per the re-use, recycle, or disposal.

Lastly, not every study may naturally consider some of the elements of infrastructure and capital equipment—information, although it is considered in some studies, particularly long-term or high-resolution studies. This could be taking into account the environmental load linked with the creation and use of nanoparticle production plants, blending units, or heat exchangers where the nanofluids are employed. Information related to equipment can cover materials (e.g., stainless-steel reactors, glass vessels), the energy they consume throughout their lifetime, and the consequences of their disposal and/or recycling. Although it is commonly not included in short-life or screening-level LCAs, this data can help illustrate a more comprehensive picture of the total system effect—particularly when that infrastructure is either unique or resource-intensive [7].

8.3.3.2 Key Components of LCI Analysis

In the context of nanofluids used in process industries, the Life Cycle Inventory (LCI) analysis must extend beyond basic material and energy flow listings. Given the emerging nature of nanotechnology, it is crucial to incorporate deeper analytical considerations into the LCI phase. The following components are especially relevant.

8.3.3.2.1 Data Classification

LCI data for nanofluids should be organized into two primary categories:

- **Foreground data**: Includes specific inputs like nanoparticle concentration, base fluid volume, surfactant types, and energy used in ultrasonication or mixing. These are typically collected from experimental or pilot-scale formulations.

- **Background data**: Covers upstream datasets such as electricity generation, raw material extraction, or transport emissions, typically sourced from databases like Ecoinvent or GaBi.

This classification ensures transparency and helps isolate which parts of the inventory are directly influenced by the product system and which are external dependencies.

8.3.3.2.2 Temporal and Geographical Relevance

For nanofluids, data accuracy depends heavily on *time and location*:

- **Temporal relevance**: Lab-scale synthesis from earlier years may not represent current industrial conditions. It's important to ensure that energy use, yield, and process inputs reflect modern-day practices.
- **Geographical context**: Electricity grid mix (renewable vs. fossil-based), water availability, and regulatory disposal practices vary by region. For example, nanofluid production in India may have a higher carbon intensity compared to production in the EU due to the local grid composition.

These contextual considerations directly impact indicators such as *Global Warming Potential (GWP)* and *Water Footprint.*

8.3.3.2.3 Allocation Methods

In nanofluid systems, especially those involving co-products or byproducts (e.g., waste sludge, heat recovery), it may be necessary to allocate environmental burdens appropriately:

- *Mass-based allocation* is used when co-products are physically comparable.
- *Economic allocation* applies when one output (e.g., nanofluid) is significantly more valuable than the other (e.g., waste stream).

The chosen method should be justified and sensitivity-tested, as allocation can influence the perceived environmental advantage of nanofluids.

8.3.3.2.4 Cut-Off Criteria

Given that nanofluids involve additives in trace amounts (e.g., dispersants at <0.1%), it is tempting to exclude these from the inventory. However, some additives may have *high environmental impacts* despite their low mass. Therefore, cut-off thresholds should not exclude:

- Hazardous surfactants.
- Rare stabilizing agents.
- Solvents with known toxicity or GWP.

Including these ensures a more accurate reflection of the overall impact, especially in categories like *Human Toxicity Potential* or *Freshwater Ecotoxicity.*

8.3.3.2.5 *Uncertainty and Sensitivity Considerations*

Due to variability in nanofluid formulations and limited industrial-scale data, LCI results should be supported by *sensitivity analysis*.

For example:

- Varying nanoparticle concentration (e.g., 0.25% vs. 1%) can significantly change energy savings in the use phase.
- The lifespan of a nanofluid (e.g., 3 months vs. 1 year) affects how often it is replaced, thereby altering cumulative environmental impact.

These analyses help identify which parameters most influence environmental performance and support robust interpretation in the later LCA phases.

8.3.3.3 Special Considerations for Nanofluids in LCI Analysis

Although the traditional organization of the Life Cycle Inventory (LCI) analysis is universally usable throughout the various systems in the chemicals and industries, nanofluids present distinct challenges and complications that require specialized handling. These problems are caused by the way materials behave at the nanoscale level, the fact that synthesis and formulation technologies are still in their early stages of development, and the fact that there isn't a lot of industrial-scale data available. Nanofluids are not typical working fluids—the nanofluids are not only different in composition but also in their functionalities, life cycle behaviors, and response to the environment. The following considerations will be necessary to make sure that the LCI part of a nanofluid-oriented LCA is both meaningful and methodologically reasonable.

8.3.3.3.1 *Material Complexity and Novelty*

One of the most prominent challenges in compiling an accurate inventory for nanofluids is their inherent material complexity and structural novelty. Unlike conventional fluids, the material complexity and structural novelty of nanofluids represent one of the most significant challenges when attempting to obtain an accurate inventory for LCA purposes. Nanofluids, in contrast to conventional fluids, are chemically and thermodynamically stable engineered colloidal systems whose nanoscale content is made up of particles with dimensions of less than 100nm in diameter. The surface area to volume ratios are very large in these nanoparticles, which makes them behave differently; for example, they become more reactive, much more energetic where the surfaces are, and mobile, than those observed in their bulk counterparts.

Although the mass concentration of nanoparticles in a nanofluid is typically low (usually lying in the range of 0.01 to 2 vol.%), nanoparticles can have a disproportionately elevated environmental burden per mass. This is given the fact that nanoparticle synthesis processes are usually characterized by extensive energy-consuming steps, high-priced chemical precursors, and poisons or scarce additives. Moreover, they could be rather reactive, which could add to being toxic in biological/aquatic systems, even in low concentrations. Thus, the LCI of nanofluids should include not only the macro inputs (such as base fluids and container materials) but also the microscale flows of nanomaterials to account that the major environmental impact of tinier, apparently nonsignificant, constituents would not be underestimated [8].

8.3.3.3.2 Process Variability

The synthesis of nanofluids is not a standard procedure; it widely depends on the type of nanoparticle, application requirements, and industrial capability. This means that there is much variation in the way processes are undertaken, which requires different energy use, chemical use, and emissions. As an example, the sol-gel route can include intermediate temperatures and solvent chemistry, whereas in hydrothermal synthesis, a long duration heating in pressurized autoclaves is required, increasing energy consumption enormously. Chemical vapor deposition (CVD) is, in contrast, carried out in gas-phase high-temperature reactions and is likely to contain harmful byproducts vaporized during the process, such as volatile organic compounds (VOCs) or NO_x.

This kind of variation cannot be described in a single LCI dataset or in a generalized model. Rather, the LCI should be process specific by explaining the synthesis procedure adopted and measuring its distinct number of resources as well as emissions. The inability to discriminate between methods can result in severe error in impact assessment, especially with respect to groups such as global warming potential, cumulative energy demand, and human toxicity potential. In addition, as these methods keep getting refined with every new study, which usually involves a tradeoff between efficiency and sustainability, the LCI should be dynamic, as well, taking into account the latest advancements in the processes.

8.3.3.3.3 Stability and Degradation

One of the primitive parameters of the nanofluids' performance is their dispersion stability, which influences the long-term performance in thermal systems. Nanoparticles, with time, are inclined to agglomerate, settle, or be oxidized on the surface, which causes a reduction in thermal conductivity and fluid homogeneity. This kind of degradation not only reduces the efficiency of energy utilization but can also cause corrosion, abrasion, or plugging of heat exchangers and other industrial devices. Losses in performance are frequently the cause of replacement or further processing of the nanofluid, thus adding more flow of materials and energy to the system.

In the context of LCI, it implies that nanofluid systems can be associated with shorter operational lives than regular fluids, just in case of formulation quality and operating conditions. Hence, the data concerning inventory should contain the replacement frequencies, the quantity of loss or wastage of materials, and other resources needed to sustain maintenance activities. As an example, when a nanofluid needs to be changed after three months (because of the sedimentation), the resources will triple per year, and the environmental score will be directly negatively affected. By considering these degradation processes, it is ensured that the LCA will portray the actual operational footprint of the nanofluids through their life cycle.

8.3.3.3.4 Data Gaps and Estimation

A lack of reliable primary data is also common due to the relatively infantile commercialization of nanofluid technologies, particularly on huge scales of industrial production. The majority of the available data refers to the laboratory-scale studies that cannot be directly translated to an industrial practice due to the variance in the yield efficiency, process energy intensity, and waste production. Consequently, LCI practitioners are often forced to interpolate the data, make aspirational assumptions, or use proxy data sets of more relevant materials or previous technology.

Though modeling method of such sort is required, it presents an uncertainty in the inventory, and they need to be well documented and justified. As an example, the energy consumption record of a 100 mL batch of nanofluid synthesized through sonication will not reveal the behavior of a 1000 L batch prepared at an industrial scale with a homogenizer. Consequently, assumptions should be clearly written, and, where possible, undergo sensitivity analysis in order to determine the effect made by alterations in the assumptions on the actual findings. The handling of data gaps in an honest and transparent manner improves valuation, increases the trustworthiness of the LCA, and facilitates improved comparability and reproducibility between studies [9].

8.4 CASE STUDY 1: LIFE CYCLE ASSESSMENT OF AL_2O_3/WATER NANOFLUID IN A SHELL-AND-TUBE HEAT EXCHANGER

8.4.1 Step 1: Goal and Scope Definition

Goal:
The goal of this LCA is to assess and compare the environmental impacts associated with using Al_2O_3/water nanofluid in a shell-and-tube heat exchanger (STHE) for process cooling applications, relative to conventional water-based systems. The assessment aims to identify key environmental hotspots and determine whether the enhanced heat transfer benefits of nanofluids justify their environmental costs.
Scope:

- Functional unit: Transfer of 1 MJ of heat in the STHE over a one-year operational period.
- System boundary: Cradle-to-grave, including:
  - This LCA adopts a cradle-to-grave system boundary, which includes every phase of the nanofluid's life cycle:
  - Cradle (raw material extraction): Mining and processing of aluminum ore for Al_2O_3 nanoparticle production.
  - Nanoparticle synthesis and nanofluid formulation: Energy, materials, and emissions associated with producing and dispersing nanoparticles into the base fluid.
  - Packaging and transportation: Energy use and emissions during storage and delivery to the industrial site.

- Use phase (operational performance): Enhanced thermal performance and reduced energy demand during heat exchanger operation.
- Maintenance and replacement: Cleaning and periodic replacement of nanofluid due to degradation or fouling.
- End-of-life (EoL): Disposal or treatment of spent nanofluid, including potential environmental risks of nanoparticle release and water recovery processes.

- Geographical scope: India
- Temporal scope: Based on data from 2022

8.4.2 Step 2: Life Cycle Inventory (LCI) Analysis

Raw material acquisition:

- Aluminum oxide (Al_2O_3) nanoparticles produced via sol-gel method.
- Base fluid: Deionized water.
- Energy input:
    - 15 kWh/kg of Al_2O_3 (lab-to-pilot scale).
- Materials used:
    - Aluminum nitrate: 2.5 kg per kg Al_2O_3
    - Ethanol solvent: 4.2 L/kg Al_2O_3
    - Deionized water: 5 L per functional unit

Nanoparticle synthesis

- Sol-gel synthesis under ambient pressure, 80–100 °C.
- Waste solvents are partially recovered.
- Emissions: 0.25 kg CO_2-eq/kg Al_2O_3 from heating and drying processes.

Nanofluid formulation

- 0.75 vol.% Al_2O_3 nanoparticles dispersed in water.
- Ultrasonication energy: ~1.2 kWh per liter of nanofluid.
- Stabilizers: 0.05 wt.% sodium dodecyl sulfate (SDS)
- Total volume used per year: 1000 L

Packaging and transportation

- HDPE drums used for storage and delivery.
- Transport distance: 400 km (by diesel truck).
- CO_2 emissions: ~0.15 kg CO_2-eq/L of nanofluid transported.

Use phase (Heat exchanger operation)

- Heat exchanger transfers 1 MJ of heat per cycle.
- Use of nanofluid improves the heat transfer coefficient by 28% vs. water.

- Operational electricity saving: 7.5 kWh/year
- Energy savings result in avoided emissions of ~5.4 kg CO_2-eq/year.

Maintenance and replacement

- Nanofluid replacement cycle: once per year due to agglomeration.
- Cleaning chemical: Citric acid (0.5 L per cleaning)
- Filter disposal: 2 kg/year of sludge containing metal oxides.

End-of-Life

- Nanoparticle sludge treated as hazardous waste.
- 80% of water recovered and reused.
- Energy used in disposal: ~0.5 kWh per functional unit.
- Estimated EoL emissions: 0.3 kg CO_2-eq/year.

8.4.3 Step 3. Life Cycle Impact Assessment (LCIA)

Using ReCiPe Midpoint Method, impacts were assessed across relevant categories as shown in Table 8.2.

8.4.4 Step 4. Interpretation

Hotspot identification

- The nanoparticle synthesis stage (especially heating and solvent use) contributed most to energy use and GWP.
- Use phase offered substantial environmental savings (~40% reduction in operational CO_2 emissions) due to improved heat exchanger efficiency.
- End-of-life impacts were moderate but manageable through controlled disposal and water recycling.

TABLE 8.2
Life Cycle Impact Assessment (LCIA) of Al_2O_3/Water Nanofluid

Impact Category	Al_2O_3/Water Nanofluid	Conventional Water
Global Warming Potential (GWP)	6.1 kg CO_2-eq	10.9 kg CO_2-eq
Cumulative Energy Demand (CED)	92 MJ	110 MJ
Human Toxicity Potential (HTP)	0.021 CTU-human	0.012 CTU-human
Freshwater Ecotoxicity	0.015 CTU-eco	0.008 CTU-eco
Water Use	11.3 L	9.4 L
Particulate Matter Formation	0.005 kg PM_{10}-eq	0.008 kg PM_{10}-eq

Trade-off analysis

- While toxicity indicators (HTP and ecotoxicity) are higher for the nanofluid due to the presence of engineered nanoparticles and surfactants, the overall system performance shows environmental benefits in terms of energy use and GWP.

Sensitivity analysis

- Varying nanoparticle concentration from 0.5% to 1% increases environmental burden but yields diminishing heat transfer returns beyond 0.75 vol.%.
- Use of renewable electricity in nanoparticle synthesis reduces GWP by ~25%.

Improvement suggestions

- Switching to green synthesis routes (e.g., plant-based sol-gel precursors).
- Optimizing nanofluid lifespan by improving dispersion stability to reduce replacement frequency.
- Incorporating closed-loop filtration systems to recover nanoparticles and reduce EoL toxicity.

8.5 CASE STUDY 2: LIFE CYCLE ASSESSMENT AND OPTIMIZATION OF TiO_2–SiC/WATER HYBRID NANOFLUID IN A SHELL-AND-TUBE HEAT EXCHANGER

8.5.1 Background and Objectives

This case study presents a detailed analysis of the integration of a TiO_2–SiC/water hybrid nanofluid into a shell-and-tube heat exchanger (STHE) system. The primary goal was to enhance thermal performance while evaluating environmental and economic sustainability using Life Cycle Assessment (LCA) and Artificial Neural Network (ANN)-based optimization. This hybrid nanofluid was investigated to replace traditional coolants (e.g., water) in chemical process industries, where higher thermal conductivity and heat transfer rates are required.

Nanofluid Preparation and Characterization

A two-step ultrasonication method was employed to prepare the hybrid nanofluid:

- Nanoparticles used
  - TiO_2 (particle size < ±20 nm, purity 95%)
  - SiC (particle size < ±30 nm, purity 99%)
- Stability and dispersion
  - Ultrasonication ensured stable mono-dispersion (confirmed by low PDI < 0.2).

- Zeta potential remained ~43 mV after five days, indicating excellent electrostatic stability.
- FESEM and DLS confirmed uniform TiO_2 coating on SiC nanofibers.
- Thermophysical properties (for 0.5 vol% hybrid nanofluid)
  - Thermal conductivity increased to 0.65 W/mK from 0.61 W/mK (water).
  - Density increased to 1168.1 kg/m³, with improved specific heat capacity.

Experimental setup

A 1-pass STHE was fabricated using stainless steel with the following specifications:

- Shell diameter: 0.13 m outer, 0.125 m inner
- Tube dimensions: 8 tubes, each 0.6 m long, with inner diameter of 0.0107 m
- Tube layout: Inline, triangular pitch, 4 segmental baffles (25% cut)
- Working fluids
  - Hot fluid (shell): Water at 65 °C
  - Cold fluid (tube): Nanofluid at 25–42 °C
  - Reynolds number (N_{Re}): 1300–18000 (tube side)

Performance outcomes

- Convective heat transfer coefficient
  The hybrid nanofluid showed a marked increase in convective heat transfer with increasing Reynolds number. Enhanced thermal conductivity, stable dispersion, and nanoparticle motion were attributed to this effect.
- Nusselt number
  - Water: Baseline
  - TiO_2 nanofluid: ~25% increase
  - SiC nanofluid: ~43% increase
  - TiO_2–SiC hybrid nanofluid: ~65% enhancement over water
- Temperature profile
  Improved outlet temperatures for the cold side (tube) with hybrid nanofluid, confirming better energy absorption even at low flow rates.
- Pressure drop
  Slight increase (~8%) due to higher viscosity but acceptable, considering the substantial gains in heat transfer.

Life Cycle Assessment (LCA)

An LCA was conducted using the IdematLightLCA 2.8.6 software, assessing five key phases:

1. Material Extraction
2. Production (Synthesis and Formulation)
3. Transportation
4. Usage (Operation for 25 years)
5. End-of-Life (Recycling and Disposal)

Findings

- Usage phase dominated energy use (89.5%) and CO_2 emissions (75.7%).
- Cost breakdown over 25 years:
  - Materials: 20%
  - Production: 4.5%
  - Transportation: 3%
  - Operation: 72.5%
- Even with $150 additional capital cost and $50 maintenance, the system achieves net energy savings due to enhanced heat transfer.
- A 7% reduction in heat exchanger surface area was possible due to higher thermal efficiency.

8.5.2 ANN-Based Cost and Ecological Optimization

To evaluate the economic and ecological trade-offs, a feedforward backpropagation ANN was trained using system parameters as inputs:

- **Design parameters**: Shell/tube diameters, tube length, and number of tubes
- **Output targets**: Ecological function (EF) and total cost

Optimization Results

- ANN achieved a correlation coefficient $R^2 = 0.9998$, indicating strong prediction accuracy.
- Increasing tube length had the most significant impact on both EF and cost.
- Increasing the number of tubes increased the cost without significantly improving EF at constant length.

Example

- For a system with 0.8 m shell diameter, 0.0055 m tube diameter, and 0.55 m tube length (9 tubes), the total cost was optimized to $820.6/year with an ecological function of ~39.99 kW.

8.6 CONCLUSION

The use of nanofluids in process industries is an innovative way to improve thermal responses, energy efficiency, and contribute to the intensification of processes. This chapter has discussed in detail the life cycle assessment (LCA) framework as adopted to nanofluids, which are capable of giving an insight into not only the technological benefits of adopting them but also the environmental impacts of such adoption. This cradle-to-grave examination, which takes into consideration sequential evaluation of every step involved in the process of raw material extraction, synthesis of nanoparticles, usage in heat exchange systems, and disposal of the nanoparticles at the end of the life of the product, illustrates the utmost significance of viewing every product or service through the lens of full life cycle. The results of the case studies, especially concerning Al_2O_3/water and TiO_2-SiC/water nanofluids, will indicate that, although nanofluids require more energy and material sumptuary gone-by synthesis, all the benefits that they can bring to the process through energy saving and minimized equipment over time can overbalance the environmental costs they incur in the early days of existence.

Sustainability of nanofluids however is extremely context dependent. Factors such as nanoparticle stability, the methods of synthesis, formulation tactics, and disposal activities somehow determine the overall environmental performance. Moreover, underlying the debate is the insufficiency of complete toxicological information and the doubts about the long-term effects of nanomaterials toward a cautious and knowledgeable implementation course. The value of inserting sustainability indicators, Global Warming Potential (GWP), Cumulative Energy Demand (CED), and toxic indicators into the assessment has also been illustrated in this chapter. These metrics, assisted by highly effective optimization processes such as artificial neural networks (ANN), can inform the traditional design and policy options to achieve the deployment of nanofluid in a more sustainable manner.

REFERENCES

1. Thakur, P., Sonawane, S. S., Sonawane, S. H., & Bhanvase, B. A. (2020). Nanofluids-based delivery system, encapsulation of nanoparticles for stability to make stable nanofluids. In *Encapsulation of active molecules and their delivery system* (pp. 141–152). Elsevier.
2. Bethi, B., Sonawane, S. H., Bhanvase, B. A., & Sonawane, S. S. (2021). Textile industry wastewater treatment by cavitation combined with Fenton and ceramic nanofiltration membrane. *Chemical Engineering and Processing-Process Intensification, 168*, 108540.
3. Malika, M., & Sonawane, S. S. (2020). Effect of nanoparticle mixed ratio on stability and thermo-physical properties of CuO-ZnO/water-based hybrid nanofluid. *Journal of the Indian Chemical Society, 97*(3), 414–419.
4. Dunmade, I. S. (2020). A linear assignment based conceptual lifecycle assessment method for selecting optimal agri-industrial materials production pathway: A case study on Nigerian yam value chain. *Agronomy Research, 18*(3), 2004–2021.
5. Parate, V. R., Kawadkar, D. K., & Sonawane, S. S. (2011). Study of whey protein concentrate fortification in cookies variety biscuits. *International Journal of Food Engineering, 7*(2), 1–12.
6. Sonawane, S., Thakur, P., & Paul, R. (2020). Study on thermal property enhancement of MWCNT based polypropylene (PP) nanocomposites. *Materials Today: Proceedings, 27*, 550–555.
7. Malika, M. M., & Sonawane, S. S. (2019). Review on application of nanofluid/nano particle as water disinfectant. *Journal of Indian Association for Environmental Management (JIAEM)*, 39(1–4), 21–24.
8. Malika, M., Bhad, R., & Sonawane, S. S. (2021). ANSYS simulation study of a low volume fraction CuO–ZnO/water hybrid nanofluid in a shell and tube heat exchanger. *Journal of the Indian Chemical Society, 98*(11), 100200.

9. Sonawane, A. V., Rikame, S., Sonawane, S. H., Gaikwad, M., Bhanvase, B., Sonawane, S. S., . . . & Gaikwad, R. (2024). A review of microbial fuel cell and its diversification in the development of green energy technology. *Chemosphere*, *350*, 141127.
10. Malika, M., and Sonawane, S. S. (2024). Ecological optimization and LCA of TiO2-SiC/water hybrid nanofluid in a shell and tube heat exchanger by ANN. *Proceedings of the Institution of Mechanical Engineers, Part E: Journal of Process Mechanical Engineering*, *238*(1), 45–55.

9 Nanofluids in Thermal Management and Heat Recovery Systems

9.1 INTRODUCTION TO APPLICATION-ORIENTED USE OF NANOFLUIDS

The constant growth in the need for energy globally, coupled with the downsizing nature of the miniaturization of electronics and the intensification of industrial operations, has burdened heat management technologies. The most important aspect of the energy conversion, storage, and utilization systems is efficient heat transfer. The commonly used heat transfer fluids (water, ethylene glycol, and mineral oil), however, have low intrinsic thermal conductivities that restrict the capabilities and efficiencies of the thermal systems. The modern scientific breakthrough is nanofluids as an innovative technology that can control matter at the atomic and molecular levels through nanotechnology. Nanofluids are the suspensions of nanoparticles (1 100 nm) in typical coolants. These fluids have the capability of giving great thermal performance with the addition of high-conductivity nanoparticles. In the past 20 years, nanofluids have attracted attention as a potential group of heat transfer fluids in a number of application-specific thermal systems (i.e., thermal systems, electronics cooling, renewable energy systems, HVAC, transportation, and industrial heat exchangers).

This section explores the *need for enhanced heat transfer in energy systems*, followed by the *critical role nanofluids play in modern thermal management and heat recovery* processes.

9.2 NEED FOR ENHANCED HEAT TRANSFER IN ENERGY SYSTEMS

9.2.1 Rising Global Energy Demand and Efficiency Challenges

The switch to more sustainable energy sources and the trend to the electrification of systems cause higher complexity and density of thermal loads. Be it power plants, renewable energy, or electric vehicles, the capability to effectively transfer and dissipate heat has a direct impact on the performance of the system, fuel, and reliability. Conventional fluids cannot satisfy modern requirements of higher thermal conductivity and stability, necessitating advanced solutions.

DOI: 10.1201/9781003628071-9

9.2.2 Miniaturization and High-Performance Requirements

Extremely high heat fluxes have been achieved as a result of the drive to compact high-performance systems, which include microprocessors, microreactors, and compact heat exchangers. The microscale requirements are that the fluids use high thermal characteristics to enable a high cooling performance in a small space. Conventional fluids, such as water and oil, can hardly handle these thermal loads, and as a result they can cause overheating and system failure.

9.2.3 Renewable Energy and Waste Heat Recovery Systems

A significant level of heat transfer is also needed in solar thermal systems, geothermal units, as well as waste heat recovery installations, where efficient heat transfer is essential for improving energy conversion rates. The total efficiency can also be enhanced and the carbon footprint lowered by improving heat exchange in such systems. Nanofluids and their tunable thermophysical attributes hold the promise of performance enhancement of energy harvesting and recovery systems.

9.2.4 Industrial Process Intensification

The latest industries, particularly chemical and petrochemical industries, are considering the process intensification methods—process strategies that substantially increase the rates of heat and mass transfer in order to minimize the required energy and the size of equipment. In this case, nanofluids are used to achieve optimal heat transfer without the additional complexity of the system or high-pressure drop.

9.3 ROLE OF NANOFLUIDS IN MODERN THERMAL MANAGEMENT AND HEAT RECOVERY

Nanofluids exhibit significantly higher thermal conductivities compared to their base fluids. This enhancement is primarily due to:

- *Brownian motion* of nanoparticles, which improves micro-convection.
- *Thermal conduction* through high-conductivity nanoparticles.
- *Improved surface area*, allowing better interaction with heated surfaces.
- *Nanoparticle clustering and percolation*, which provides enhanced thermal pathways.

Additionally, the thermophysical properties—such as viscosity, specific heat capacity, and density—can be tuned through nanoparticle material selection, concentration, and functionalization, offering application-specific customization.

9.4 APPLICATIONS IN ELECTRONICS COOLING

Due to the development of more and more electronic technologies, the size of devices has become reduced, and they work faster and more efficiently. This, nevertheless, is at a price of heat generation. As electronic devices are miniaturized (smartphones, laptops, power electronics, in electric cars), there is a corresponding increase in power density and operating temperatures. Poor thermal management not only leads to low energy efficiency but also reduces the number of working hours of parts. The traditional cooling mechanism approaches are becoming less effective as applied to the compact systems to control the high heat fluxes in such systems, which requires the development of new cooling techniques that could guarantee reliable performance.

Engineered nanofluids, composed of nanoparticles in base fluids, provide improved thermophysical characteristics, including high thermal conductivity, and convey heat transfer coefficients. They are being used in high-performance electronics with advanced cooling systems, microchannel heat sinks with spray cooling systems, and dielectric liquid-based immersion cooling. Energy Al_2O_3, CuO, TiO_2, and carbon nanotubes (CNTs) have proved to enhance heat transfer significantly when added as a new nanoparticle to an electronic cooling system.

Cooling technologies can also be roughly divided into passive and active methods, depending on whether they require no outside force (therefore relying entirely on natural forces such as surface tension and gravity) or require an active force (mechanical in nature through the use of pumps or fans). The major cooling practices that have developed involving the incorporation of nanofluids are elaborated as follows:

9.4.1 Thermosyphons

The thermosyphons are a type of two-phase, gravity-driven device working on the principles of phase change—liquid evaporates at the point of heating and condenses at the point of cooling. They are discrete and passive, thus used in electronics. The nanofluids' alleviation improves the CHF and boiling heat transfer efficiency. For instance, the thermosyphons with titanium oxide and copper oxide nanofluids showed smaller thermal resistance or higher heat transfer coefficient, especially at the optimal concentrations of nanoparticles.

Case Study: Thermosyphon-Based Anti-Freezing Systems for Water Meters: Case Study by Bahar et al. (2025)

This paper evaluates how thermosyphon systems with nanofluids can be utilized to avoid the cases of freezing water meters in cold regions, such as Tabriz, Iran. Based on a two-dimensional numerical model with experimental data as its validation, the study will use a volume of fluid (VOF) method in ANSYS Fluent to generate a two-phase heat transfer problem that includes the role of the phase change mechanisms. The experimental arrangement comprises a copper thermosyphon fitted vertically into a water meter chamber, and the design is such that it makes use of underlying soil heat (Figure 9.1). The system is comprised of evaporator, adiabatic, and condenser, and analyzed with different nanofluids, such as Al_2O_3, CuO, and Ag, of different concentrations. The

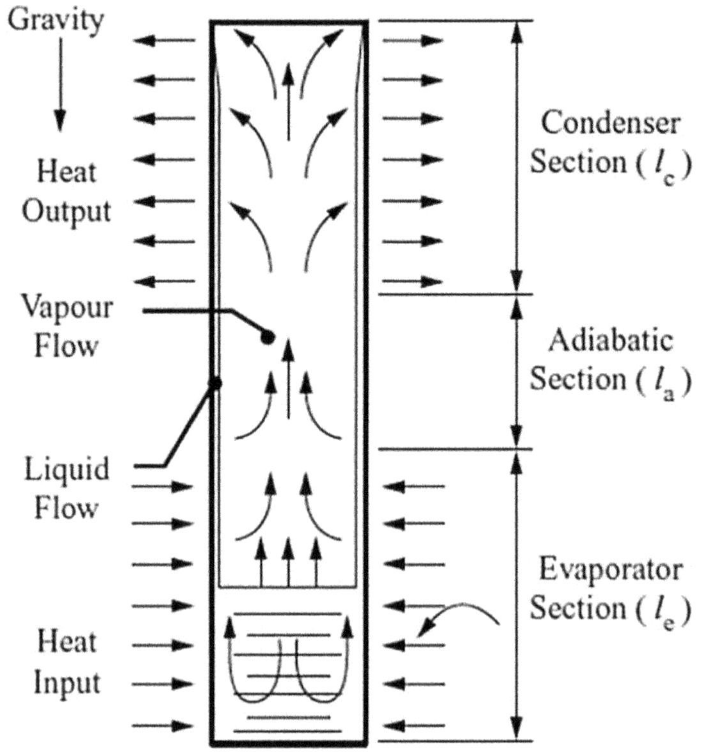

FIGURE 9.1 Experimental setup for the thermosyphon system [1].

simulation parameters were based on the real meteorological ones, providing practical value. The heat source was soil at a depth of 1 meter, and the average temperatures were maintained above freezing temperatures even though the environment was very cold.

Nanofluids were demonstrated to perform greatly in the transfer of heat when compared to conventional coolant like water. Their high thermal conductivity improved the evaporation-condensation process, leading to high levels of heat flow and a high temperature control within the chamber. For instance, the enhancement of the nanoparticle concentration saw a significant rise in the sense of thermal conductivity, of which silver nanofluids exhibited the best results, followed by CuO and Al_2O_3. Nanofluids were critical in improving thermosyphon performance by lowering thermal resistance and increasing the rate of transfer of energy. Their importance is emphasized by the practical potential of the findings to be used in energy systems and thermal management systems, and represents a sustainable, low-maintenance solution to the problem of preventing freezing of parts of the utility infrastructure.

9.4.2 Heat Pipes

Heat pipes are essential components in modern thermal management systems due to their ability to efficiently transfer heat over long distances with minimal temperature loss. They involve no external source of power nor any moving components, and they operate passively using the evaporation and condensation cycle, therefore, being extremely reliable and maintenance free. Their outstanding thermal conductivity, which is frequently hundreds of times more than solid metal materials, such as copper, allows high rates of heat dissipation, which is essential in ensuring temperature continuity and eliminating hotspots within delicate systems, including CPUs, batteries, and photovoltaic cells, among others. Heat pipes are compact and lightweight, making them suitable for applications in constrained spaces such as electronics, aircraft, and electric cars. Besides, nanofluids integration in heat pipes has also served to increase their effectiveness furthermore with regard to heat transfer values, wherein they can be greatly appropriate in high heat flux applications. On the whole, heat pipes are essential to assure efficiency, stability, and performance of advanced thermal systems.

Experimental Analysis of Graphene Nanofluid in Heat Pipe-Based Solar Collectors: Case Study by Sathish et al. (2025)

The study presents an innovative investigation into improving the thermal performance of heat pipe-based solar tube collectors (STCs) using graphene-based nanofluids. Graphene nanoparticles were synthesized through electrochemical exfoliation and dispersed in deionized water to obtain stable nanofluids in three volume concentrations: 0.1%, 0.2% and 0.3%. Ultrasonication was done to maintain uniform dispersion of the nanofluids and to prevent the nanoparticles from agglomeration. In order to simulate real-time solar collection conditions, a closed-loop experimental setup (Figure 9.2) was established at a solar testing facility in India. The installation had a nanofluid storage tank, two pumps to circulate nanofluid, a heat exchanger, a flow control valve, and a pyranometer to measure the solar irradiance (instantaneous power per unit area received from the sun (W/m^2). A pyranometer measures solar irradiance). The STC was subjected to variability in flow (1.5, 3.0, and 4.5 L/min), and comparison of some of the most critical performance parameters was made relative to temperature gain, thermal efficiency, and useful heat discharge. The findings showed the apparent improvement in performance on using graphene nanofluids. The nanofluids with deionized water showed an increase of up to 52.94% in heat gain and 41.3% increase in thermal efficiency, especially at a fluid concentration of 0.3% and a flow rate of 4.5 L/min. Long-term stability of nanofluid by zeta potential was confirmed as no formation of sediment was found after 30 days, and thus heat transfer performance stays constant. The results can highlight that the graphene nanofluids have great potential for greater thermal efficiency of the solar energy systems. Maximizing the concentration of nanoparticles and the flow rate were

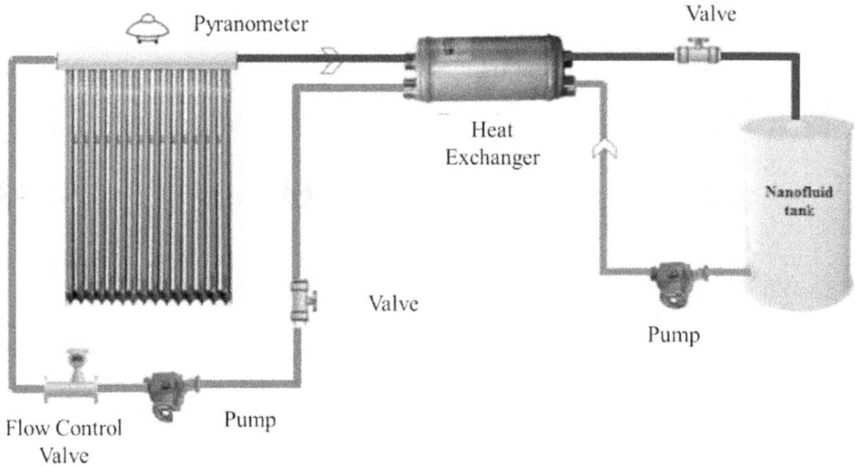

FIGURE 9.2 Experimental setup for evaluating graphene nanofluid in heat pipe-based solar thermal collectors [2].

identified as core considerations in expressing higher rates of heat absorption and energy transfer, which advances the use of future solar thermal technologies toward a more sustainable and efficient process.

The findings underscore the significant potential of graphene nanofluids in enhancing the thermal efficiency of solar energy systems. Optimizing nanoparticle concentration and flow rate proved essential in maximizing heat absorption and energy transfer, paving the way for more sustainable and efficient solar thermal technologies.

9.4.3 Microchannel-Based Forced Convection

In high-end cooling systems (heat sinks), microchannels are highly popular because they allow effective cooling of dissipative heat in small high-power electronic devices. Their small passages greatly augment the surface area in which heat can be transferred, and small designs can handle large heat fluxes. But the problem is that the thermal characteristics of such common working fluids as air or water usually constrain the efficacy of microchannels. Nanofluids like Al_2O_3, SiO_2, TiO_2, and CNT dispersion are added in microchannel systems to overcome this shortcoming. Nanofluids improve the total thermal efficiency of a fluid by using a high thermal conductivity nanoparticle suspended in base fluids. At such low concentrations (e.g., 0.1 vol.%), alumina nanofluids have demonstrated a 22% reduction in CPU temperatures, resulting in stable dispersion. The need to reduce the temperature has seen graphene nanoplatelet-based fluids reduce temperatures by up to 10°C because of their superior thermal conductivity. With application in microchannels, they enhance convective heat transfer, lowering thermal resistance and enabling more control over temperature of systems with large circulating thermal loads. Nanofluids are therefore incorporated in the microchannels to realize their potential in cooling applications and achieve the requirements of future electronic and thermal gadgets.

Heat Transfer and Pressure Drop Characteristics of Microchannel Cold Plate in Commercial CPU-Package Cooling System: Case Study by Zi-Xing Wang and Wen-Quan Tao (2025)

The current paper reports an experimental study on the thermal and hydraulic behavior of multi-layered microchannel cold plate with the specific design of a cooling commercial CPU package (CPcCPU) at high heat flux. The study indicates that the development of enhanced liquid cooling has been a priority, as liquid cooling is now essential to keep modern high-performance CPUs cool since their excess heat reaches above 100 W/cm2. To counter this, the authors suggest an integrated cold plate and CPU-package model and consider various inlet/outlet manifold shapes—I-type, Z-type, and U-type in addition to the impact of the microchannel layer number (MCLN) increase of 1–9. The I-type manifold was more evident and superseded the rest of the tested designs because it had better uniformity in flow, less pressure drop, and low thermal resistance. Again, adding more layers of microchannels had a big impact on cooling and reducing pressure drop, but at some point, the subsequent layers would give fewer returns in terms of maximum CPU power dissipation (MPCPU).

The fifth version of the cold plate, a modified nine-layer cold plate (I-type-9L-CUT1), was presented, saving poorly performing channels to achieve optimal flow and decrease energy dissipation. This was tested and managed to reach the maximum chip heat flux of 206.5 W/cm^2, minimal thermal resistance of 0.1727 K/W, and CPU loading of 336.70 W when indium was used as the thermal interface material (TIM 2) since it is a very conductive material. Optimized TIM and a multi-layer channel configuration gave substantial improvement in cooling performance at low pumping power (3.337 mW) and satisfactory pressure drop (523.5 Pa).

The originality of the study consists in the fact that a new method of realistic modeling of a complex CB-structure of the commercial sample of the CPU-type connected to the cold plate was created, since previous works based on the hypothesis about the unity of the heat flow denied the initial hypothesis. Such global analysis is more realistic to predict than local ones that will be applied in practice in CPU thermal management, hence very applicable in future high-power electronic cooling systems [3].

CPU Heat Sink Cooled by Alumina-Water Nanofluid: Case Study by Ghasemi et al (2021)

This study experiments on the improvement of CPU heat sink efficiency with the use of alumina (Al_2O_3)-water nanofluid as the coolant. A closed-loop test system was made, in which aluminum minichannel heat sinks with different numbers and diameters of channels were included. The two-step ultrasonic homogenization nanofluid was prepared without the addition of surfactant in order to remain stable. Thermal performance was determined with inlet and outlet temperature measurements.

The objective of the experiment was to test the heat transfer capabilities of a CPU heat sink with the help of alumina-water nanofluid as a coolant. The arrangement of the closed-loop flow circuit was planned (Figure 9.3), which was a surrounding area of a reservoir, a pump, a flow meter, and an aluminum heat sink that tested the sections. Temperature was measured by thermocouples at the inlet and outlet, and the data were recorded via a data acquisition system.

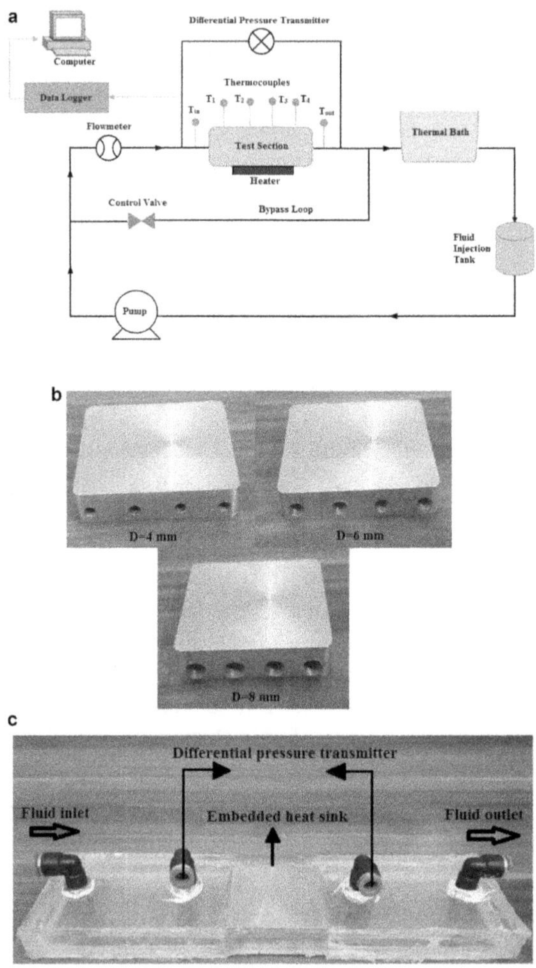

FIGURE 9.3 (a) Schematic diagram of the experimental setup for nanofluid-cooled CPU heat sink, (b) Geometric configuration of the microchannel heat sink with variable channel diameter and number, and (c) Flow loop system used for circulating alumina-water nanofluid in heat sink testing [4].

There was no addition of surfactants, but alumina nanoparticles (90 nm) were dispersed in deionized water at concentrations up to 0.1% and stabilized through uniform ultrasonication. The effect of the number of channels and the hydraulic diameter on heat dissipation was analyzed by testing three heat sinks, with four, six, or eight straight channels of 4 mm, 5 mm, or 6 mm dia. The flow and heat load at the CPU level were modulated to represent real operating conditions, and efficiency determined when it was driven under varying configurations.

The outcomes revealed that in comparison to water, the application of Al_2O_3 nanofluid has effectively amplified heat shift. Particularly, the convective heat transfer coefficient rose by as much as 25% and the increment was linked to the fact that the nanofluid used had an elevated thermal conductivity, and the ability of the nanofluid to affect thermal diffusion through Brownian motion was accentuated. It was also discovered that small sizes of channel diameters (4 mm) and the increased flow rate sped up the cooling and augmented the pressure drop. The paper used Response Surface Methodology (RSM) in order to optimize several objectives, which are concerned with balancing the thermal and the hydraulic aspects.

The most effective cooling with the manageable pressure loss was reached by the best configuration: four channels of 4 mm diameter and 0.1% of the concentration of nanofluid. This shows how geometry and working fluid optimization are important to advanced CPU cooling.

9.4.4 Thermoelectric Cooling (TEC)

The thermoelectric devices include thermoelectric generators (TEGs) and thermoelectric coolers (TECs), which can represent solid-state solutions to thermo-management and energy conversion. Their use lies in their ability to harness the Seebeck effect into waste heat generation to produce electrical energy; hence, their potential application in automotive, aerospace, and industrial applications. TECs instead work based on the Peltier effect, allowing finer and more accurate, vibration free and compact cooling of sensitive electronics, medical equipment, and portable devices. The two technologies will appeal on the basis of their reliability, the fact that they have no moving parts, are silent, and scale well. They have been however limited by a somewhat low conversion efficiency and thermal performance, which relies highly on the temperature differential across the thermoelectric module.

Thermoelectric generators (TEGs) and thermoelectric coolers (TECs) have potential as solid-state devices to accomplish energy harvesting and cooling tasks, but their applications are not widespread because of low efficiency, their ineffective heat transfer properties, and the necessity (due to their low efficiency) to use big heat sinks. Such devices are useful only when they can maintain a large temperature difference between their hot and cold sides but with low-conductivity heat transfer fluids, such as water or air, it is hard to get to this level. Consequently, TEGs provide weak electrical conversion efficiency, whereas TECs tend to cause excessive power expenditure and local heat sources. As a way of circumventing these handicaps, nanofluids (suspension of high-conductivity nanoparticles within base fluids) are being used to increase the thermal capabilities of the system. Nanofluids can greatly enhance the thermal conductivity and convective heat transfer, which could make the TECs work better by keeping cold-side temperatures low; it can also extract the heat from TEGs more effectively and increase their coefficient of performance (COP) and electrical output, respectively.

Case Study: Performance Enhancement of Thermoelectric Devices Using Nanofluid-Coupled Heat Exchangers: Case Study by Ruan et al. (2021)

This study investigates the performance improvements in thermoelectric generators (TEGs) and thermoelectric coolers (TECs) using graphene nanoplatelet (GNP) nanofluids as enhanced heat transfer media. A three-dimensional fluid-thermal-electric multiphysics model was developed in COMSOL Multiphysics to simulate the combined thermodynamic and electrical behavior of these systems under varying operational parameters. The experimental setup/modeling framework was as shown in Figure 9.4. The simulated system comprises 20

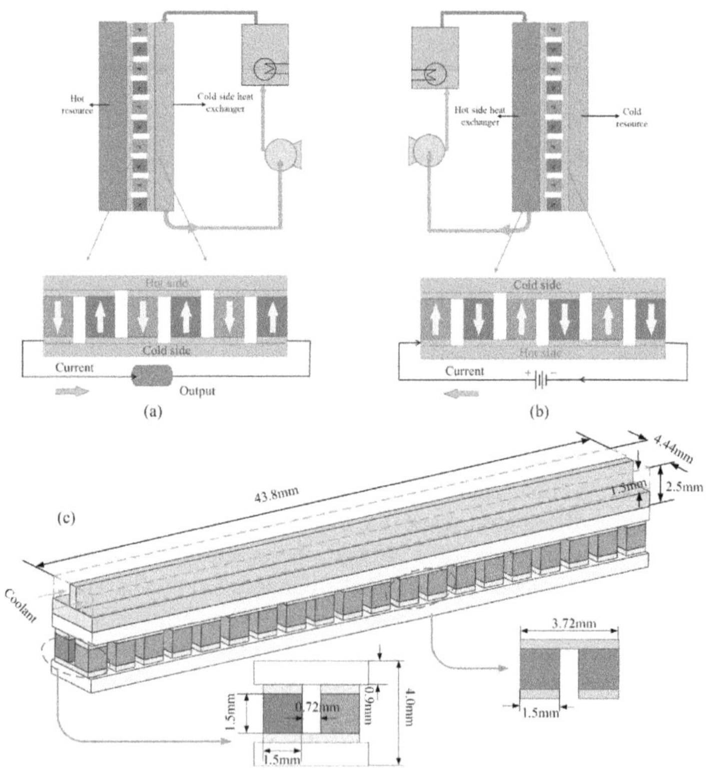

FIGURE 9.4 (a) Schematic diagram of the thermoelectric module configuration, (b) Cross-sectional view of the TEG/TEC integrated with nanofluid-cooled heat exchanger, (c) Simulated geometry of the thermoelectric system for multiphysics modeling [5].

pairs of thermoelectric legs (Bi_2Te_3-based), sandwiched between heat exchangers and hot/cold sources. A plate heat exchanger with nanofluid as the working fluid is integrated to remove excess heat efficiently. In the study the Reynolds numbers were varied between 100 and 2000 and nanoparticle weight concentrations (0–0.1 wt.%) to assess their impact on cooling and power generation. Three nanofluids—GNPs, Al_2O_3, and CuO—were compared for their effectiveness.

The numerical simulations demonstrated that nanofluids significantly improve thermal and electrical performance, especially at low Reynolds numbers. For a GNP nanofluid at 0.1 wt.% and Re = 100:

- TEG output power increased by 11.38% and conversion efficiency by 5.7%.
- TEC cooling power rose by 12.13% with a 9.1% improvement in coefficient of performance (COP).

As Reynolds number increased, the enhancement effect diminished, stabilizing beyond Re = 1200. Among the nanofluids tested, GNPs outperformed both Al_2O_3 and CuO in enhancing thermoelectric behavior, despite a slightly higher pressure drop.

Lithium Ion Battery Cooling with Hybrid Nanofluids: Case Study by Kanti et al. (2025)

In a significant 2025 experimental study, Kanti et al. investigated the thermal management of lithium-ion batteries (LIBs) using both mono (Al_2O_3, CuO) and hybrid (Al_2O_3-CuO) nanofluids. Using a 0.5% volume concentration of nanoparticles in water, they tested cooling performance across flow rates of 150–350 ml/min and discharge rates of 0.5°C and 1°C. The hybrid nanofluid outperformed both mono and base fluids, reducing battery temperature by 54.23% at the highest flow rate.

Thermal conductivity of the hybrid fluid exceeded that of CuO and Al_2O_3 by 35.26% and 29.1%, respectively, at 60°C. To complement experimental insights, machine learning models were developed using Random Forest (RF) and Linear Regression (LR) techniques. The RF model achieved 98.4% prediction accuracy, significantly surpassing the LR model's 80.82%, thereby providing a robust predictive framework for battery temperature control.

This study not only emphasizes the superior cooling capacity of hybrid nanofluids but also demonstrates the synergy of AI-based modeling with experimental approaches for designing efficient Thermal Management Systems (TMS) for batteries and electronics. The integration of nanofluids with machine learning offers a cost-effective path to optimize future thermal systems [6].

9.4.5 Natural (Free) Convection

The mechanism of natural convection is one of the basic concepts of heat transfer that is driven by buoyancy forces produced in fluids between a heavy-dense fluid and a light fluid with a high power of density due to different temperatures. It finds important applications in passive cooling systems where reliability, simplicity, and energy efficiency are important factors, for example, in building ventilation, solar collectors, cooling of electronic equipment, safety assemblies in nuclear reactors. The performance of the systems using pure natural convection, however, can be restricted by the low thermal conductivity of the working fluid, normally water, air, oils, and insufficient heat removal capabilities of such systems when applied to high-flux geometries or compact systems as the thermal resistances in a natural convection-dominated system are comparatively high.

In order to surmount this deficiency, nanofluids are added into natural convection systems. Such nanoparticles with high thermal conductivity increase the thermo-physical properties like thermal conductivity and the specific heat capacity of the fluid, resulting in an overall increased convective heat transfer. Nanofluids also lead to enhanced temperature homogeneity, postponement of boundary layer splintering, and possible decimation of thermal resistance against warm surfaces. Also, they help to overcome the thermal stratification and promote thermal dispersion because of the increased Brownian motion. This means that the implementation of nanofluids across the systems that rely on natural convection makes the systems much more efficient and responsive than they were before, as well as better suited to the demands of thermal management in contemporary environments without undermining the traditional properties of passive and energy-saving processes. Recently, it has been observed that, in studies conducted under different enclosure geometries (rectangular, semicircular, and trapezoidal) and experimental setups, it has been demonstrated that nanofluids do have a significant effect on heat transfer, and performance levels depend on the concentration of the particles used and the geometry of the enclosure.

Enhanced Natural Convection Heat Transfer Using Hybrid Nanofluid: Case Study by Vinod Kumar et al. (2025)

A new passive cooling strategy to enhance natural convection heat transfer in horizontal annular systems is examined as a case study; that is, the enhancement mechanism by means of a passive cooling strategy is studied with a corresponding simulation study. The main drawback of classical annular heat exchangers is the occurrence of stagnant areas of fluid around hot surfaces and poor turbulence, which has the property to limit convective heat transfer. In order to overcome this, researchers suggested the use of an integrated augmentation shear technique, that is, to use a Diamond Fin Vortex Generator (DFVG) and a hybrid nanofluid comprising Multi-Walled Carbon Nanotubes (MWCNTs) and Magnesium Oxide (MgO) nanoparticles dispersed in demineralized water.

ANSYS Fluent was used to execute the simulation, which represented a concentric elliptical annular region in which natural convection occurred under the influence of thermal buoyant forces. The inner elliptical cylinder was considered to be the heat source and outer elliptical cylinder to be the heat sink (Figure 9.5). The DFVG was placed at the bottom of the inner cylinder to induce the formation of the vortex that increased the turbulence and reduced the stagnant areas.

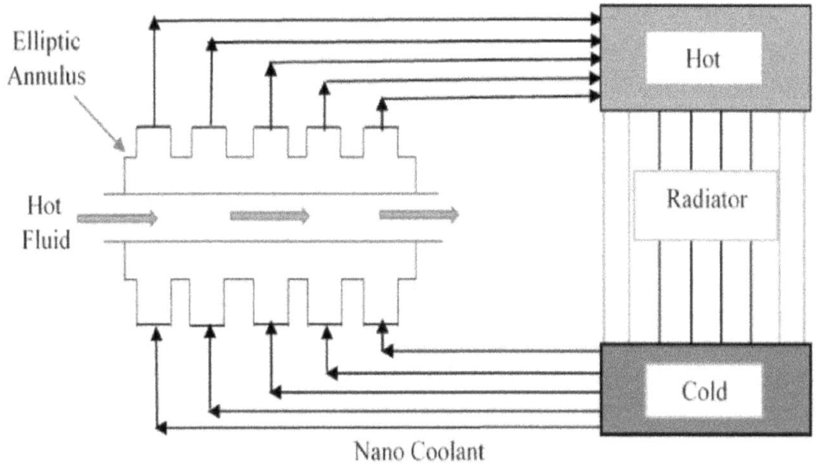

FIGURE 9.5 Geometric configuration of the experimental setup for natural convection simulation [7].

The demineralized water was selected as a base fluid due to its high thermal capacity and compatibility with nanoparticles. In order to avoid an increase in viscosity and the formation of scales, the hybrid nanofluid was prepared in the absence of surfactants. MWCNTs added high thermal conductivity (up to 3000 W/mK), whereas MgO nanoparticles provided thermal stability and corrosion resistance and oxidation.

Key Results

- **Thermal conductivity**: The hybrid nanofluid was able to attain a thermal conductivity of 0.278 kW/mK, which is superior to other nanofluids that were tested, including the TiO_2-based nanofluids.
- **Heat transfer coefficient**: The suggested DFVG enhanced ellipse annulus achieved a peak convective heat transfer coefficient of 858 W/m^2 °C at a heat flux of 3000 W/m^2, which is higher compared to the adopted conventional geometry, such as square, triangular, and circular shapes.
- **Temperature drop**: The inside cylinder surface temperature was at 25.8°C (relative to 26.4–27.5°C in the conventional geometries) with an amount of 1000 W/m^2 heat flux that exemplifies enhanced cooling.
- **Nusselt number**: The maximum Nusselt number of the experiment was observed when the nanoparticle concentration was 0.5% and provided a Nusselt number of 9.4, thus showing superior convective characteristics provided by the better mixing of a fluid and the thermal diffusion.
- **Optimization insight**: Across the range of fin heights (3 mm, 5 mm, and 7 mm), the 7 mm height provided the best overall (based on the efficiency measure 3 = 2.4902, which must balance the heat transfer advantage over the higher flow resistance disadvantage).

9.5 SIGNIFICANCE OF NANOFLUIDS

The addition of the MWCNT–MgO hybrid nanofluids was paramount to the enhancement of the rate of convective heat transfer. The tubular nanostructure of MWCNTs fulfilled the functions of better thermal conductivity channels, whereas MgO nanoparticles added stability and thermal resilience to the structure. This blend of theirs allowed achieving greater rates of energy transportation, with no use of chemical surfactants due to which complications like particle agglomeration and fluid instability were not a part of the scene.

The combinatory effect shows that the synergistic interaction of novel focus-enhancing geometry (such as the DFVG) and an exceptional heat transfer fluid (hybrid nanofluids) is one of the most promising in terms of boosting the thermal potential of naturally convective systems, as a suitable measure to use in energy systems, electronic systems, and heat exchangers.

9.5.1 Phase Change Material (PCM)-Based Cooling

Phase Change Materials (PCMs) are thermal storage media capable of absorbing and releasing significant amounts of heat during their phase transitions—typically from solid to liquid and vice versa. Their high latent heat storage capacity allows them to stabilize temperature fluctuations, making them highly effective in applications such as solar water heating, thermal insulation, electronics cooling, and energy-efficient buildings. PCMs can store solar energy during the day and release it during the night or cloudy periods, extending the operational effectiveness of renewable energy systems without active control mechanisms.

However, while PCMs are excellent at storing thermal energy, they exhibit certain limitations, such as low thermal conductivity and limited heat transfer rates. These properties can slow down the charging and discharging cycles, reducing the overall efficiency of the system. In large-scale applications, this limitation becomes particularly critical where quick response times and uniform temperature distribution are essential. To address the low thermal conductivity issue of PCMs, researchers have started incorporating nanofluids.

Enhancing Solar Water Heating System Performance Using Nanofluid and PCM Integration: Case Study by Chen et al. (2025)

This case study puts emphasis on combining nanofluids and phase change materials (PCMs) to promote thermal performance of a solar water heating system. The main goal was to enhance the efficiency of thermal storage and heat retention because these attributes remain low in conventional systems due to the low thermal conductivity of PCMs and the lack of proper heat transfer in the base liquids, such as water.

The system included a flat-plate solar collector and a hot water storage tank (similar to the implementation as shown in Figure 9.6), where the PCM was embedded in the form of vertical PCM tubes, and a closed circulation loop of CuO nanofluid with water. The ANSYS Fluent numerical tool was utilized to

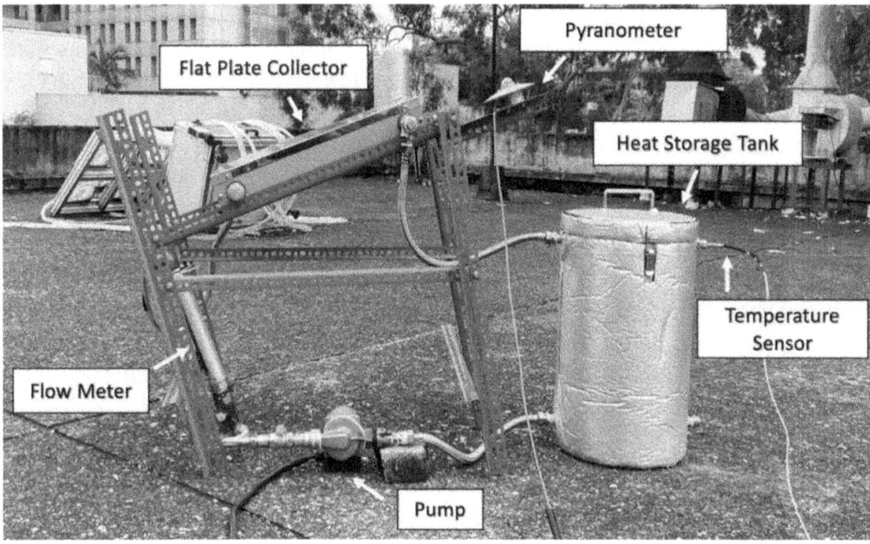

FIGURE 9.6 PCM and CuO-water nanofluid integrated solar collector experimental setup [8].

perform numerical simulations in order to compare different configurations, changing the number of PCM tubes (which ranges between 10 and 16) and PCM volume ratio (10–30%). CuO-water nanofluid was chosen because of its increased thermal conductivity and stability, and the concentration of the nanoparticle of CuO used was 1% in terms of volume. The outcomes showed that the most appropriate combination was 14 PCM tubes with a 20% PCM volume ratio, having a 94.2% thermal storage efficiency and 31.7 hours as hold-up time. This was a 31.9% improvement in retention compared to that of a system that had water only. The nanofluid made sure that PCM took a shorter/lesser time to charge and discharge and had a quicker melting and solidification property—in the view that a decrease (reduction) in thermal resistance is highly ensured. In general, a combination of nanofluids and PCMs together had a great effect on the performance of the solar water heater, which is suitable and effective in long-term thermal energy storage in the residential and industrial sectors.

9.5.2 Jet Impingement Cooling

Jet impingement cooling (JIC), also known as jet impingement and often thought of as sprayed cooling, is a very effective technique in thermal management of surfaces to dissipate large quantities of heat by jetting a high-velocity fluid jet (most commonly air, water, or other coolant) onto a specific heated surface. It is especially applicable for cooling small zones of high-temperature twist electronic compact devices, turbine blades, high-power laser systems, and semiconductors. The principal benefit of jet impingement is that the thermal boundary layer is thin and that convective heat transfer is intense, notably in the stagnation region (where impact actually takes place). Jet impingement cooling performance is also a function of various parameters, which include nozzle type, size, jet-to-surface distance, Reynolds number, flow rate, and the physical properties of their respective working fluids. Nevertheless, despite optimized arrangements, small thermal conductivity has a limit to the cooling capacity of conventional fluids even in the case of high heat flux flow.

In jet impingement systems, nanofluids lead to an increase in the local heat transfer coefficient as well as enhanced critical heat flux (CHF). The introduction of the nanoparticles enhances convective mixing and thermal diffusion, taking into consideration that Brownian motion and thermal conductivity are intensified.

Enhancing Jet Impingement Cooling Performance Using Hybrid Nanofluids: Case Study by Atofarati et al. (2025)

The case study is examining the synergistic attribute of nanoparticle size and pulsating jet flow on the thermal performance of hybrid nanofluids concerning jet impingement cooling of solar PV systems to manage heat. The study combines both machine learning and experimental processes to determine the impact of both particle and flow parameters on how well heat is transferred. Figure 9.7 shows the characteristics of a special test rig that was designed to simulate the thermal environment of a solar photovoltaic (PV) backplate. A copper cylindrical heat sink formed the main content that was warmed up by 6,100 W cartridge heaters placed in the block at temperatures that were below. To complete a closed-loop system, the Lechler circular nozzle (3 mm diameter) was used to redirect the jet after impingement, a heat exchanger to control the temperature of the coolant (29–31°C), and a pulsating unit controlled through the use of a function generator and by solenoid valves.

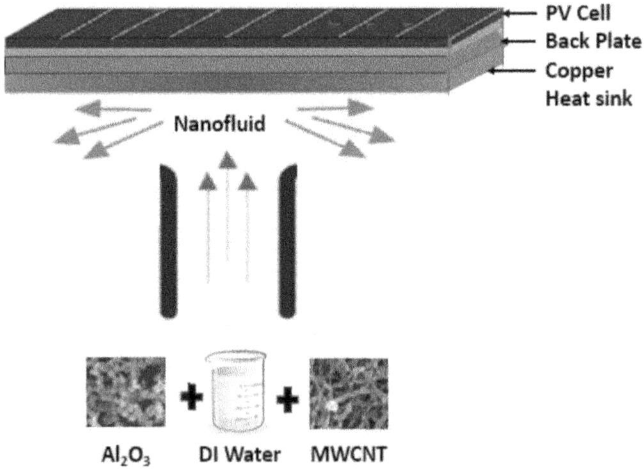

FIGURE 9.7 Experimental setup for the PV cell test section [9].

The nanofluid synthesis was done by subjecting 60:40 mixing ratio by weight of MWCNT-γ-Al_2O_3 and deionized water of different concentrations (0.05–0.3 vol %) to the effect of radiation size. Preparation involved magnetic stirring and ultrasonic homogenization. Viscosity, thermal conductivity, and stability were determined and verified. The embedded thermocouples were used to track the surface temperatures of the copper block, whereas the concentric thermocouple was used to track the exit coolant temperatures above the impingement zone. The tested system was able to test steady as well as pulsating jet regimes where the waveforms were under control (sine, F = 0.2 Hz, A = 8 Vp, J = 2) and Reynolds number 15,000–40,000.

Key Findings

- **Nanoparticle size effect:** Nanofluids with larger MWCNT sizes (30–50 nm) showed superior heat transfer due to enhanced thermal conductivity. Sample D (Al_2O_3: 5 nm, MWCNT: 30–50 nm) consistently exhibited the best performance under continuous and pulsating flow.
- **Jet type comparison:** Pulsating jets outperformed continuous jets at higher Reynolds numbers, improving the Nusselt number by up to *79.06%,* compared to water.
- **Concentration influence:** A higher nanofluid concentration (0.3 vol%) yielded the greatest heat transfer enhancement, though performance gains plateaued at certain levels.

9.5.3 Absorption Refrigeration Systems (ARS)

ARS frequently employing ammonia-water solutions does not need mechanical compressors and instead uses heat energy to perform the refrigeration cycle. Nanofluid additives, in particular ones that contain titania, increase thermal conductivity and decrease surface tension, increasing the rate of absorption and regeneration. Closed-coupled hybrid systems, such as those including ARS and nanofluid-cooled liquid loops, have reached cooling frequencies of 540 W and are being tested as options to local chip-level cooling in environmentally aware designs.

9.5.4 Spray Cooling

Spray cooling is an emerging method of thermal management that uses atomized fluid sprayed onto a hot body. The impinging droplets spread by evaporating and causing the cool heat removal on the surface by *convective and evaporative* mechanisms. This method is very effective in handling large heat fluxes (up to 100 W/cm^2 or beyond) and thus it can be applied in cooling high-performance electronics, aerospace equipment, laser diodes, and data center equipment. Spray cooling exhibits one of the greatest benefits, which is the possibility to keep the surface temperature low enough below the boiling point of the coolant, even with very high thermal loads present, because of the evaporative cooling effect. It is also temperature uniform, has no hot spots, and is compact and scalable. Nevertheless, its parameter is erratic to a range of things, such as droplet size, nozzle configuration, spray angle, proximity to the surface, and coolant characteristics, which are the viscosity, thermal conductivity, and surface tension.

One of the main reasons that allows significantly improving the efficiency of spray cooling by using nanofluids is their more effective thermal transport behavior than conventional types of base fluids. In the spray cooling, the reduced volume fraction of nanoparticles changes multiple interfacial and fluid dynamic features, providing enhanced heat removal when nanofluid droplets strike the heated surface. The deposition technique applied through repeated droplet impingement and subsequent evaporation is one of the major improvement mechanisms of the nanoparticles onto the target surface. This also changes the surface morphology to enhance wettability by lowering contact angles, and this leads to improved liquid spreading and rewetting in the boiling case. As a consequence, the system gets more nucleation site density, improves boiling heat transfer, and delays the formation of film boiling, thus increasing *critical heat flux (CHF).*

Furthermore, the effective high rate of thermal conductivity of nanofluids facilitates the absorption of energy conspicuously slower than the surface and aids in improving the coefficient of convective heat transfer.

Jet Impingement Cooling of High-Power LEDs Using Al_2O_3-Water Nanofluid: Case Study by Li et al. (2025)

The case study examines an advanced cooling method of high-power LEDs through a multi-nozzle array jet cooling system, which is targeting numerical simulation of Al_2O_3-water nanofluid as the working medium. This study aims to overcome the thermal management problems of LEDs, as the electro-optical efficiency is low (only ~20%). The LEDs produce a lot of heat, and this may present problems to their thermal management and ultimately to their performance.

EXPERIMENTAL SETUP

The simulation represented a 100 W LED array (100 chips, 1 W each), mounted over a drilled jet cavity through a 31 3 nozzle grid as in the case of Figure 9.8. The nozzles helped blow coolant jets on the LED surface in order to absorb heat. The most important geometric variables that were tested were the nozzle diameter (1–3 mm), cavity height (15–25 mm), and nozzle spacing (4–10 mm). They also designed a closed-loop experimental validation system, which included a water tank, pump, flowmeter, nozzles, thermocouples, and data-acquiring system.

A volume of fluid (VOF) method was applied to the base fluid (water) and nanofluid (Al_2O_3-water, up to 6 vol %). Non-dimensional boundary conditions also were used, such as the velocity inlet (3.48 m/s), constant point heat source of 4×10^9 W/m³ at each chip.

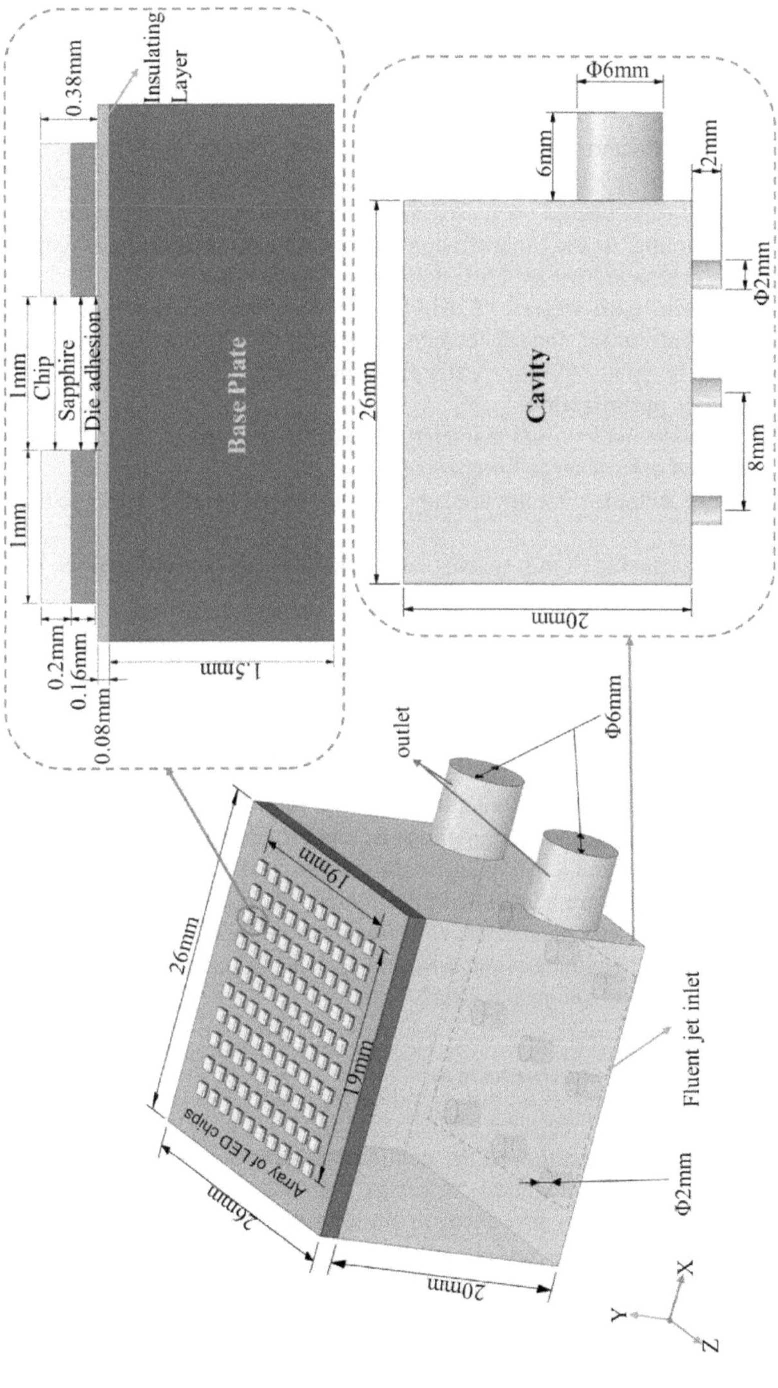

FIGURE 9.8 Schematic diagram of multi-nozzle jet impingement cooling system for LED module using nanofluid [10].

Key Findings

i. **Reynolds number effect:** Increasing the Reynolds number from 6000 to 8000 enhanced convective mixing and reduced LED surface temperatures by up to 2.5°C. Optimal jet dynamics and vortex intensity were observed at Re = 8000.
ii. **Nanofluid concentration:** A nanofluid volume fraction of 2–3% yielded the best cooling performance and thermal uniformity. Higher concentrations (4–6%) caused increased viscosity, leading to recirculation and uneven cooling. At 4% concentration, the nanofluid lowered the LED chip's maximum temperature by 4.5°C compared to pure water.
iii. **Comparison with water:** 4% Al_2O_3-water nanofluid offered better temperature distribution and reduced thermal hotspots. The cooling efficiency improved by up to 7.9% over water at optimal volume fractions.
iv. **Geometric optimization:**
   a. Nozzle diameter: 1 mm diameter yielded the strongest local cooling but required a dense array for coverage.
   b. **Cavity height:** 15 mm was optimal, as larger heights weakened jet impact.
   c. **Nozzle spacing:** 4 mm spacing enhanced thermal coupling between jets and ensured uniform heat removal.

9.6 PRACTICAL CONSIDERATIONS AND LIMITATIONS

Use of nanofluids in different thermal systems is a huge leap toward enhancing heat transfer. These fluids have demonstrated positive improvements in thermal conductivity, convective heat transfer coefficients, and critical heat flux. In spite of the aforementioned considerable advantages, the commercialization and broad implementation of nanofluids are not limited by a number of practical factors and limitations. This discussion will present the most important of these, and they include the issue of stability and agglomeration, cost and scalability, compatibility and corrosion, and environmental and health issues.

9.6.1 Stability and Agglomeration

Achieving and keeping long-term colloidal stability is one of the major difficulties of the practical application of the nanofluids. Nanoparticles tend to undergo agglomeration (clumping together) depending on the van der Waals forces of intermolecular forces that cause them to clump together and ultimately settle in the suspension. The result is particle sedimentation, loss of effective surface area, and reduction of thermal performance. This instability may, in some systems like microchannel heat exchangers and jet impingement cooling systems lead to clogging in the channels and unequal flow distribution, along with high losses in heat transfer efficiency.

To combat these effects, several stabilization techniques are employed:

- **Surfactants and dispersants**: These additives adhere to the surface of nanoparticles, creating a barrier that reduces attractive forces between particles. However, their effectiveness can degrade over time or at elevated temperatures, leading to eventual destabilization.
- **pH control**: Adjusting the pH of the nanofluid alters the surface charge of the particles, enhancing electrostatic repulsion and preventing aggregation. This method is highly effective for specific types of nanoparticles but may not be universally applicable.
- **Surface functionalization**: Chemical functional groups can be grafted onto nanoparticle surfaces to improve their compatibility with the base fluid. This method enhances dispersion stability and can improve thermal performance but adds complexity and cost to nanoparticle synthesis.
- **Ultrasonication**: A mechanical method used during nanofluid preparation to disrupt agglomerates and ensure a uniform dispersion. While effective in laboratory settings, prolonged sonication is not practical for industrial-scale or continuous-flow systems.

Even with these techniques, maintaining long-term stability under operational conditions such as fluctuating temperatures, flow rates, and pressure remains a challenge. For instance, in jet impingement cooling applications using hybrid nanofluids like Al_2O_3-MWCNT, significant performance drops can occur over time if the fluid begins to separate or agglomerate.

9.6.2 Cost and Scalability

A major constraint to the large-scale application of nanofluids is their economic feasibility. Good nanoparticles are frequently synthesized by well-established procedures, which include sol-gel techniques, chemical vapor deposition (CVD), and hydrothermal. Such methods are expensive, and they might not be feasible to be produced at a large scale without a huge investment.

Quality control is also hindered by the fact that it is hard to maintain the same size, morphology, and purity of the particle in batch production. Use of nanofluid systems in industrial processes requires a high level of consistency regarding thermophysical parameters such as viscosity, thermal conductivity, and stability.

The better performance that nanofluids bring in most areas of industry must reckon with their cost of production and maintenance. As an illustration, a small improvement in the heat transfer efficiency will not justify the incremental cost unless it results in energy savings, a reduction in size of the system, or an increase in the service life of its operation. Such a trade-off between cost and performance is of primary concern in the realization of the usefulness of the nanofluids.

Scalability also remains problematic. Although laboratory investigations are encouraging, scaling the same to practical systems with high volume and intricate geometries has not been straightforward. No common practices on bulk preparation, stabilization, and testing of nanofluids are available. Furthermore, long-term storage and shelf-life tests are often not stated in the literature, and this makes them less reliable when implemented in real life.

9.6.3 Compatibility and Corrosion

Nanofluids must be chemically and mechanically compatible with all components of the thermal system, including heat exchangers, piping, pumps, and seals. However, certain nanoparticles can be reactive or abrasive, leading to corrosion, erosion, and fouling over time.

For example:

- **CuO and Fe_3O_4 nanoparticles**: In aqueous environments, these can catalyze redox reactions that accelerate the corrosion of metallic surfaces.
- **Carbon-based nanoparticles**: Though thermally efficient, materials like carbon nanotubes (CNTs) may pose abrasive wear risks in high-velocity flow environments.

To mitigate such issues, several strategies are employed:

- **Use of corrosion inhibitors**: These chemical additives protect metal surfaces by forming a passive film that reduces reactivity.
- **Coatings**: Applying polymer or ceramic coatings to internal system surfaces can enhance corrosion resistance.
- **Material selection**: Choosing corrosion-resistant metals like stainless steel or using polymeric components in contact with nanofluids can help reduce degradation.

Although these precautions are taken, there is inadequate knowledge about the long-term behavior of nanofluids in a complex thermal system. The majority of the experiment studies are carried out over short periods, and it is an open issue whether there were cumulative effects like biofouling, fatigue of the material to use, and formation of deposits. Hence, intensive endurance testing in genuine working cycles is needed to determine the life performance of nanofluid-based systems.

9.6.4 Environmental and Health Concerns

The increasing application of engineered nanomaterials raises legitimate concerns about their environmental and biological impacts. Due to their small size and high reactivity, nanoparticles can interact with biological systems in potentially hazardous ways.

- **Health risks**: Nanoparticles can enter the body through inhalation, ingestion, or skin contact during handling and preparation. Inhalation is particularly risky during the mixing of dry powders, as particles can penetrate deep into the lungs, leading to respiratory and systemic effects.

- **Environmental impact**: Accidental spills, leakage, or improper disposal of nanofluids can lead to contamination of water sources and ecosystems. The long-term ecological effects, including bioaccumulation and toxicity to aquatic organisms, are still under investigation.

To address these concerns, regulatory agencies are formulating guidelines for the safe use and disposal of nanomaterials. Furthermore, researchers are actively exploring environmentally friendly alternatives:

- **Silica and alumina-based nanoparticles**: These are generally considered to be less toxic and more inert.
- **Green synthesis methods**: Using plant extracts or biopolymers to produce nanoparticles reduces chemical waste and improves biodegradability.
- **Biodegradable base fluids**: Replacing conventional glycols and synthetic oils with vegetable oils, glycerol, or bio-based esters can reduce environmental risks.

Still, comprehensive life cycle assessments (LCA), including cradle-to-grave analyses, are essential before large-scale adoption. These assessments must consider raw material extraction, nanoparticle synthesis, operational usage, and end-of-life disposal to fully quantify environmental sustainability.

9.6.5 Summary and Future Outlook

Although nanofluids represent a potentially revolutionary advancement in promoting heat transfers and improving the design of small thermal systems, practical use is constrained by unaddressed challenges. Key among these are:

- Long-term colloidal stability in operating conditions.
- Achieving cost-effective and scalable production processes.
- Demonstrating chemical and mechanical compatibility with system components.
- Mitigating environmental and occupational health risks.

To solve such problems, alliances of materials science, thermal engineering, chemistry, and environmental sciences will need to be proposed. Potential ways forward are smart self-healing nanofluids, machine-learning-based optimization of nanofluid formulations, and robust methods of real-time monitoring.

Moreover, the adoption process in the industry will be enhanced with the creation of nanofluid synthesis and characterization states, as well as the test states for applications. Responsible deployment will also be enhanced through government policies and safety regulations as aligned with the objectives of sustainability.

To sum up, although work to achieve the large-scale adoption of nanofluids remains ongoing, research and development activity stimulated by performance criteria, as well as sustainability indicators, has been advancing the adoption path of nanofluids in future mainstream energy and cooling systems technology.

9.7 INDUSTRIAL CASE STUDIES AND EMERGING TRENDS

9.7.1 Case: Nanofluid-Enhanced District Cooling in UAE

In a groundbreaking pilot project, UAE-based district cooling leader Tabreed, in collaboration with Ireland's HT Materials Science (HTMS), has successfully tested a nanofluid-based heat transfer technology called Maxwell in one of its cooling plants in Khalifah City, Abu Dhabi. This experimental study is considered the first of its kind globally and highlights the significant energy efficiency potential of nanofluids in district cooling systems. The Maxwell nanofluid—an engineered suspension of sub-micron aluminum oxide particles in water or a water/glycol mix—was added to the chilled water system at a 2% concentration. During the summer 2023 trial, it cooled a 55,742 m^2 educational campus and achieved chiller efficiency improvements of 13.6% and an overall plant efficiency gain of 9%, compared to performance in the same period of the previous year.

The enhanced thermal properties of Maxwell improved the heat transfer capability of the chilled water, thereby allowing components like chillers and pumps to operate more efficiently and consume less electricity. If scaled across Tabreed's 89-plant portfolio, this could lead to annual carbon emissions reductions of up to 200,000 metric tons, equivalent to removing approximately 43,500 cars from the roads, and saving 335 million kWh of electricity—enough to power 20,000 homes. Maxwell is non-toxic, non-corrosive, and fully recyclable, making it a sustainable solution for HVAC systems. Encouraged by the success of the pilot, Tabreed and HTMS are now exploring the broader implementation of this nanofluid across more facilities. Given that cooling accounts for 70% of energy use in the Gulf Cooperation Council (GCC) and demand is set to triple globally by 2050, this innovation could significantly contribute to carbon reduction goals and support the UAE's Net-Zero 2050 strategy.

9.7.2 Trina Solar's Record-Breaking 26.58% Efficiency in i-TOPCon PV Cells

Trina Solar, one of the global leaders in solar energy innovation, has achieved a groundbreaking milestone in the field of photovoltaic (PV) technology by developing an n-type i-TOPCon solar cell with a record-setting power conversion efficiency (PCE) of 26.58%. This achievement marks a significant leap in solar cell performance and solidifies the company's position at the forefront of next-generation solar technologies.

9.7.2.1 What Is i-TOPCon Technology?

i-TOPCon stands for *industrial Tunnel Oxide Passivated Contact* solar cell technology. It is an advanced form of *n-type crystalline silicon* technology. The core idea is to reduce energy losses by improving surface passivation and minimizing recombination of charge carriers. i-TOPCon cells use:

- A thin tunnel oxide layer to passivate the silicon surface.
- A polycrystalline silicon (poly-Si) layer that helps transport electrons more efficiently.

This architecture boosts both *open-circuit voltage (Voc)* and *fill factor*, resulting in *higher overall efficiency.*

9.7.2.2 Trina Solar's Innovation Details

Trina's 26.58% PCE milestone is not just incremental—it's revolutionary. Here's how they achieved it:

9.7.2.3 Material Choice—N-Type Silicon Wafers

The base material used is *n-type silicon*, specifically *phosphorus-doped wafers.* Compared to traditional p-type wafers, n-type silicon offers:

- Higher minority carrier lifetimes.
- Lower susceptibility to light-induced degradation (LID).
- Better resistance to impurities.

This leads to longer-lasting and more efficient solar cells. *Cell Size—210 mm × 182 mm*

The large-area wafers allow for a greater surface area for light absorption and electricity generation, helping scale the efficiency in actual modules, not just in lab prototypes.

9.7.2.4 Quantum Tunneling Passivated Contact

Trina used a *proprietary quantum tunneling technology* in the passivated contact layers:

- This approach minimizes recombination at the rear surface of the cell.
- It allows better carrier selectivity, improving current flow and overall efficiency.

9.7.2.5 Ultrafine Line Printing and Optic Trapping

They implemented *ultrafine metallization techniques* that reduce shading on the solar cell surface while maintaining excellent electrical conductivity.

- Optic trapping further ensures more sunlight is absorbed by the silicon, maximizing photon utilization.

9.7.2.6 Verification and Impact

The record efficiency was *independently verified by the Institute for Solar Energy Research in Hamelin (ISFH CalTeC)* in Germany, giving credibility and scientific validation to Trina Solar's claim. This new benchmark exceeds the previous record of 25.9% by a margin of 0.5%, set just a month prior, showcasing the *rapid pace of development.*

9.7.2.7 Why Is This Important?

i. **Efficiency breakthroughs lower costs**
   a. Higher efficiency means *more power per square meter*, reducing the area and material required for installations.

   b. This drives down the *levelized cost of electricity (LCOE)*, making solar power even more affordable and competitive.

ii. **Industrial viability**
   a. The breakthrough is not just academic—it uses *industrial-scale wafers* and fabrication techniques.
   b. This means it is readily adaptable for *mass production.*

iii. **Boosts renewable energy transition**
   a. With global goals to phase out fossil fuels, high-efficiency solar cells accelerate the shift toward a *low-carbon economy.*
   b. Improved PV efficiency is essential for *energy-dense installations* in urban areas or land-constrained regions.

9.7.2.8 Trina's Vision and R&D Leadership

Trina Solar's CEO, *Gao Jifan*, emphasized that this success reflects the company's *long-term commitment to R&D.* The ongoing investment in *second-generation i-TOPCon cell development* not only enhances performance but also:

- Protects *intellectual property (IP).*
- Improves *technological competitiveness.*
- Sets a clear pathway for *future renewable energy innovations.*

This record is not just a technological accomplishment but also a strategic milestone for energy sustainability, aligning with global efforts to meet *net-zero carbon goals.*

9.7.2.9 Future Outlook

With the efficiency of *n-type TOPCon cells now breaking the 26% barrier*, the PV industry is expected to see:

- Rapid adoption of *n-type technologies.*
- A gradual phasing out of older p-type PERC cells.
- Continued R&D to reach *27%+ efficiencies*, possibly integrating *hetero-junction* or *IBC (Interdigitated Back Contact)* technologies.
- Emerging trend: AI-optimized nanofluid formulation for dynamic systems.

9.8 CONCLUSION

Nanofluids have emerged as a highly promising class of thermal fluids capable of significantly enhancing heat transfer performance across a broad spectrum of thermal management and energy recovery systems. Their unique nanoscale properties, including improved thermal conductivity, increased surface area, and enhanced convective behavior, make them particularly effective in compact and high-flux environments where traditional fluids fall short. The integration of nanofluids into technologies such as microchannel heat sinks, thermosyphons, PCM-based systems, and thermoelectric modules has led to measurable improvements in heat dissipation, energy efficiency, and operational reliability.

Experimental case studies—ranging from CPU cooling and lithium-ion battery thermal management to district cooling and solar thermal applications—demonstrate the versatility and effectiveness of nanofluids. Notably, hybrid nanofluids, which combine multiple nanoparticle types, offer enhanced thermophysical properties and superior performance under varying operating conditions.

However, practical limitations persist. Issues related to long-term colloidal stability, cost-effective large-scale synthesis, material compatibility, and environmental safety must be addressed to facilitate widespread industrial adoption. The potential risks of agglomeration, corrosion, and nanoparticle toxicity also require comprehensive life cycle and toxicological assessments.

To overcome these challenges, future research should prioritize:

- Standardization of nanofluid formulation and testing protocols.
- Development of eco-friendly and biodegradable nanofluids.
- Advanced modeling techniques (e.g., machine learning and AI) for performance prediction and optimization.
- Smart, self-healing, and adaptive nanofluid systems for real-time applications.

Moreover, policy frameworks and regulatory guidelines must evolve in tandem with technological advancements to ensure safe, scalable, and sustainable deployment.

REFERENCES

1. Bahar, A., Ghafouri, J., & Zamzamian, K. (2025). A numerical investigation into the application of thermosyphons for preventing the freezing of water meters, utilizing pure water and different nanofluids. *International Journal of Thermofluids*, *26*, 101088.
2. Sathish, T., Giri, J., Allehaibi, M., Hourani, A. O., & Anderson, A. (2025). Performance enhancement of heat pipe-based solar tube collectors using graphene nanofluids: Experimental analysis and thermal optimization. *Case Studies in Thermal Engineering*, 106460.
3. Wang, Z.-X., & Tao, W. Q. (2025). Heat transfer and pressure drop characteristics of microchannel cold plate in commercial CPU-package cooling system. *International Journal of Heat and Mass Transfer*, *246*, 127060.
4. Ghasemi, S. E., Ranjbar, A. A., Hoseini, M. J., & Mohsenian, S. (2021). Design optimization and experimental investigation of CPU heat sink cooled by alumina-water nanofluid. *Journal of Materials Research and Technology*, *15*, 2276–2286.
5. Ruan, H., Xie, H., Wang, J., Liao, J., Sun, L., Gao, M., & Li, C. (2021). Numerical investigation and comparative analysis of nanofluid cooling enhancement for TEG and TEC systems. *Case Studies in Thermal Engineering*, *27*, 101331.
6. Kanti, P. K., Yang, E. S. J., Wanatasanappan, V. V., Sharma, P., & Said, N. M. (2024). Impact of hybrid and mono nanofluids on the cooling performance of lithium-ion batteries: Experimental and machine learning insights. *Journal of Energy Storage*, *101*, 113613.
7. Kumar, V. M., Vinod, S., Kumar, A., & Vikas, R. (2025). Enhanced natural convective heat transfer in a horizontal elliptic annulus using a diamond fin vortex generator and MWCNT-MgO nanofluid. *Case Studies in Thermal Engineering*, *71*, 106099.
8. Chen, G.-R., Liao, T.-W., Hsieh, C.-C., Barman, J., Huang, C.-Y., & Kuo, C.-F. J. (2025). Using the Taguchi method and grey relational analysis to optimize the parameter design

of flat-plate collectors with nanofluids, and phase change materials in an integrated solar water heating system. *Energy Conversion and Management: X*, *26*, 100910.
9. Atofarati, E. O., Sharifpur, M., Huan, Z., Awe, O. O., & Meyer, J. P. (2025). Experimental and machine learning study on the influence of nanoparticle size and pulsating flow on heat transfer performance in nanofluid-jet impingement cooling. *Applied Thermal Engineering*, *258*, 124631.
10. Li, Z., Xin, W., Ma, D., & Jian, W. (2025). Numerical investigation of jet cooling for high-power LEDs using A12O3-water nanofluid with a multi-nozzle array. *Case Studies in Thermal Engineering*, *70*, 106083.
11. New material improves cooling performance in electronics. (2024, March 21). *Gulf Daily News*. https://www.gdnonline.com/Details/1295366. Accessed 12 July 2025.

10 Nanofluids in Intensification of Process Industry

10.1 INTRODUCTION

10.1.1 Definition of Process Intensification (PI)

Process Intensification (PI) is a forward-looking and transformative approach in chemical engineering and tries to enhance efficiency of chemical, biochemical, and physical processes by major modifications of the standards of efficiency, compactness, and sustainability. Essentially, PI is not just about changing existing processes but thinking about how the process can be carried out and based on essentially different principles, new materials, and innovations in equipment design. It is a philosophy aimed to place processes in a cleaner, faster, safer, or more economical manifold by increasing the fundamental transport and reaction phenomena.

The conventional design of a chemical plant has relied on high-volume unit functions, each specialized in a specific job that may need a great deal of space, capital, and energy. Although the method has served the industry well over the past decades, it is gradually being threatened by the requirements of the contemporary manufacturing industry, which requires greater environmental controls, pressurized cost of energy, decentralized flexible, and sustainable production systems. Process Intensification seeks to overcome these challenges by targeting integration or reconfiguration of unit operations to drastically minimize the equipment size, operational expenditures, and environmental burden, even as it enhances the quality of output and throughput. This is among the most common definitions of PI as proposed by Stankiewicz and Moulijn (2000), who referred to it as a collection of strategies resulting in much smaller, cleaner processes and energy-efficient ones [1]. With this definition, PI has two goals: first, dramatic improvement in performance and, second, decreasing the environmental impact of chemical processes (Figure 10.1). In contrast to incremental improvements, PI aims at step-change progress, meaning it may result in order-of-magnitude increases in process efficiency. In an overview presented in this paper, the chief principles of process intensification, as well as the common means of implementing them in industry, are also discussed.

The main objectives of process intensification may be grouped into five major results. The initial one is the downsizing of the equipment and plant size. The necessity of having large and numerous equipment can be avoided by increasing the rate of heat and mass transfer or integrating several process steps in a single and intensified one (e.g., reactive distillation or membrane reactors). This reduces the size and modularity of the processes designed, and the latter is particularly useful in space-sensitive

DOI: 10.1201/9781003628071-10

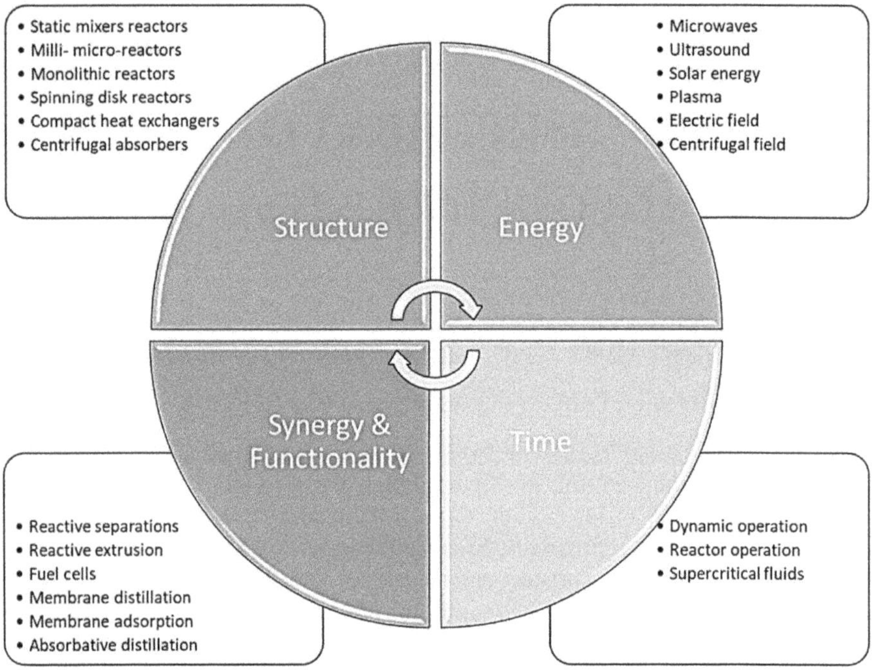

FIGURE 10.1 Core principles of process intensification and representative industrial applications [2].

industries such as offshore oil platforms, mobile refineries, or remote pharmaceutical manufacturing. Second, another significant cause of PI is the enhancement of the throughput of the processes. Reactions and separations are also made faster in intensified systems and therefore lead to increased productivity on a per unit of volume basis. This would be attained via enhanced phase contacting, elevated interfacial space, minimal diffusion hindrances, and more effective utilization of energy. As an example, mixing and heat transfer in microreactors and spinning disk reactors can be very rapid (several orders of magnitude faster) than in standard reactors and can make reactions possible that would otherwise be restricted.

Third, PI seeks to develop *energy efficiency* in the entire supply chain. Due to reduced temperature and pressure gradients, and making more productive use of waste heat or renewable energy sources, intensified processes have much lower energy consumption. As an example, compact heat exchangers that are integrated with process units reduce heat losses, and, since the temperature difference is lower, they permit heat recovery. In the same manner, application of solar nanofluids in thermal collectors constitutes an enhanced system integrating the heat transportation with environment-friendly energy creation.

The fourth goal is the *improvement of the rate of heat and mass transfer*. Faster reactions, enhanced separations, and short residence times are possible with efficient transfer of both heat and mass. The properties of materials and geometry of equipment restrict such transfers in traditional units. PI methods enhance these processes because they are facilitated by making use of highly conductive materials, new flow configurations (oscillatory or swirling flows), micro-structured surfaces, or indeed new working fluids, for example, nanofluids. With the improvement, there is an increased control capacity, reduced energy requirements, and an improved continuity of products.

Finally, PI contributes to the *minimization of environmental impact.* With the minimization of the size of equipment and efficiency in using resources, it is possible to use less energy and produce less emissions. In addition, there are numerous PI technologies to which greener methods are added, including solvent-free reactions, utilization of waste heat, use of renewable feedstocks, or catalysts. An increased eco-efficient and more sustainable chemical process is the outcome, compliant with international environmental goals like those envisaged by the Paris Agreement and the UN Sustainable Development Goals (SDGs).

Some enabling technologies and design strategies lie at the core of PI. These are *compact multifunctional reactors, microreaction technology, intensified separation systems, and alternative energy inputs* (microwave, ultrasound, or electric fields). There are also *novel materials*, such as nanostructured catalysts, superhydrophobic surfaces, and stimuli-responsive membranes, that add to PI, due to their increased activity, selectivity, and effectiveness, compared to their traditional analogs. Another strong weapon in the PI arsenal is *process coupling*, such as is found in membrane reactors or reactive absorption, in which reaction and separation are accomplished in one step.

New computational modeling, *artificial intelligence (AI)*, and *machine learning* capabilities also find use in process intensification, to empower predictive process design and optimization. They enable a virtual test of process configurations, which saves cost and time in the development. This flexibility and control of intensified processes are further augmented by growing Industry 4.0 integration, in the form of digital twins and real-time data analysis. The prosperity of PI can be observed in many parts of the chemical process industry. PI has been used in continuous manufacturing, fast scale-up, and personalized medicine in the pharmaceutical industry. Compact heat exchangers and intensified distillation units are useful to make the petrochemical sector more energy-efficient and to minimize emissions. The PI techniques are used in the food industry to enhance drying, mixing, and pasteurization costs. The PI can be applied in environmental engineering, which allows effective removal of pollution and capture of carbon and decontamination of water using new adsorbents and catalytic mechanisms.

10.1.2 Role of Heat and Mass Transfer in Chemical Industries

In chemical and process industries, *heat and mass transfer* serve as the fundamental pillars that govern the efficiency, performance, and safety of nearly all unit operations. From reactors and heat exchangers to distillation columns and absorbers, the movement of energy (as heat) and matter (as mass) is central to the transformation and separation of chemical species. An in-depth understanding of these transport phenomena is therefore essential not only for the design and scale-up of individual equipment but also for achieving overall process optimization and intensification.

10.1.2.1 Heat Transfer: Central to Thermal Management and Reaction Control

Heat transfer encompasses the movement of heat energy as a result of a temperature difference and is mainly divided into three modes: that is, conduction, convection, and radiation. The most frequently met with in industrial processes are conduction and convection. Thermal management is essential in chemical plants as most of the processes, particularly those involving chemicals, transitions, and separations, are temperature-sensitive.

An example of this can be seen in chemical reactors, particularly when they perform exothermic or endothermic reactions, where control of the temperature is important to support reaction kinetics and selectivities, and safety. In an exothermic reaction, like in the nitration of benzene, thermally uncontrolled heat accumulation can result in a dangerous situation called thermal runaway. Heat transfer systems that are maximized are thus incorporated into the design of a reactor to ensure a constant removal or addition of the required heat to a reactor to ensure control of isothermal or optimal working temperatures. Heat transfer is the main role of heat exchangers, one of the most widely used equipment in the chemical process industry (Figure 10.2). These gadgets allow the movement of heat between two or more fluids without any direct contact between them. They are critical in recapturing energy and preliminarily heating streams of feed and regulated yields. *Heat exchanger* performance is normally measured in relation to the total heat transfer coefficient (U_0-value), which is a factor of the thermal conductivity of the fluids, fluid velocities (influencing convective coefficients), flow scheme (counter-current or co-current), and fouling resistance.

The drawback of traditional fluids like water, ethylene glycol, or oils is that due to their low thermal conductivities, they have limited values of the maximum attainable heat flux. Enhanced thermal conductivity is of particular importance in compact systems, or microscale heat exchangers, where the area is generally limited to begin with. Beating these heat transfer constraints is among the best reasons why advanced fluids as nanofluids that exhibit much superior thermal conductivities can be incorporated.

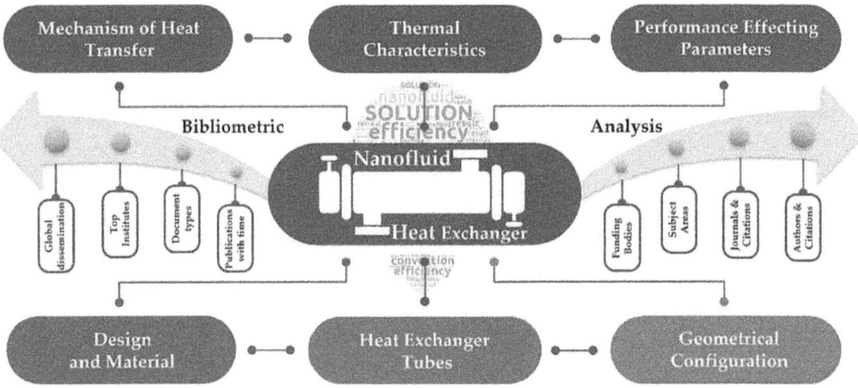

FIGURE 10.2 Application of nanofluids in the field of heat transfer [3].

10.1.2.2 Mass Transfer: Governing Separation and Reaction Processes

Meanwhile, mass transfer is the movement of mass (molecules or species) from one place to another and is also forced by concentration gradients. As with heat transfer, it can also take place in several modes: molecular diffusion convection and interfacial mass transfer (like gas–liquid or liquid–liquid exchange). In the separation process transfer of mass is pivotal and to any system that needs mixing or interaction of phases.

A simple example of a mass transfer is described in *distillation*, where a liquid mixture is separated into its individual components based on differences in their volatilities. The effectiveness of this operation is mostly influenced by the mass transfer rates between the vapor phase and the liquid phase on every tray or in every packing layer. In *absorption and stripping columns*, a liquid (or a gas) is absorbed into a liquid (or a gas), the speed of the mass transfer at the gas–liquid interface deciding the size and efficiency of the column.

The process of *liquid–liquid extraction* is controlled by mass transfer, which controls solute distribution between two immiscible phases. There are generally two ways of achieving this: the enhancement of the interfacial area and the decrease in the resistance of the boundary layers. In solid–liquid systems (leaching or adsorptions), solute movement in porous media and concentration gradient impact the rate and efficiency of separation kinetics.

Mass transfer also plays a very important role in *heterogeneous catalysis reactions*, in which the reactants need to be brought to the catalyst surface and products easily removed. In a typical gas-phase catalytic reactor applied to make ammonia or methanol, inadequate mass transfer may cause a concentration gradient to form, which lowers the overall reaction rate, as well as catalyst activity. In this case, to increase productivity, it is important to maximize the rate of mass transfer through turbulence, optimal geometry of the reactor, or use of surface modification.

10.1.2.3 Interdependence of Heat and Mass Transfer

It is important to note that in practice, heat and mass transport are often intertwined phenomena. An example is the vaporization that occurs in the evaporators, where heat is added to a liquid, making it vaporize, and both heat and mass are transferred. Likewise, in condenser, latent heat removal as well as mass movement of the condensing vapor should be taken into account. Diffusion of moisture away through vapor phase and heat addition to surface happen in a coordinated manner in dryers, and this controls the evaporation of moisture at a solid surface. Therefore, enhancement in any type of transfer tends to drive another form, which is mostly applicable in process intensification.

10.1.2.4 Barriers to Efficient Transport and the Need for Intensification

While heat and mass transfer are well understood and widely applied, they are not always efficient in practice. Several factors limit their effectiveness in industrial operations:

- *Low thermal conductivity* of conventional fluids
- *Laminar flow conditions*, especially at small scales, that limit convective heat/mass transfer

- *Large boundary layer resistances* in poorly mixed or poorly designed systems
- *Fouling and scaling*, which increase thermal resistance and reduce effective surface area
- *Inefficient interfacial contact* between phases in multiphase operations

These limitations result in higher energy consumption, larger equipment, longer processing times, and reduced product quality. Therefore, improving heat and mass transfer is not just a performance goal but a necessity for meeting today's demands for compact, energy-efficient, and sustainable processing.

This is where *Process Intensification (PI)* makes its most substantial contribution. PI seeks to *maximize heat and mass transfer rates* through:

- The use of structured surfaces or microchannels that promote turbulence.
- The integration of mixing-enhancing internals like static mixers or baffles.
- The design of multifunctional equipment that combines heat and mass transfer in a single unit (e.g., reactive distillation).
- The adoption of *high-performance working fluids*, such as nanofluids, that exhibit superior thermophysical properties.

For instance, in microreactors used for fine chemical synthesis, the *high surface-area-to-volume ratio* enhances both heat removal and species transport, enabling isothermal operation and faster reactions. In rotating packed beds (HiGee systems), *centrifugal forces* enhance mass transfer far beyond conventional gravity-driven systems. These intensified configurations show how deliberate design and smart materials can overcome traditional transport limitations.

10.1.2.5 Emerging Role of Nanofluids in Transport Phenomena

Nanofluids are engineered mixtures of nanoparticle suspensions in a fundamental fluid. In recent years, engineers and scientists have found nanofluids to be promising heat transfer enhancement media. Their rheological properties (especially tailored to control heat and mass transfer) in conjunction with their high thermal conductivity allow them to outperform in heat sinks, cooling components, and even some mass transfer. To take one example, in *membrane-based separation* systems, nanofluids may help to uniformize temperature, minimize fouling, and possibly even enhance diffusivity because of micro-convection driven by Brownian motion of the nanoparticles.

While research on mass transfer enhancement using nanofluids is still emerging, early studies suggest that improved mixing, reduced boundary layer thickness, and better temperature control can indirectly benefit mass transfer efficiency as well.

10.1.3 Emergence of Nanofluids as a Novel Heat Transfer Medium

In recent decades, the demand for high-efficiency thermal management systems has grown significantly across various industrial sectors, including chemical processing, power generation, electronics cooling, automotive, and renewable energy systems.

Widely used heat transfer fluids include traditional fluids like water, ethylene glycol, engine oil, and refrigerants, and have the advantages of their availability, non-toxicity, and their ability to be compatible with existing infrastructure. Nevertheless, these traditional fluids tend to have, in many cases, bad thermal conductivity, which restricts their ability to match the increasing thermal performance requirements of more recent technologies to a drastic extent. In an effort to address these limitations, researchers have developed nanofluids, a new type of engineered fluid that shows great enhancements with regard to thermal transport properties as compared to the base fluids.

As discussed in earlier chapters, nanofluids are nanosuspensions, that is, colloids that consist of nanometer-sized particles (<100 nm) suspended in familiar heat transfer fluids. Such nanoparticles (possibly metallic ones, e.g., Cu, Ag; metal oxides, e.g., Al_2O_3, TiO_2, ZnO; carbon-based, e.g., CNTs; graphene; or hybrid composites) can increase the thermal conductivity and convective heat transfer characteristics of the parent fluid at low concentrations (typically <5 vol.%), in some cases exponentially. Choi and Eastman (1995), who have become known as the first to coin the term nanofluid, showed that by introducing copper nanoparticle solutions into water, thermal conductivity could be brought to a significantly high level, heralding the current revolutionary study.

10.1.3.1 The Emergence of Nanofluids Can Be Attributed to Several Converging Factors

10.1.3.1.1 Nanotechnology Advancements

The rapid evolution of nanotechnology has enabled the synthesis of a wide variety of nanoparticles with controlled size, shape, surface chemistry, and dispersion characteristics. These advancements allow the precise tailoring of nanofluid properties to meet specific application requirements. Techniques such as chemical vapor deposition, sol-gel synthesis, hydrothermal methods, and laser ablation have facilitated the production of stable and highly thermally conductive nanoparticles.

10.1.3.1.2 Enhanced Thermal Conductivity

One of the most compelling reasons for the rise of nanofluids is their ability to significantly increase the effective thermal conductivity of the base fluid. Experimental studies have shown thermal conductivity enhancements ranging from 10% to over 50%, depending on the type, size, shape, and concentration of nanoparticles used. This makes nanofluids ideal candidates for heat exchanger applications, microchannel cooling systems, and other process intensification technologies where enhanced heat transfer is critical.

10.1.3.1.3 Improved Convective Heat Transfer

Beyond conductivity, nanofluids also influence convective heat transfer. The presence of nanoparticles increases the effective surface area for energy exchange, promotes micro-convection due to Brownian motion, and may modify boundary layer behavior. Together, these effects lead to higher Nusselt numbers and better heat transfer coefficients, making nanofluids suitable for both laminar and turbulent flow regimes commonly encountered in industrial operations.

10.1.3.1.4 Compatibility with Existing Systems

Despite their advanced performance, nanofluids can often be integrated into existing thermal systems with minimal modification. This feature lowers the barrier to adoption in industries where retrofitting costs can be a major concern. For example, in compact heat exchangers or reactor cooling loops, nanofluids can replace water or glycol-based fluids without requiring complete system redesign.

10.1.3.1.5 Multiphase Heat Transfer Applications

Nanofluids are increasingly being investigated for boiling, condensation, and phase change heat transfer processes. Their ability to alter bubble dynamics, reduce thermal resistance at interfaces, and enhance critical heat flux makes them promising candidates for applications such as refrigeration, desalination, and high-performance boiling systems in process plants.

10.1.3.1.6 Sustainability and Energy Efficiency

The deployment of nanofluids contributes to energy conservation and sustainability goals by improving process efficiency and reducing the energy input required for heating or cooling. Their use in renewable energy systems such as solar thermal collectors and geothermal heat exchangers further underscores their potential in supporting low-carbon technologies.

10.1.3.1.7 Applications in Process Intensification

The unique properties of nanofluids align strongly with the objectives of process intensification (PI)—namely, achieving higher energy efficiency, compact design, and enhanced mass and heat transfer. In microreactors, plate-fin heat exchangers, and absorption chillers, nanofluids serve as effective working media that contribute to the miniaturization and optimization of the equipment.

10.1.4 Motivation for Using Nanofluids in Process Intensification

The motivation for using nanofluids in process intensification is due to their better thermophysical properties as well as their capability to improve heat and mass transfer in different industrial processes. Nanofluids, suspensions of nanoparticles in base fluids (e.g., water or ethylene glycol), have much greater thermal conductivity and convective heat transfer coefficients than more common fluids. This facilitates the transportation of thermal energy in a more speedy and efficient manner, and the size of the equipment can be reduced, besides enhancing the capabilities of the heat exchangers as well as the control in a reactor or a unit of separation. In addition, nanofluids may enhance the mass transfer across multiphase systems via enlarged interfacial surface and micro-mixing due to nanoparticle presence. Their abilities to fit small and innovative devices such as microreactors, oscillatory baffled reactors, and membrane systems enhance their contribution in attaining the prime objectives of process intensification, which include energy efficiency, equipment miniaturization, and sustainable processes. The ability to tailor the nano-particle type, volume density, and surface modification of nanofluids also gives flexibility to tune their properties based on a specific application. They can be used under high heat flux and support waste heat recovery, besides helping to produce greener processing, which makes nanofluids an exciting and potential medium toward improving process intensification with traditional energy-based and renewable energy integrated industrial systems.

10.2 PRINCIPLES OF PROCESS INTENSIFICATION (PI)

10.2.1 Overview of Process Intensification Goals and the Tools and Techniques Employed

Process Intensification (PI) is a transformative approach that strives to dramatically enhance the efficiency, sustainability, and economic feasibility of chemical and process industries. In contrast to more conventional approaches to developing a process where one tends to scale up particular unit operations to form a larger process, usually with the concept of modifying a conventional established unit operation, PI aims at reconsidering and reengineering parts of the process, leading to a much superior performance. This covers broadly integrating several operations in one location, incorporation of new materials and designs, and utilizing the latest technologies to lower energy consumption, operational base, and environmental degradation. The reason behind PI is a strong motive within itself, and it lies in the necessity to provide more compact, safer, energy-efficient, and cost-effective production processes. This has been enhanced by the growing pressure to adhere to the strict environmental laws and the rise in energy and material storage costs, which has made the use of PI methodologies. The center part explains that the principal objectives of PI are energy efficiency, miniaturization of equipment, and throughput. These objectives are then discussed in great detail using many examples of tools and techniques that can assist in achieving these objectives.

10.2.2 Goals of Process Intensification

Process intensification is built upon achieving certain core objectives that reflect the evolving demands of modern industries. These goals are not isolated; rather, they are interlinked and collectively contribute to the broader transformation of the chemical and process sectors.

10.2.2.1 Energy Efficiency

One of the most compelling reasons for embracing PI is its potential to substantially reduce energy consumption. Conventional processes are often energy-intensive and involve significant thermal losses. PI addresses this by:

- Introducing integrated heat exchange mechanisms (e.g., heat-integrated distillation columns, recuperative reactors) to recycle and reuse energy within the system.
- Utilizing alternative energy sources, such as microwave, ultrasonic, or plasma energy, which enable faster and more selective reactions with lower energy input.

- Replacing energy-wasteful separation techniques (like distillation) with more efficient alternatives such as membrane separation or reactive distillation.

By minimizing unnecessary energy dissipation, PI not only reduces operational costs but also contributes to environmental sustainability by lowering greenhouse gas emissions.

10.2.2.2 Equipment Miniaturization

PI aims to condense large-scale operations into smaller, more efficient units without compromising on output or performance. This principle is achieved through:

- Employing microreactors or oscillatory baffled reactors that offer high surface-area-to-volume ratios, thereby enhancing heat and mass transfer rates.
- Utilizing modular process systems, which are compact, portable, and can be scaled out (rather than up) to meet production demands.
- Integrating multiple process steps—such as reaction, separation, and heat exchange—into multifunctional equipment to eliminate the need for separate units.

Miniaturized equipment results in reduced capital investment, smaller plant footprints, and improved safety due to lower inventories of hazardous materials.

10.2.2.3 Increased Throughput and Productivity

By improving reaction kinetics, enhancing transport phenomena, and reducing processing times, PI systems are capable of significantly increasing throughput. This is particularly evident in:

- Continuous flow processes that replace batch operations, allowing for uninterrupted production with improved control over product quality and consistency.
- High-intensity reactors that facilitate fast reactions under optimal conditions, often enabled by better mixing, temperature control, and catalyst performance.
- Process intensification via hybrid operations, such as reactive distillation, which combines reaction and separation in one unit, reducing cycle times and increasing yield.

Ultimately, PI enables industries to meet higher production demands without a proportional increase in equipment size or resource consumption.

10.2.3 Tools and Techniques in Process Intensification

To achieve the goals outlined earlier, process intensification relies on a wide array of innovative tools and techniques that span the fields of chemistry, materials science, thermodynamics, transport phenomena, and systems engineering. The following are some of the most influential PI strategies.

10.2.3.1 Novel Reactor Designs

Innovative reactor geometries and configurations lie at the heart of PI. These designs aim to enhance mixing, improve heat and mass transfer, and achieve better control over reaction kinetics. Examples include:

- **Microreactors**: These are characterized by channels with micrometer-scale dimensions, offering high heat and mass transfer coefficients. Suitable for fast, exothermic reactions and pharmaceutical synthesis.
- **Oscillatory baffled reactors (OBRs)**: These enhance mixing through oscillatory motion, making them ideal for slow or multiphase reactions.
- **Spinning disc reactors**: Thin liquid films formed on a rotating disc promote rapid heat and mass transfer, enabling high selectivity and yield.
- **Monolithic and structured reactors**: Use of honeycomb or foam structures with catalytic surfaces reduces pressure drop and increases contact efficiency.

10.2.3.2 Multifunctional Process Units

Multifunctional reactors aim to integrate multiple steps (reaction, separation, heat transfer) into a single unit. This reduces capital costs and improves process synergy. Key examples include:

- **Reactive distillation**: Simultaneous reaction and separation of products increase conversion and reduce equipment.
- **Membrane reactors**: Catalytic reaction is coupled with selective removal of products through membranes to drive equilibrium reactions forward.
- **Heat exchanger reactors**: Reactions occur in heat exchanger surfaces, providing precise temperature control, especially for highly exothermic reactions.

10.2.3.3 Enhanced Heat and Mass Transfer Devices

Efficient transport of heat and mass is crucial in intensified systems. Tools to achieve this include:

- **Compact heat exchangers**: Such as plate-fin, microchannel, or printed circuit heat exchangers, which offer high surface areas for rapid thermal exchange.
- **Static mixers**: Provide efficient mixing in continuous processes, eliminating the need for moving parts.
- **Microchannel devices**: Enable very fast transport processes with low pressure drops, used in cooling of electronics and high-intensity reactors.

10.2.3.4 Use of Alternative Energy Sources

PI leverages non-conventional energy forms to accelerate processes, reduce temperatures, and achieve unique product qualities. Examples include:

- **Microwave-assisted processing**: Provides volumetric heating and fast activation of polar molecules, useful in drying, pyrolysis, or synthesis.

- **Ultrasound-assisted extraction**: Enhances cell wall disruption and mass transfer during extraction.
- **Plasma reactors**: Generate reactive radicals for advanced oxidation or gas treatment processes.

10.2.3.5 Intensified Separation Techniques

Separations often account for a major share of energy use. PI offers alternative techniques such as:

- **Membrane technologies**: Including nanofiltration, pervaporation, and membrane distillation, which offer energy-efficient alternatives to distillation.
- **Hybrid separation**: Combining separation mechanisms, such as membrane-adsorption or crystallization-extraction, for better selectivity and energy savings.
- **Chromatographic and centrifugal separations**: Suitable for fine chemicals and pharmaceuticals, where precision and purity are critical.

10.2.3.6 Process Integration and Heat Recovery

Integration strategies help in minimizing utility requirements and waste generation. These include:

- **Pinch analysis**: A systematic method to identify the most efficient heat recovery network.
- **Process-to-process heat integration**: Using waste heat from one unit as energy input for another.
- **Combined heat and power (CHP) systems**: Simultaneous generation of electricity and usable heat increases overall efficiency.

10.2.3.7 Process Simulation and Optimization Tools

Modern simulation software supports the design and optimization of intensified systems:

- **Computational fluid dynamics (CFD)**: Used to analyze flow, mixing, and reaction phenomena in complex geometries.
- **Process simulation tools** (e.g., Aspen Plus, COMSOL): Help predict performance, perform sensitivity analysis, and test integration schemes.
- **Design of experiments (DOE)**: Facilitates optimization of process parameters with minimal experimental runs.

10.2.3.8 Materials and Catalysts for Intensification

Material innovations can significantly enhance the performance of PI systems:

- **Nanocatalysts**: Offer high surface areas and improved reactivity.
- **Phase change materials (PCMs)**: Employed in thermal management for energy storage and heat transfer.

- **Functionalized membranes**: Enable selective separations and coupling of reactions.

10.2.4 Process Intensification 4.0 (PI4.0): The Digital Evolution of Process Engineering

Process Intensification 4.0 (PI4.0) is a combination of the historical process intensification (PI) efforts and the potential of Industry 4.0 technologies (artificial intelligence (AI), machine learning (ML), Internet of Things (IoT), advanced sensors, cyber-physical systems, and digital twins) to radically transform process industries. This will transform design, control, optimization, and operation of chemical/process systems by uniting real-time data-driven intelligence with intensified processes.

10.2.4.1 What Is PI4.0?

PI4.0 builds upon the foundational goals of traditional PI—such as reducing equipment size, improving energy efficiency, maximizing throughput, and enhancing sustainability—but leverages *digital technologies* to achieve these objectives faster, smarter, and more flexibly. It introduces an adaptive, intelligent framework where process equipment, control systems, and operators are interconnected and communicate in real time to make informed decisions.

10.2.4.2 Core Components of PI4.0

1. **Smart Sensing and IoT**
   - Real-time monitoring of temperature, pressure, flow rate, and composition via advanced sensors.
   - IoT devices facilitate continuous data collection from all process units for predictive control and optimization.
2. **Artificial Intelligence and Machine Learning**
   - Algorithms trained on historical and real-time data can predict optimal operating conditions, detect anomalies, and recommend adjustments.
   - Enables closed-loop control with learning-based feedback mechanisms for process optimization.
3. **Digital Twins**
   - A digital replica of the physical process or equipment that can simulate performance under varying conditions.
   - Useful for scenario analysis, troubleshooting, and virtual experimentation without physical trial-and-error.
4. **Advanced Process Control (APC)**
   - Model Predictive Control (MPC), real-time optimization, and self-tuning controllers allow for dynamic response to changes in feed conditions or disturbances.
5. **Process Analytical Technology (PAT)**
   - Integrated spectroscopy, chromatography, and chemometrics enable real-time analysis of product quality and process parameters, ensuring consistency and regulatory compliance.

6. **Edge and Cloud Computing**
   - Enables storage, processing, and analysis of large datasets from multiple units, allowing seamless data integration across the plant.

10.2.4.3 Advantages of PI4.0 in Process Industries

- **Increased Process Agility and Responsiveness**
  Systems can autonomously adapt to changes in raw material quality, product demand, or environmental conditions.
- **Enhanced Energy and Resource Efficiency**
  Real-time analytics and control allow for minimized waste, reduced energy consumption, and optimal resource utilization.
- **Reduced Downtime and Maintenance**
  Predictive maintenance algorithms help forecast equipment failures and schedule maintenance without interrupting production.
- **Safety and Environmental Benefits**
  Continuous monitoring ensures operation within safe limits, and early fault detection can prevent accidents or emissions.
- **Accelerated Innovation and Customization**
  Rapid prototyping and simulation using digital twins enable faster development of novel intensified processes.

It introduces an adaptive, intelligent framework where process equipment, control systems, and operators are interconnected and communicate in real time to make informed decisions. A visual representation of the core dimensions and domains of PI4.0 is presented in Figure 10.3.

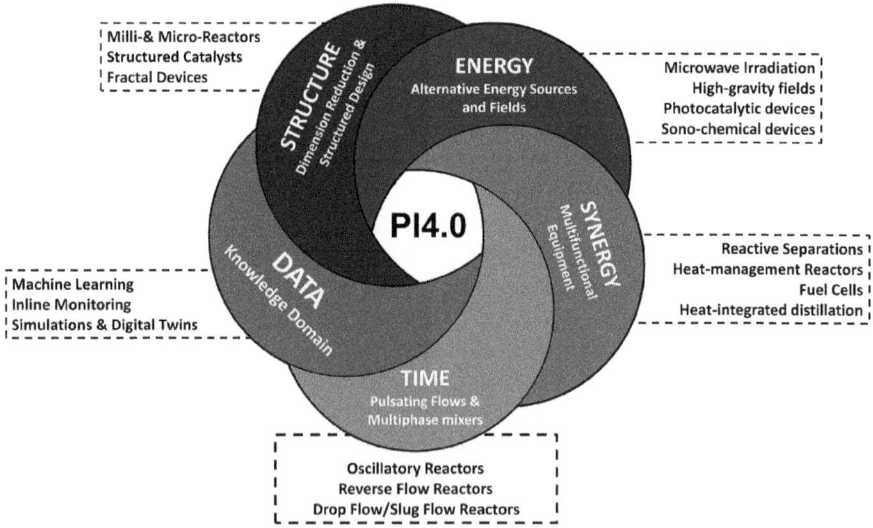

FIGURE 10.3 Key dimensions of Process Intensification 4.0 (PI4.0) [4].

TABLE 10.1
Integration of Process Intensification Techniques with Industry 4.0 Enhancements in Various Unit Operations

Application Area	PI Technique	Industry 4.0 Enhancement
Heat exchanger networks	Compact, high-efficiency heat exchangers	Real-time fouling detection using IoT-based smart sensors
Continuous flow reactors	Microreactors, oscillatory baffled reactors	Machine learning for optimizing residence time and kinetics
Separation units	Membrane distillation, reactive separations	Digital twin modeling for fault detection and feed variation
Energy systems	Heat recovery loops, thermosyphons	AI-enabled process scheduling and load forecasting

Rapid prototyping and simulation using digital twins enable faster development of novel intensified processes. Table 10.1 presents illustrative examples of PI4.0 applications, demonstrating how traditional PI strategies are enhanced using smart technologies.

10.2.4.4 Challenges and Future Perspectives

While PI4.0 offers transformative potential, several challenges must be addressed:

- **Integration complexity:** Combining legacy equipment with smart technologies requires robust interfacing.
- **Data security and privacy:** Ensuring cyber-physical system protection is vital.
- **Skill gap:** Workforce must be trained to handle AI, big data, and smart systems.
- **Cost:** High initial investment in sensors, infrastructure, and computing platforms is required.

Future prospects include fully autonomous plants capable of self-learning, self-healing, and self-optimizing functions—essentially realizing the vision of *"lights-out" manufacturing* in the chemical process industries.

10.3 INTEGRATION OF NANOFLUIDS IN PROCESS UNITS

Applying nanofluids in process units forms one of the crucial entries into contemporary chemical engineering and heat systems. Due to their high thermophysical properties and increased heat and mass transport potential, nanofluids are becoming a strategic working fluid in the process intensification context. They are adopted in line with the energy efficiency, compactness, and optimal performance in diverse industrial sectors. Substitution of nanofluids as a technology not only assists in traditional process design but also allows creative redesigning of unit operations, that is, heat exchangers, reactors, cooling towers, evaporators, condensers, and solar thermal systems.

10.3.1 Reactors

10.3.1.1 Nanofluid-Assisted Temperature Control

Nanofluids have demonstrated some potential toward enhancing efficiency and control capabilities of chemical reactors through the ability to achieve great heat and mass transfer. Proper control over the temperature, along with diffusion of designed heat, is essential in the use of reactors; more so in the case of highly sensitive or energy-intensive reactions. Due to the peculiarities of thermal behavior of nanofluids, namely very high thermal conductivity, enhanced convective heat transfer, and customized optical absorption, they can easily be incorporated into a variety of reactor technologies, like catalytic, photochemical, and biochemical reactors.

10.3.1.1.1 Nanofluid-Assisted Temperature Control in Reactors

Temperature control has been one of the most important aspects in chemical processing industries, which contribute to the rate, selectivity, and safety of chemical reactions. Regardless of whether a system is experiencing an endothermic or exothermic reaction, a stable and optimal temperature profile will allow exciting the desired reaction pathways with minimal risk of unwanted side reactions, thermal degradation of the materials, and catalyst losses. Traditional heat transfer fluids such as water, mineral oils, or ethylene glycol are generally limited with respect to their thermal conductivity, resulting in an inefficient heat transfer process and limitation of heat removal or supply within an intensive reaction. Conversely, nanofluids, which are designed by the incorporation of finer-length solid particles in a base fluid, reveal a greater level of thermal conductivity, convective heat transfer coefficients, and specific heat capacities. Thus, these nanofluids are a better option in the thermal management of the reactor.

Nanofluids are highly efficient in absorption, transportation, and dispersion of heat throughout the reactor volume and hence lessen the possibility of local hot spots, uneven heat distribution, and/or temperature changes. This uniform thermal pattern is particularly critical in delicate chemical systems where small changes in temperature can alter reaction equilibria, as well as impact product yield and, in some cases, induce runaway reactions. Nanofluids can directly produce a safer process, better products, and energy savings by promoting more accurate temperature control. One real-world nanofluid situation is in thermo control, where tubular and jacketed batch reactors are common in pharmaceutical, specialty chemical, and petrochemical industries. The heat transfer between the wall of the reactor and the fluid or process occurs in such type of systems in order to control the required reaction temperature. Upon the utilization of nanofluids in the outer jackets of such reactors, the high thermal conductivity will increase the rate of heat transfer. The implications are less time to stability during initial start-up and shutdown events, increased startup/shutdown rates, as well as a more adaptable control system to the challenges of variable heat loads during transient processes.

Similarly, in continuous stirred tank reactors (CSTRs), which are used extensively in bulk chemical production and in biochemical processing as well, heating control can be a very tricky problem when faced with extremely exothermic reactions. The abrupt release of heat in the reactions will require a heat removal system that will act in real-time to the varying thermal state. In this instance, nanofluids that work in internal cooling coils and heat exchangers are very effective. They have the capacity to soak up and discharge huge quantities of heat very fast, a factor that assists the reactor to experience stable internal temperature with absence of risks of overshoot and thermal instability. This gives the advantage of steady performance of the reactor at prolonged periods, which minimizes the reverse effects of catalyst fouling or oxidative damage of vulnerable reactants. Besides, thermophysical behavior of nanofluids may also be customized through the choice of the right type of nanoparticles and concentrations. This flexibility enables the engineers to maximize the heat transfer properties by correct reactor design, heat load, and fluid dynamics. As an example, a low viscosity base fluid like water or ethylene glycol can be used, and nanoparticles highly conductive at high temperature like titanium oxide can be added to the base fluid, bringing the performance of the nanofluid to maximum levels of convective heat transfer without necessitating large jumps in pressure drop or power demands of the pump.

The low thermal conductivity of conventional fluids can impose a limitation on responsiveness in systems in which rapid changes in temperature are required, typical examples being batch polymerization, oxidation, or nitration processes. The dynamic thermal response may be improved with nanofluids due to their high thermal diffusivity, so extreme heat control over the onset and end of a proposed reaction is feasible. This is especially helpful in multi-reaction series schemes or cascading reactors, where each stage could need different thermo regimes. Moreover, the use of nanofluids in temperature management of reactors also blends in the objectives of process intensification, that is, to produce more with little and more efficient machinery. Nanofluids can be used to enhance thermal management to facilitate higher throughputs, higher reaction yields per unit volume, and reduced reactor footprint, contributing to compact and energy-efficient designs that are consistent with sustainability in processing.

Case Study 1: High-Temperature Spray Cooling Using Cu-Zn-Al LDH Nanofluid for Steel Quenching

Chakraborty et al. (2018) performed an extensive experimental study of thermo-physical conduct of a Cu-Zn-Al-based layered double hydroxide (LDH) nanofluid and its usefulness in high-temperature cooling, mainly in the form of the spray cooling of hot steel plates. The synthesis of the nanofluid was accomplished with nanoparticles of the rod shape with an average particle size of around 49 nm, utilizing three metals with high thermal conductivity, copper, zinc, and aluminum. The nanofluid reached a notable increase in thermal conductivity (13.9%) at a trivial concentration of 160 ppm as compared to pure water. Besides, it had a maximum cooling rate of 158.4°C/s, which is

18.5% faster than water. Thus, it effectively improved the process of quenching metals, namely steel, to enhance their final mechanical character. The paper also emphasized that there are two components promoted by nano thermal transport and deposition of the nanoparticles on the target surface that serve as new nucleation sites and hasten heat conductivity. Nevertheless, the phenomenon was noted that when the optimal nanofluid concentration is surpassed, the cooling performance and the thermal conductivity diminish. This drop in suspension stability is attributed to the increase in suspension agglomeration or attachment and the formation of a nanoparticle layer, which prevents the cooling fluid from making contact with the metal surface. Such results support the necessity of having a suitable nanoparticle loading to counter the potential advantage of using thermal enhancement against the operational safety of a spray cooling system [5].

10.3.1.2 Use in Exothermic Reaction Systems

The introduction of nanofluids into chemical reactors processing exothermic or endothermic reactions also becomes the paradigm shift in developing thermal management and reliable performance. Such reactions can either release heat or absorb heat, and in order to maintain the optimal reaction kinetics, avoid safety issues, and material consistency, very careful thermal control is necessary.

Large amounts of intense heat are usually released during the exothermic reaction, nitration, hydrogenation, oxidation, or polymerization. Otherwise, this heat accumulation may result in localized temperature above ambient (hotspots), creating undesirable side reactions, thermo-degradation of substances, or catalyst sintering or even dangerous thermal runaway conditions. The traditional cooling mediums that are used, such as water, brine, or organic oils, have relatively lower heat capacities and thermal conductivities and impede the heat removal process, increasing the response time of a dynamic system.

These limitations were resolved with nanofluids, which is a nanotechnology-engineered liquid containing nanoparticles of thermally conductive materials dispersed in a base fluid and have greatly improved thermal properties. In practice, on utilization in reactor jackets, heat exchangers, or internal cooling coils, they have yielded significant improvements in uniformity of temperature and rates of heat removal. To illustrate, in fixed-bed catalytic reactors that carry out exothermic partial oxidation of hydrocarbons, nanofluids have been utilized to ensure uniform distribution in both radial and axial directions in temperature, thereby alleviating catalyst hotspotting, which enhances life of the catalyst.

Moreover, nanofluids could be implemented into microreactor systems, which are characterized by their large surface area-to-volume ratios and thus could be used to manage fast heat produced. Use of nanofluids in microchannels leads to ultra-fast cooling that is essential in time-sensitive work such as flash polymerization or large throughput synthesis in pharmaceutical industries. This is especially useful in continuous processing, where any thermal transient should be blocked out within a finite time or the process will be disrupted.

Case Study: Application of Nanofluids in Reactor Jackets Using SiO_2-Water Nanofluid

In many industrial-scale chemical reactors, *jacketed vessels* are used to regulate the temperature of the reaction medium. The thermal fluid circulating in the jacket must provide efficient heat exchange—either to remove the heat generated during exothermic reactions or to supply the necessary energy for endothermic processes. The use of nanofluids, as explored in the study by Azimi et al. (2016), presents a significant advancement in improving the thermal performance of such jacketed systems.

The authors investigated the effect of using a *SiO_2-water nanofluid* in a jacketed vessel through both *experimental measurements* and *computational fluid dynamics (CFD) simulations*. Their study aimed to understand the heat transfer characteristics and temperature uniformity in the reactor using different nanofluid concentrations and flow conditions [6].

10.3.1.2.1 Key Findings and Importance

a. **Enhanced Heat Transfer in Reactor Jackets**
The experimental results demonstrated that the inclusion of SiO_2 nanoparticles in water significantly enhanced the overall heat transfer coefficient in the jacketed vessel. Specifically, the convective heat transfer rate increased with nanoparticle volume concentration, with improvements up to 14.8% at a 0.2 vol% concentration of SiO_2. This indicates that even at low concentrations, nanofluids can outperform traditional heat transfer fluids in jacketed reactors.
b. **Improved Temperature Uniformity Inside the Vessel**
Uniform temperature distribution is critical in chemical reactors to avoid hotspots, ensure reaction selectivity, and maintain product quality. The study showed that nanofluids led to a more homogeneous temperature field within the reactor. This is particularly beneficial in batch and semi-batch processes where precise thermal control directly affects reaction kinetics.
c. **Faster Heating and Cooling Cycles**
Due to their superior thermal conductivity, nanofluids enabled faster heating and cooling responses in the jacketed vessel. This dynamic responsiveness is important for processes requiring quick thermal adjustments—such as those in pharmaceutical synthesis or polymerization—thus reducing batch cycle time and increasing productivity.
d. **Numerical Validation through CFD**
The authors validated their experimental findings using CFD simulations incorporating the two-phase mixture model. The simulated temperature contours and heat flux distributions were in good agreement with the experimental data, confirming that nanofluid behavior can be effectively modeled and predicted for design optimization. This enables process engineers to simulate jacket performance before physical implementation.

e. **Low Viscosity Penalty and Stability**
Importantly, the SiO_2 nanofluids used in the study showed minimal increase in viscosity compared to the base fluid, thereby avoiding significant pumping power penalties. The nanofluid also exhibited good stability, with no noticeable sedimentation over the experimental time, which is essential for industrial scalability.

10.3.1.2.2 Industrial Significance

The use of nanofluids like SiO_2-water in reactor jackets offers several advantages over conventional thermal fluids:

- **Energy efficiency**: Faster thermal response translates to reduced heating/cooling durations and energy savings.
- **Compact design**: Enhanced heat transfer allows for smaller jacket sizes or thinner walls, aligning with the goals of process intensification.
- **Process safety**: Improved thermal management helps mitigate the risk of thermal runaway in exothermic systems.
- **Scale-up potential**: The positive results at lab scale can be translated into pilot or industrial scales with appropriate design adaptation.

10.3.1.3 Nanofluids in Endothermic Reaction Systems

Conversely, endothermic reactions, for example, steam reforming of methane, biomass pyrolysis, ammonia synthesis (reverse Haber-Bosch), and thermal cracking of hydrocarbons all require a flow of energy to maintain the equilibrium of the reaction and force product formation. The conversion degree and the final yield are determining factors for the effectiveness of heat delivery in such systems. Poor thermal transfer of traditional heating fluids, possible in the high-temperature gradient conditions, may result in the cold regions or under-heated reaction areas, lowering the conversion efficiency.

In this scenario, nanofluids are better alternatives to heat delivery vectors, guaranteeing high rates of thermal penetration and stable heating of the mixture of reactants. Their use on packed bed reformers, as an example, increases the stability of their thermal envelope around catalyst particles, allowing a more uniform endothermic reaction and less carbon formation as a result of cooling holes. In nanofluids, optical nanofluids (such as plasmonic nanoparticles or carbon-based nanofluids) can improve solar thermochemical system performance by increasing solar energy absorption and thermal capacity, using concentrated solar energy to realize high-temperature reactions. This duality increases the efficiency of transferring heat and is also useful in terms of integrating sustainable and renewable power.

Additionally, in fluidized bed reactors (e.g., biomass gasification or thermal decomposition), the internal coils or plates of a heat exchanger are cooled by nanofluid to facilitate quick temperature ramp-up and hence reduce start-up time and enhance fuel-to-gas conversion efficiency. The increased properties of convective heat transfer of nanofluids give the chance of better thermal matching of the reaction zone and the heat source.

A promising advancement in the field is the utilization of magnetically responsive nanofluids (ferrofluids), which have the capability of localized induction heating, with the application of alternating magnetic fields. This enables spatial control of heating in reactors, which is especially useful when there is a need in the supply of different heats to various parts of the reactor in such systems where there are batch or plug flow reactors. This degree of control can be revolutionary in endothermic reactions where the reaction intermediates or products are heat sensitive.

Case Study: Enhancing Solar Thermal Collector Performance Using Ionic Liquid–Graphene Oxide Nanofluids [7]

BACKGROUND AND MOTIVATION

Solar energy systems are increasingly used to combat climate change and reduce reliance on fossil fuels. However, traditional fluids like water and ethylene glycol used in solar collectors suffer from limitations such as high vapor pressure and low thermal stability. Recent advancements have highlighted the potential of nanofluids—fluids engineered by dispersing nanoparticles in base fluids—to overcome these challenges.

In this context, the study investigates the use of *graphene oxide (GO)* dispersed in various combinations of *ionic liquids (IL)* and *water* to form *ionic nanofluids (INF)* for flat plate solar collectors. The goal is to enhance thermal efficiency under the influence of nonlinear thermal radiation and convection while considering velocity slip and temperature jump at boundaries.

OBJECTIVE

To evaluate and compare the thermal performance and heat transfer characteristics of four nanofluids:

- Pure IL/GO
- IL(75%)-H_2O(25%)/GO
- IL(50%)-H_2O(50%)/GO
- H_2O/GO

These are assessed under various heat transfer mechanisms including the following:

- Nonlinear thermal radiation
- Nonlinear free convection
- Heat absorption effects
- Slip and jump boundary conditions

METHODOLOGY

- **Model Development:** A 2D mathematical model of nanofluid flow was developed in the XY-plane representing a solar panel surface

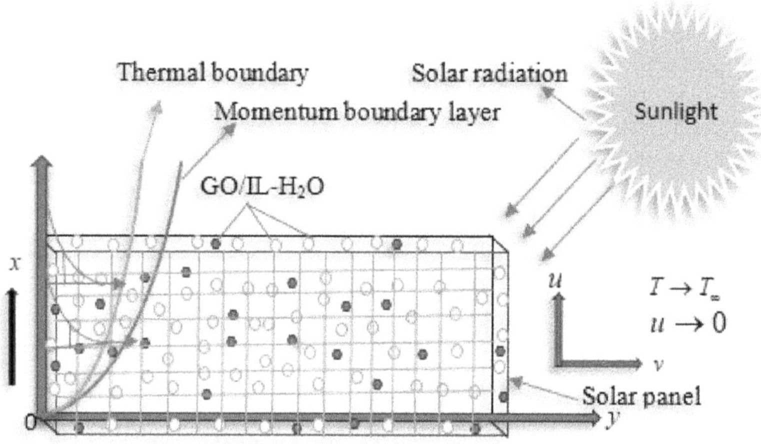

FIGURE 10.4 Physical and geometrical configuration of nanofluid flow in a solar thermal collector [7].

(**Figure 10.4**). The model included the effects of solar radiation, heat absorption, velocity slip, and temperature jump.

- **Solution Technique:** The *Galerkin Weighted Residual Method (GWRM)* was used to solve the nonlinear partial differential equations. The thermophysical properties were sourced from experimental data at 20°C.
- **Experimental Inputs:** Parameters like density, specific heat capacity, thermal conductivity, and viscosity for each nanofluid were measured experimentally (see Table 3 of the paper).
- **Performance Metrics:** Skin friction coefficient and Nusselt number were evaluated to compare the thermal transport behavior.

KEY FINDINGS

- **Thermal efficiency**
  - IL/GO nanofluid showed the highest thermal energy efficiency, outperforming H_2O/GO by 72.8%, IL(50%)/GO by 36.8%, and IL(75%)/GO by 9.001%.
  - The results matched experimental data, validating the model.
- **Effect of free convection** (λ_1)
  - Increased λ_1 led to enhanced buoyancy-driven convection, optimizing heat removal and improving thermal stability.
- **Impact of heat absorption (Q_0)**
  - Higher Q_0 improved uniform heat distribution, reducing thermal stress and hotspots in the collector.
- **Role of radiation parameter (Nr)**
  - Higher Nr values enhanced nanofluid heat absorption from solar radiation, though at extreme values, radiative losses increased, affecting Nusselt number.

- **Biot number** (βt)
  - An increased Biot number promoted more effective boundary heat transfer, enhancing the bulk fluid temperature rise.
- **Velocity slip and temperature jump**
  - Moderate slip and jump conditions were found to optimize flow resistance and heat transfer at the nanofluid–panel interface.
- **Streamline and isotherm behavior**
  - IL/GO displayed dense, uniform streamline and temperature contours, indicating stable and effective heat transfer performance compared to other fluids.

APPLICATIONS

- High-intensity solar systems: IL/GO is ideal due to its superior thermal conductivity and stability.
- Moderate conditions: H_2O/GO or IL(50%)/GO mixtures are preferable to avoid overheating.

PRACTICAL CONSIDERATIONS AND DESIGN IMPLICATIONS

Though some theoretical benefits of nanofluid in exotherms as well as endotherms are well confirmed, some design and operational considerations are essential to be made in practice. Nanofluids should be optimized to overcome the problems of clogging in narrow reactor channels or fouling of the heat transfer surfaces through viscosity, stability, and flow behavior. Furthermore, the type, size, and surface modification of the nanoparticle and compatibility with the type of base fluids used have to agree with the chemistry of the reactor to avoid contamination or reduction.

Design-wise, nanofluids may be used in *secondary loops* of thermal control systems, as a transporter of heat between the reactor and an adjacent heat exchanger. The strategy ensures that the risk of nanoparticle contamination of the process stream is minimized, yet the potential increase in the level of heat transfer is capitalized.

Also, in order to project reactor performance during transients, *dynamic simulation models* are also designed in which the thermophysical characteristics of nanofluids are integrated. These models prove useful in the optimization of the size of heat exchangers, flow rate, and temperature setpoints that enabling process engineers to achieve smaller, quicker response thermal management systems.

10.3.1.4 Nanofluid in Catalytic and Photochemical Reactors

Nanofluids have emerged as promising multifunctional media in chemical reactors, particularly catalytic and photochemical systems, where both thermal and reaction kinetics play crucial roles in determining efficiency. Their ability to enhance heat and mass transfer, along with the functional properties of the nanoparticles themselves, offers considerable advantages in improving reaction rates, selectivity, and process intensification.

10.3.1.4.1 Nanofluids in Catalytic Reactors

Moreover, in heterogeneous catalytic applications, particularly slurry reactors, nanofluids are a combination of nanoparticles of metal oxides with an underlying fluid chosen as the base fluid (usually water, alcohols, or ionic liquids), which fulfills two functions to increase thermal conductivity and act as active or co-catalysts. The nanoparticles enhance the heat transfer behavior by maximizing the areas and paths of heat in the reactor to reduce the temperature gradients and hotspots during an exothermic or endothermic reaction.

More so, in cases where the nanoparticles are catalytically active (i.e., TiO_2, CeO_2, CuO, or Fe_3O_4), then they can also take part directly in the chemical reaction. It is especially valuable when used in Fischer–Tropsch synthesis, methanol synthesis, hydrogenation, and oxidation reactions, where there would be a more effective dispersion of catalysts, and provision of more active sites and decreased intraparticle diffusion restrictions, due to uniform suspension of catalyst particles in the nanofluid form during reaction.

For example, in Fischer–Tropsch reactors, in which long-chain hydrocarbons are produced on a catalyst using syngas (CO + H_2) new products, uniform temperature profile, and increased mass transfer caused by nanofluids result in lower hot spots, sintering of the catalyst particles, and improved selectivity of the products and conversion levels. In a similar way, nanofluids support effective and quick heat removal, reduce concentration polarization, and enhance the rate of reaction due to ease of catalyst access in slurry-phase hydrogenation of organic compounds.

Case Study: High-Efficiency Electrochemical Ammonia Removal Using Pt/NbC Nanofluid Membrane Reactor (Ma et al. 2022)

BACKGROUND

Wastewater containing ammonia pollutants poses serious threats to the health of human beings and the ecosystems. The conventional technologies of ammonia removal, such as biological treatment and chemical oxidation, are characterized by slow retention time, low rate of conversion, or formerly unacceptable byproducts, helping to form nitrates. Electrochemical oxidation has come out as a feasible alternative with the potential of breaking down pollutants without the introduction of external chemicals. Nevertheless, it has been a challenge to achieve high rates of conversion and nitrogen selectivity at the same time [8].

INNOVATION IN REACTOR DESIGN

To overcome these limitations, researchers developed a novel electrochemical reactor that incorporates *platinum (Pt)* as a catalytic layer on a *niobium carbide (NbC) nanoporous membrane.* This design leverages *nanofluidic behavior* within nanochannels to significantly enhance mass transport and catalytic efficiency.

Key features of the reactor include:

- *Controllable pore sizes* between *200 nm and 8.3 nm*, precisely engineered using a *dip-coating technique.*

- *Electrochemical oxidation occurring within nanochannels*, significantly reducing the diffusion distance for reactants.
- *Integration of Pt catalyst on NbC membrane*, allowing for high catalytic activity with minimal loss of surface area.

EXPERIMENTAL SETUP AND PERFORMANCE

The reactor was tested for ammonia removal under varied operating conditions, including pore size and applied potential. Notable results include:

- *Complete removal (100%) of ammonium (NH_4^+)* was achieved at an applied potential of *4 V vs. SCE* using membranes with the smallest pore size (8.3 nm).
- The process exhibited a *nitrogen gas selectivity of 88.09%*, with *nitrate formation limited to 11.91%*, indicating controlled oxidation without excess nitrification.
- Remarkably, these results were obtained under a *short hydraulic retention time (HRT) of 249 seconds*, demonstrating the feasibility of rapid treatment.
- Further tuning of the operating voltage and membrane pore size enabled *selectivity toward nitrogen to reach 100%*, though this came at the cost of reduced ammonia conversion.

MECHANISTIC INSIGHTS

Using *Density Functional Theory (DFT)* simulations, the team proposed a mechanistic pathway for the oxidation process on the Pt (111) surface:

$$NH_4^+ \rightarrow NH_3 \rightarrow NH_2 \rightarrow N \rightarrow N_2$$

This stepwise dehydrogenation and recombination pathway elucidates how nitrogen gas is selectively formed, offering valuable insights into catalyst design and surface chemistry for electrochemical wastewater treatment.

IMPLICATIONS AND APPLICATIONS

This membrane-based nanofluid reactor system offers several advantages:

- *Efficient ammonia removal* without the need for added chemical oxidants.
- *Rapid treatment cycles*, making it suitable for decentralized or compact treatment systems.
- *High nitrogen selectivity*, mitigating the formation of nitrates, which can pose secondary environmental hazards.
- *Scalability* through the modular nature of membrane fabrication and reactor assembly.

The technology holds promise for deployment in *municipal wastewater treatment*, *industrial effluent polishing*, and *aquaculture wastewater remediation*, where ammonia management is critical.

NANOFLUIDS IN PHOTOCHEMICAL REACTORS

Nanofluids are particularly effective in *photochemical and photocatalytic reactors*, where light absorption, heat generation, and mass transfer must be simultaneously managed. In these systems, nanofluids offer *optical*, *photocatalytic*, and *thermal advantages* that make them ideal for solar-driven or UV-assisted chemical processes.

Metal oxide nanoparticles such as *titanium dioxide (TiO_2)*, *zinc oxide (ZnO)*, and *iron oxide (Fe_2O_3)* exhibit strong photocatalytic activity under UV or visible light. When suspended in a suitable base fluid, these nanofluids can:

- Enhance *light absorption* by scattering and trapping incident photons.
- Generate *reactive oxygen species* (e.g., hydroxyl radicals, superoxide ions) through photoactivation.
- Facilitate *heat removal* from the illuminated zone, preventing thermal degradation of reactants or deactivation of photocatalysts.

A typical application is the *photocatalytic degradation of organic pollutants* in water treatment. TiO_2 nanofluids, when irradiated with UV light, initiate oxidation reactions that break down complex organic molecules. The enhanced heat transfer maintains system stability and allows operation at optimal conditions for catalyst activity.

In *solar photoreactors*, nanofluids are also used in *hydrogen generation via water splitting*. Here, the nanofluid absorbs solar energy efficiently, initiates the photochemical reaction, and rapidly transports heat away from the reactor interface. The simultaneous improvement in *solar absorption* and *thermal regulation* leads to a significantly higher *solar-to-chemical energy conversion efficiency*.

DUAL FUNCTIONAL ROLE AND PROCESS INTENSIFICATION

The unique advantage of nanofluids in these systems lies in their *dual functional capability*—acting both as *reaction media* and *catalytic entities*. This leads to intensified processes where:

- The *reactor volume* can be reduced due to faster kinetics.
- *Heat exchangers* may become obsolete due to intrinsic thermal management.
- *Catalyst recovery* can be simplified using magnetically active nanofluids (e.g., Fe_3O_4).
- Reaction *selectivity and yield* are improved by mitigating diffusion and thermal limitations.

Moreover, the flexibility in tuning nanofluid composition allows tailoring their physicochemical properties (e.g., viscosity, thermal conductivity, optical absorption) to match the requirements of specific catalytic or photochemical reactions.

10.4 FUTURE PROSPECTS

The field of nanofluids has reached a critical juncture where foundational research must evolve into scalable, application-driven innovations. As process industries increasingly adopt energy-efficient and sustainable technologies, nanofluids are expected to play a pivotal role in shaping next-generation process intensification strategies [9]. The following future directions outline key areas of advancement.

10.4.1 Development of Sustainable and Biodegradable Nanofluids

Conventional nanofluids often involve synthetic base fluids (e.g., ethylene glycol) and chemically synthesized nanoparticles (e.g., Al_2O_3, CuO), which may pose risks to human health and ecosystems. There is a growing research push toward *green nanofluids*, which utilize:

- *Biodegradable base fluids* such as vegetable oils, water–glycerol mixtures, or ionic liquids derived from natural compounds.
- *Eco-friendly nanoparticles* synthesized from plant extracts, biowastes, or naturally occurring minerals like clays and carbon-based materials (e.g., graphene from biomass sources).

Sustainable nanofluids would not only reduce environmental toxicity but also comply with regulations on wastewater discharge, occupational safety, and life cycle impact. The transition toward green synthesis and renewable materials aligns with the broader principles of circular economy and responsible manufacturing [10–13].

10.4.2 Role in Achieving Net-Zero Emission Targets

To meet global climate objectives such as those defined by the *Paris Agreement* or national *net-zero emission roadmaps*, industries must drastically reduce their carbon footprint. Nanofluids can contribute to this through:

- *Enhanced energy efficiency* in thermal processes such as distillation, condensation, and evaporation.
- *Lower fuel consumption* due to faster heat transfer and reduced thermal resistance in heat exchangers.
- *Reduced reliance on auxiliary cooling systems*, especially in exothermic chemical reactors.

By improving heat recovery and minimizing process losses, nanofluids can contribute significantly to *Scope 1 and Scope 2 emission reductions* in energy-intensive sectors such as petrochemicals, pharmaceuticals, and metal processing.

10.4.3 Integration with Renewable Energy Systems

The high thermal conductivity and photothermal properties of certain nanofluids make them ideal candidates for use in *solar thermal collectors, geothermal heat pumps, and hybrid PV-T systems.* Key applications include:

- **Solar-driven water heaters and desalination systems**: Nanofluids with materials like TiO_2, ZnO, or carbon nanotubes can increase solar absorption, enhance thermal gradient development, and improve phase change behavior in solar stills.
- **Thermal Energy Storage (TES):** Nano-enhanced phase change materials (NEPCMs) offer higher heat storage capacity, faster charging/discharging cycles, and stability in latent heat applications.
- **Geothermal energy loops:** By improving the convective heat transfer inside subsurface loops, nanofluids can increase geothermal system efficiency even in low-temperature reservoirs.

Such integration supports *clean energy transition*, reduces dependency on fossil-based utilities, and strengthens the decarbonization of industrial energy systems.

10.4.4 Smart Nanofluids with Tunable and Responsive Properties

Next-generation nanofluids are being engineered to exhibit *stimuli-responsive behaviors*, enabling *adaptive process control* and *real-time optimization*. These *smart nanofluids* can change their properties in response to:

- **Temperature**: Thermo-responsive fluids that modify viscosity or conductivity at specific thresholds, useful in automatic thermal regulation.
- **Magnetic or electric fields**: Magneto-rheological or electro-rheological nanofluids for precision control in fluidic systems and separation units.
- **pH and chemical environment**: pH-sensitive nanofluids for drug delivery, reactive separation, or targeted catalysis.

These functionalities will be integral in developing *self-optimizing reactors*, *heat exchangers with dynamic flow control*, and *digital twin-enabled process loops*. In conjunction with AI and machine learning, smart nanofluids will form the backbone of *autonomous process intensification systems*.

10.4.5 Scalability, Economic Viability, and Regulatory Standards

Although laboratory studies show promising results, real-world applications of nanofluids are limited by challenges in:

- *Long-term stability and dispersion* in turbulent and high-shear environments.
- *Economic feasibility* related to nanoparticle synthesis, recycling, and maintenance of fluid performance.
- *Standardization and safety regulations*, especially in food, pharmaceutical, and water treatment sectors.

Future research must focus on *techno-economic analysis*, *life cycle assessment (LCA)*, and *regulatory compliance* to transition nanofluids from laboratory prototypes to industrial mainstream.

10.5 SUMMARY

Nanofluids have proven to be game-changers in the drive toward process intensification, offering superior thermal and mass transport properties that enable higher performance across a wide range of applications. This chapter has highlighted how nanofluids improve energy efficiency, reduce equipment footprint, and support greener operations. Their use in catalytic, photochemical, and renewable energy-integrated systems underscores their versatility and potential in revolutionizing industrial operations. As challenges related to stability, cost, and environmental impact are addressed, the future of nanofluids appears promising—especially in the context of smart manufacturing and sustainable process engineering. The transition from conventional to intensified processes will be significantly shaped by the evolution and application of next-generation nanofluids.

REFERENCES

1. Stankiewicz, A., & Moulijn, J. A. (2000). Process intensification: Transforming chemical engineering. *Chemical Engineering Progress*, *96*, 22–34.
2. Haase, S., Tolvanen, P., & Russo, V. (2022). Process intensification in chemical reaction engineering. *Processes*, *10*(1), 99. https://doi.org/10.3390/pr10010099.
3. Babar, H., Wu, H., Zhang, W., Shah, T. R., McCluskey, D., & Zhou, C. (2024). The promise of nanofluids: A bibliometric journey through advanced heat transfer fluids in heat exchanger tubes. *Advances in Colloid and Interface Science*, *325*, 103112.
4. López-Guajardo, E. A., Delgado-Licona, F., Álvarez, A. J., Nigam, K. D. P., Montesinos-Castellanos, A., & Morales-Menendez, R. (2022). Process intensification 4.0: A new approach for attaining new, sustainable and circular processes enabled by machine learning. *Chemical Engineering and Processing-Process Intensification*, *180*, 108671.
5. Chakraborty, S., Sarkar, I., Ashok, A., Sengupta, I., Pal, S. K., & Chakraborty, S. (2018). Thermo-physical properties of Cu-Zn-Al LDH nanofluid and its application in spray cooling. *Applied Thermal Engineering*, *141*, 339–351.
6. Hafezisefat, P., Esfahany, M. N., & Jafari, M. (2017). An experimental and numerical study of heat transfer in jacketed vessels by SiO2 nanofluid. *Heat and Mass Transfer*, *53*(7), 2395–2405.
7. Yusuf, A., Bhatti, M. M., & Ellahi, R. (2025). Study of ionic water/graphene nanofluids in solar panels under the effects of thermal radiation and slip conditions using experimental data. *International Communications in Heat and Mass Transfer*, *164*, 108845.
8. Ma, J., Wei, W., Qin, G., Jin, S., Liu, S., & Jiang, L. (2021). Electrochemical oxidation of ammonia in water by Pt/NbC membrane-based catalytic nanofluid reactor. *Chemical Engineering Journal*, *486*, 150240.
9. Malika, M., Bhad, R., & Sonawane, S. S. (2021). ANSYS simulation study of a low volume fraction CuO–ZnO/water hybrid nanofluid in a shell and tube heat exchanger. *Journal of the Indian Chemical Society*, *98*(11), 100200.
10. Thakur, P., Sonawane, S., Potoroko, I., & Sonawane, S. H. (2021). Recent advances in ultrasound-assisted synthesis of nano-emulsions and their industrial applications. *Current Pharmaceutical Biotechnology*, *22*(13), 1748–1758.
11. Sonawane, A. V., Rikame, S., Sonawane, S. H., Gaikwad, M., Bhanvase, B., Sonawane, S. S., . . . & Gaikwad, R. (2024). A review of microbial fuel cell and its diversification in the development of green energy technology. *Chemosphere*, *350*, 141127.

12. Malika, M., & Sonawane, S. S. (2021). The sono-photocatalytic performance of a novel water based Ti+ 4 coated Al (OH) 3-MWCNT's hybrid nanofluid for dye fragmentation. *International Journal of Chemical Reactor Engineering*, *19*(9), 901–912.
13. Parate, V. R., Kawadkar, D. K., & Sonawane, S. S. (2011). Study of whey protein concentrate fortification in cookies variety biscuits. *International Journal of Food Engineering*, *7*(2), 1–12.

11 Nanofluids for Enhanced Solar Collector Efficiency

11.1 INTRODUCTION: OVERVIEW OF SOLAR ENERGY AND ITS IMPORTANCE IN SUSTAINABLE ENERGY SYSTEMS

One of the renewable energy sources that has the biggest supply and is the cleanest is solar energy, which is harvested by taking advantage of the radiant light and heat of the sun. Sunlight provides the earth with about 173,000 terawatts (TW) of solar energy—which is vastly more than the sum total of required energy within the entire globe—providing solar energy with the potential as an energy source that is virtually infinite in terms of use as power, heating, or industrial processes. Solar energy is quickly becoming one of the important solutions to the shift to sustainable energy systems toward low-carbon energy systems because global energy demands keep on surging despite dropping fossil fuel reserves, environmental degradation, and raging issues of climate change. Among the most important advantages of solar power is that it has a very small footprint on the environment. Solar energy systems, unlike traditional fossil fuels, do not release any greenhouse battery when running. Besides, solar technologies can decentralize energy supply, limit the exploitation of limited resources, and make energy available in remote and underdeveloped areas. There are several solar technologies, but solar thermal collectors are important in the utilization of the heat of the sun in the production of domestic hot water, industrial process heat, space heating, and the generation of power by means of solar thermal power plants.

Solar collectors use the incident solar radiation to generate thermal energy that can be used to heat an appropriate fluid that can be tapped to be utilized or stored (Figure 11.1). The efficiency of these systems is highly governed by the efficiencies of convection interactions of heat between the exposed absorber surface and the heat transfer medium. Conventional fluids such as water, ethylene glycol, and oils have been used as heat transfer media for decades. But their low thermal conductivities hamper the overall effectiveness and thermal response of solar collectors with fully powered/transient solar events. As the world quickly demands optimization of the efficiency of solar thermal systems to make them more applicable and adoptable in the application of energy infrastructures and their enhancement in the context of sustainability, there has arisen a heated research concern regarding the heat transfer fluids. Here, nanofluids have brought the answer to the first type of conventional or traditional heat transfer fluids that might not only help in breaking the performance boundaries but successfully establish a paradigm shift in solar collector technologies.

DOI: 10.1201/9781003628071-11

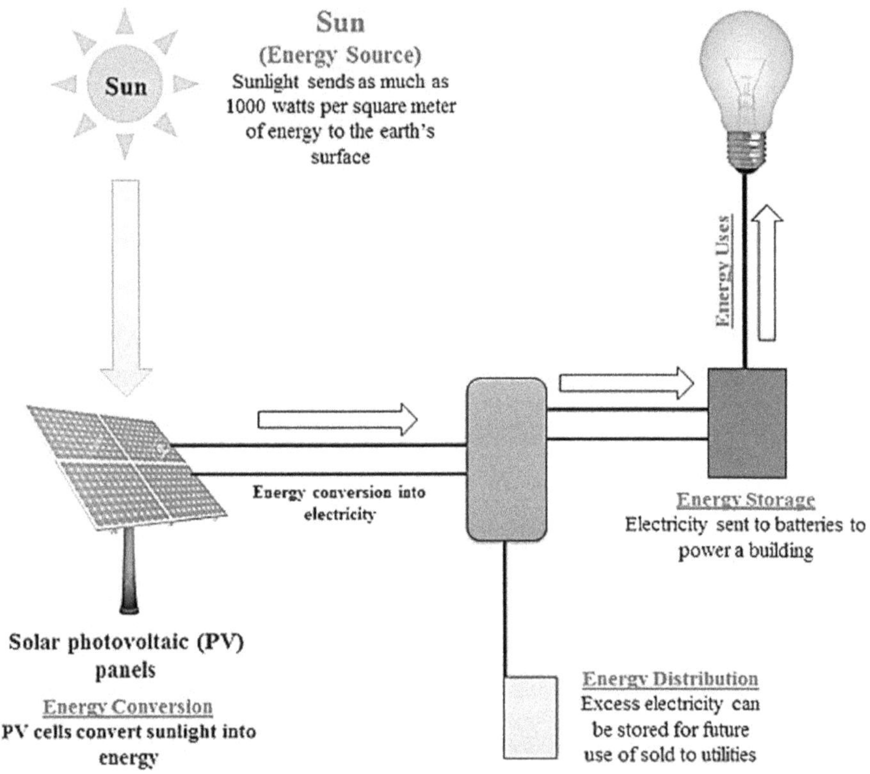

FIGURE 11.1 Schematic illustration of the conversion of solar radiation into electrical energy using a photovoltaic (PV) solar panel for various applications [1].

11.1.1 Limitations of Conventional Heat Transfer Fluids in Solar Collectors

Despite the widespread use of conventional fluids in solar thermal applications, several inherent limitations constrain their performance in modern energy systems. These limitations are particularly critical in the context of increasing energy demands, fluctuating solar availability, and the push for higher efficiency systems.

11.1.1.1 Low Thermal Conductivity

One of the primary shortcomings of traditional heat transfer fluids is their inherently low thermal conductivity. For instance, water has a thermal conductivity of approximately 0.6 W/m·K, while ethylene glycol exhibits even lower values (around 0.25 W/m·K). Such limitations reduce the rate at which thermal energy can be absorbed, transferred, and stored within the collector system, leading to larger collector sizes, longer heating times, and greater heat losses.

11.1.1.2 Limited Specific Heat and Viscosity Trade-Offs

While water boasts a relatively high specific heat capacity, other fluids such as thermal oils offer advantages in high-temperature stability but suffer from low heat capacity and higher viscosity. This results in a compromise between thermal capacity and pumping power requirements. The increased viscosity of oils and antifreeze mixtures like ethylene glycol further exacerbates energy consumption due to higher frictional losses and mechanical wear.

11.1.1.3 Thermal Stability and Degradation

At elevated temperatures, especially in concentrating solar collector systems, conventional fluids can degrade or undergo chemical changes, leading to fouling, corrosion, and reduced system reliability. Oxidative degradation and thermal cracking are common issues faced with hydrocarbon-based fluids in high-temperature regimes.

11.1.1.4 Inefficient Solar Absorption and Heat Transfer Coupling

Conventional fluids are typically transparent or weakly absorbing in the solar spectrum. As a result, solar energy must first be absorbed by a solid surface (e.g., a metal absorber plate) and then transferred to the fluid via conduction and convection. This multi-stage heat transfer process introduces thermal resistance and potential losses, thereby reducing overall system efficiency.

11.1.1.5 Environmental and Safety Concerns

Certain heat transfer fluids, such as synthetic oils and glycol-based mixtures, pose environmental and safety hazards in case of leaks or improper disposal. Their toxicity and poor biodegradability raise concerns in large-scale solar installations and environmentally sensitive areas.

These limitations collectively underline the necessity for alternative heat transfer media that can simultaneously offer high thermal conductivity, thermal stability, low viscosity, and enhanced solar absorption capabilities—criteria that nanofluids are uniquely positioned to fulfill.

11.2 INTRODUCTION TO NANOFLUIDS AND THEIR ROLE IN IMPROVING SOLAR THERMAL SYSTEM PERFORMANCE

Nanofluids are a new class of heat transfer fluids engineered by dispersing nano-sized (1–100 nm) particles into conventional base fluids. The nanoparticles used can be metals (e.g., copper, silver), metal oxides (e.g., Al_2O_3, CuO, TiO_2), carbon-based materials (e.g., graphene, carbon nanotubes), or hybrid composites. Due to their high surface area and unique thermophysical properties, these nanoparticles significantly enhance the heat transfer capabilities of the base fluid even at low concentrations.

11.2.1 Enhanced Thermal Conductivity

The most celebrated advantage of nanofluids is their markedly increased thermal conductivity compared to base fluids. Even with nanoparticle volume fractions as low as 0.1–1%, the thermal conductivity of the suspension can improve by 10–30% or more. Mechanisms responsible for this enhancement include Brownian motion, nanoparticle aggregation, liquid layering at the solid–liquid interface, and phonon transport.

11.2.2 Improved Convective Heat Transfer Coefficients

Numerous experimental and numerical studies have demonstrated that nanofluids exhibit higher convective heat transfer coefficients in both laminar and turbulent flow regimes. This improvement is attributed not only to enhanced thermal conductivity but also to increased thermal dispersion and modification of boundary layer dynamics caused by suspended nanoparticles.

11.2.3 Solar Spectral Absorption Capability

Certain nanoparticles—especially carbon-based nanomaterials and metal oxides—exhibit strong absorption in the solar spectrum. This feature enables the development of *direct absorption solar collectors (DASCs)*, where nanofluids absorb solar radiation directly rather than relying solely on an absorber plate. This eliminates intermediate heat transfer steps and significantly reduces thermal resistance and losses.

11.2.4 Tunable Thermophysical Properties

By selecting appropriate nanoparticle materials, sizes, shapes, and concentrations, researchers can tailor nanofluids to specific operating conditions and applications. Recent advances in hybrid nanofluids and functionalized nanoparticles have enabled further enhancements in stability, performance, and multifunctionality (e.g., combining heat transfer with anti-fouling or photothermal effects).

11.2.5 Real-World Impact on Solar Collector Efficiency

Studies have shown that integrating nanofluids into solar collector systems can lead to *efficiency gains ranging from 10% to 40%*, depending on the system configuration, nanoparticle type, and operating parameters. These improvements are especially pronounced under dynamic and high-flux solar conditions, making nanofluids ideal candidates for next-generation solar thermal systems [2].

11.3 OBJECTIVE AND STRUCTURE OF THE CHAPTER

This chapter aims to provide a comprehensive analysis of how nanofluids can significantly enhance the thermal efficiency and performance of solar collectors. By combining recent advances in nanotechnology, thermal engineering, and solar energy applications, the chapter seeks to:

- Explain the fundamental mechanisms through which nanofluids improve heat transfer in solar collectors.
- Evaluate different types of nanofluids and their thermophysical characteristics relevant to solar thermal systems.
- Analyze experimental, theoretical, and simulation studies on nanofluid-based solar collector performance.
- Discuss the integration of nanofluids into various collector designs, including flat plate, evacuated tube, and concentrating systems.
- Address the challenges associated with nanofluid use, such as stability, cost, environmental impact, and scalability.
- Explore future trends, including smart nanofluids, hybrid systems, and AI-optimized thermal systems using nanofluids.

11.4 FUNDAMENTALS OF SOLAR COLLECTORS

Solar collectors are devices that absorb solar radiation and convert it into thermal energy, which is then transferred to a working fluid (usually water, air, or oil) for various applications such as domestic water heating, industrial process heat, or power generation. The design, operating principle, and performance of these collectors vary depending on their intended use and environmental conditions. A fundamental understanding of different types of collectors, their working mechanisms, and efficiency metrics is essential before integrating nanofluids for performance enhancement.

11.4.1 Working Principle of a Solar Collector

A *solar collector* is a device that captures solar energy and converts it into *thermal energy (heat)*, which is then transferred to a working fluid (usually air or water). This thermal energy can be used for a variety of applications, including water heating, space heating, industrial process heat, or even electricity generation in solar thermal power plants (Table 11.1).

11.4.2 Basic Working Principle

i. **Solar Radiation Absorption**
    - Solar radiation, primarily in the *short-wave spectrum (UV and visible light)*, strikes the *absorber surface* of the solar collector.
    - This absorber is coated with a *selective surface* that has *high absorptivity* (absorbs most of the incoming solar radiation) and *low emissivity* (minimizes heat re-radiation losses).
    - The absorber converts the solar radiation into *heat energy*.

TABLE 11.1
Key Components of Solar Collector

Component	Function
Glazing	Allows solar radiation in, prevents heat loss
Absorber plate	Captures and converts solar radiation to heat
Heat transfer fluid (HTF)	Transfers the absorbed heat to storage/use
Insulation	Reduces heat loss from the collector
Manifold/Tubes	Pathways for the fluid to flow through and absorb heat

ii. **Heat Transfer to Fluid**
   - This heat is transferred to a *heat transfer fluid (HTF)*—commonly water, air, or a nanofluid—flowing through tubes or channels in contact with the absorber.
   - The heated fluid carries the thermal energy away for storage or direct use.

iii. **Minimizing Thermal Losses**
   - The collector is usually enclosed in *transparent glazing (glass or plastic)* to allow sunlight in but reduce *convective and radiative heat loss* to the environment.
   - An *insulated casing* at the back and sides further prevents heat loss.

iv. **Types of Heat Transfer**
   - Heat transfer inside the collector involves *conduction* (from absorber to fluid), *convection* (within the fluid), and *radiation* (losses to the surroundings).
   - In concentrating collectors, *mirrors or lenses* are used to *focus sunlight* onto a smaller area (e.g., a pipe), increasing the temperature significantly.

11.4.3 Energy Conversion Chain

Solar Energy (Short-Wave Radiation) → Thermal Energy at Absorber → Sensible Heat in Fluid

11.4.4 Efficiency Factors

Efficiency of a solar collector depends on:

- Solar irradiance (W/m^2).
- Collector design (type, area, orientation).
- Thermal conductivity and flow properties of the working fluid.
- Ambient conditions (temperature, wind, cloud cover).
- Heat losses via conduction, convection, and radiation.

Table 11.2 presents a comparative overview of the major types of solar thermal collectors based on their working principles, operational temperature ranges, and distinctive features. From conventional flat plate collectors used in domestic heating to high-temperature solar tower systems for industrial-scale power generation,

TABLE 11.2
Types of Solar Collectors Based on Working Principle

Type of Collector	Temperature Range	Working Principle	Key Features
Flat Plate Collector (FPC)	30–80°C	Absorbs solar radiation directly on a flat surface with glazing	Simple, low-cost, widely used for domestic water heating
Evacuated Tube Collector (ETC)	50–150°C	Uses vacuum-sealed glass tubes to reduce heat loss	High efficiency, good performance in colder climates
Compound Parabolic Collector (CPC)	60–200°C	Uses non-imaging optics to concentrate sunlight onto an absorber	Moderate concentration without tracking
Parabolic Trough Collector (PTC)	150–400°C	Parabolic mirrors focus sunlight onto a central receiver tube	Requires tracking, used in industrial process heat
Solar Tower (Central Receiver System)	250–1000°C	Heliostats reflect sunlight to a central receiver at the top of a tower	High concentration, ideal for large-scale power generation

each collector type operates on specific heat absorption and concentration mechanisms suited for different thermal applications. This classification aids in selecting the appropriate collector technology based on energy demand, climatic conditions, and intended usage in solar thermal systems.

11.5 CLASSIFICATION OF SOLAR COLLECTORS

Solar collectors are the heart of solar thermal systems. Their primary role is to *capture solar radiation* and *convert it into usable thermal energy*, which is then transported by a heat transfer fluid (HTF) to a point of application or storage. Based on their design and operational characteristics, solar collectors are *classified into two major categories* (Figure 11.2):

- **Non-concentrating (stationary) collectors**
- **Concentrating (tracking) collectors**

This section discusses the most widely used collector types within these categories, including *Flat Plate Collectors (FPCs)*, *Evacuated Tube Collectors (ETCs)*, *Parabolic Trough Collectors (PTCs)*, and *Compound Parabolic Collectors (CPCs)*—all of which differ in complexity, temperature range, efficiency, and suitability for nanofluid integration.

11.5.1 Flat Plate Collectors (FPCs)

Flat Plate Collectors (FPCs) are the most commonly used type of non-concentrating solar thermal collectors, particularly favored in residential and institutional water

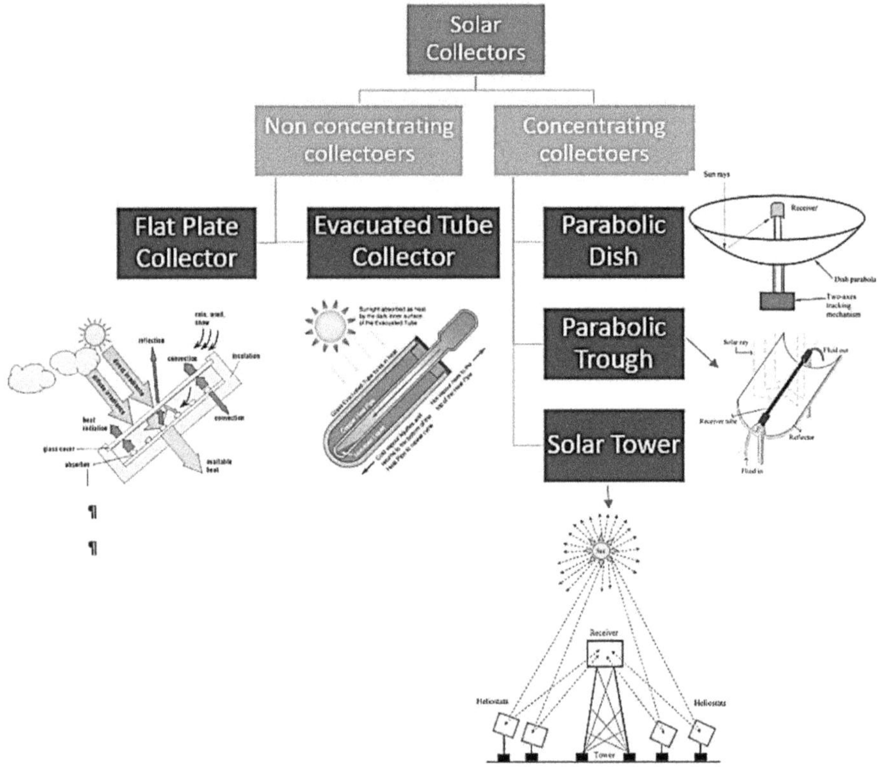

FIGURE 11.2 Schematics of common solar thermal collector designs.

heating applications. They are simple in construction, cost-effective, and capable of delivering useful heat at moderate temperatures ranging from 30°C to 80°C. These collectors are best suited for locations with ample solar radiation and mild to warm ambient conditions.

11.5.1.1 Construction and Components

An FPC typically consists of the following major components (**Figure 11.3**):

i. **Absorber Plate**
   This is the core element of the collector where solar radiation is converted into thermal energy. The plate is usually made of metals with high thermal conductivity, such as copper or aluminum, coated with a selective surface (e.g., black chrome or tin oxide) to maximize solar absorptivity while minimizing emissivity.
ii. **Transparent Cover (Glazing)**
   A sheet of glass or transparent plastic is placed above the absorber to reduce convective heat losses to the environment and to allow solar radiation to enter the collector. The glazing is often low-iron tempered glass, which has high transmissivity and durability.

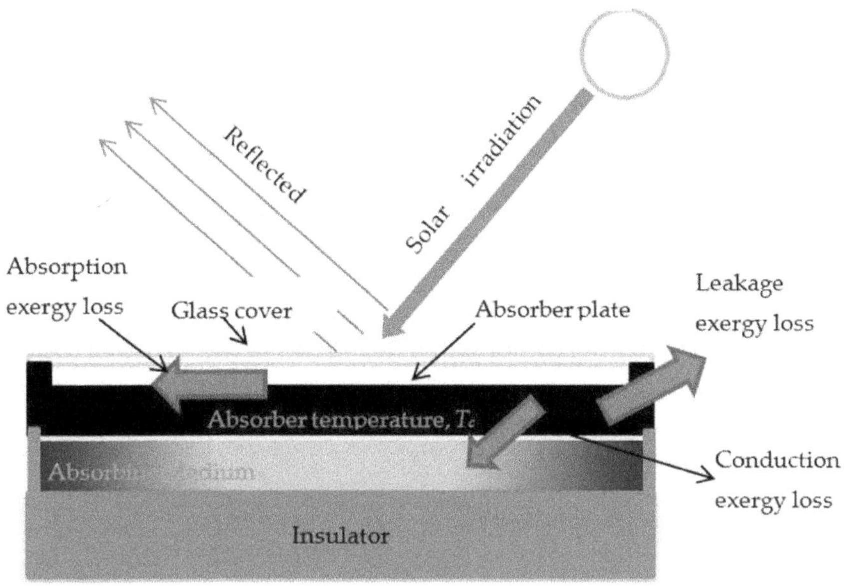

FIGURE 11.3 Schematic of energy absorption and exergy losses in a solar collector [3].

iii. **Insulation**
To minimize conductive heat losses, thermal insulation (such as fiberglass or polyurethane foam) is provided at the back and sides of the collector box.

iv. **Fluid Flow Tubes**
Tubes or channels are attached or embedded in the absorber plate through which a working fluid (usually water or glycol-based antifreeze solution) flows. These tubes collect the absorbed heat and transfer it to the fluid.

v. **Collector Housing (Casing)**
A metallic or polymer casing encloses all the components to protect the collector from dust, moisture, and mechanical damage.

11.5.1.2 Working Principle

The working of an FPC is based on the greenhouse effect and conduction-convection heat transfer mechanisms:

- Solar radiation passes through the transparent glazing and strikes the absorber plate.
- The absorber plate captures this radiation and heats up.
- Heat is transferred from the absorber to the fluid inside the tubes via conduction.
- The heated fluid is circulated to a storage tank or heat exchanger, where it releases its energy for domestic or industrial use.
- Insulation and glazing minimize heat losses due to convection and radiation to the ambient environment.

The system may be passive (natural circulation based on thermosiphon principle) or active (using pumps for forced circulation).

11.5.1.3 Energy Balance and Efficiency

The energy balance of an FPC can be expressed as:

$$Q_u = A_c\left(S - U_L\left[T_P - T_a\right]\right)$$

where:

- Q_u = useful energy gain (W).
- A_c = collector area (m²).
- S = absorbed solar radiation (W/m²).
- U_L = overall heat loss coefficient (W/m²·K).
- T_p = average absorber plate temperature (K).
- T_a = ambient temperature (K).

The thermal efficiency η is defined as:

$$\eta = \frac{Q_u}{A_c \cdot G_t}$$

where G_t is the incident solar radiation.

11.5.1.4 Advantages of Flat Plate Collectors

- **Simplicity**: Easy to design, manufacture, and maintain.
- **Cost-effective**: Cheaper compared to concentrating collectors.
- **No tracking required**: Can absorb diffuse and direct radiation.
- **Versatile**: Suitable for domestic hot water, space heating, and industrial pre-heating.

11.5.1.5 Limitations

- **Lower operating temperature**: Not suitable for high-temperature applications (>100°C).
- **Thermal losses**: Susceptible to heat losses, especially in cold or windy conditions.
- **Bulky design**: Requires larger area for installation due to low energy density.

Case Study: Performance Optimization of a Hybrid Flat Plate Solar Collector Using Phase Change Material (PCM) and Heat Sink [4]

Phase Change Materials (PCMs) are substances that absorb and release thermal energy during the process of melting and solidification (i.e., phase change). When a PCM changes from solid to liquid (melting), it absorbs heat without a significant increase in temperature. Conversely, when it changes from liquid back to solid (solidification), it releases stored heat at nearly constant temperature. This property makes PCMs ideal for thermal energy storage (TES) applications.

11.5.1.6 Technical Definition

A phase change material is a material with a high latent heat of fusion that can store and release large amounts of thermal energy by undergoing a reversible phase transition, most commonly between solid and liquid phases, at a specific transition temperature.

11.5.1.7 Key Features

- **Latent heat storage**: PCMs store energy in the form of latent heat during the phase transition.
- **Isothermal behavior**: The temperature remains nearly constant during melting or solidification.
- **Reversible**: The process can be repeated for many cycles with minimal degradation.
- **Thermal buffering**: PCMs maintain a stable temperature even when external temperatures fluctuate.

11.5.1.8 Types of PCMs

1. **Organic PCMs**
   - Examples: Paraffin wax, fatty acids
   - Advantages: Chemically stable, non-corrosive, no supercooling
   - Disadvantages: Low thermal conductivity, flammable
2. **Inorganic PCMs**
   - Examples: Salt hydrates (e.g., $CaCl_2 \cdot 6H_2O$), metallics
   - Advantages: High latent heat, higher thermal conductivity
   - Disadvantages: Corrosive, prone to phase segregation and supercooling
3. **Eutectic Mixtures**
   - Mixtures of two or more components that melt/solidify at a single temperature
   - Examples: Organic-organic or inorganic-organic mixtures

11.5.1.9 Applications

- Solar thermal energy storage
- Building thermal regulation
- Refrigeration and cold storage
- Thermal management in electronics
- Textiles (temperature-regulating clothing)

11.5.1.10 Introduction to the Case Study

Flat Plate Solar Collectors (FPSC) are widely employed for domestic and industrial thermal applications. Their performance, however, is limited by the availability of sunlight and rapid cooling after sunset. To overcome this, researchers have explored *thermal energy storage (TES)* using *Phase Change Materials (PCMs)* and passive heat transfer enhancements like *heat sinks*. This case study discusses the experimental evaluation of a *hybrid FPSC integrated with PEG/Mg(OH)$_2$-based PCM and Aluminum heat sink*, highlighting improvements in thermal efficiency and extended hot water availability.

11.5.1.11 Objective

To evaluate and optimize the performance of a flat plate solar collector (FPSC) by:

- Integrating a PEG/Mg(OH)$_2$ phase change material beneath the absorber plate.
- Employing an aluminum heat sink to enhance thermal uniformity and heat transfer.
- Comparing these hybrid designs against a conventional FPSC.

11.5.1.12 Experimental Setup

The experiments were conducted outdoors in *Abha, Saudi Arabia* (semi-arid climate) (Figure 11.4). The setup consisted of:

- A *flat plate solar collector* (absorber area: 1.5 m^2).
- Three system configurations
  1. **Case 1:** Standard FPSC (reference)
  2. **Case 2:** FPSC + PCM (PEG/Mg(OH)$_2$)
  3. **Case 3:** FPSC + PCM + aluminum heat sink

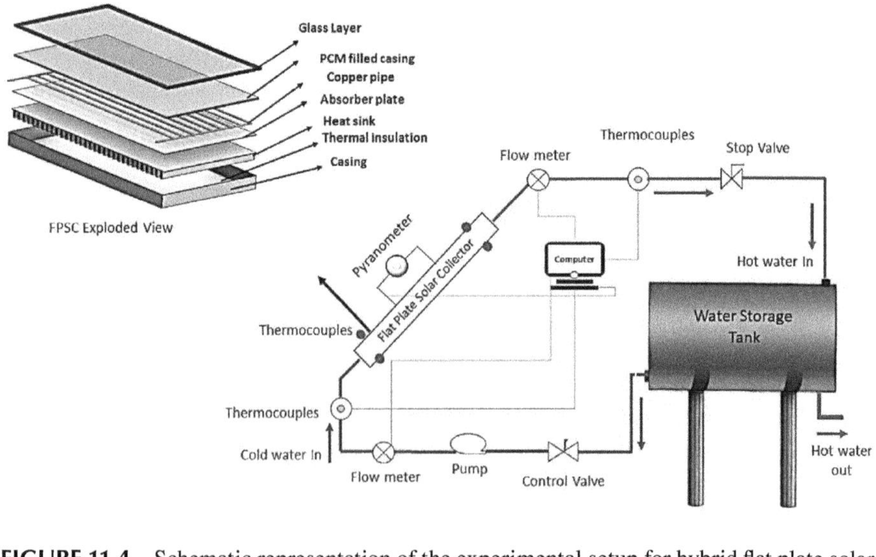

FIGURE 11.4 Schematic representation of the experimental setup for hybrid flat plate solar collector integrated with PCM and heat sink [4].

11.5.1.13 Instrumentation

- *Twelve thermocouples* (K-type) for temperature monitoring at various locations (fluid inlet/outlet, absorber plate, PCM, ambient)
- *Solarimeter* for measuring solar irradiance
- *Digital flowmeter* to regulate and measure fluid flow (fixed at *0.032 kg/s*)
- *Data logger* for continuous data acquisition
- Water as the *heat transfer fluid (HTF)*

11.5.1.14 PCM Preparation

- PEG 6000 was blended with 5% $Mg(OH)_2$ using ethanol as the dispersion medium.
- The mixture was sonicated and oven-dried at 60°C, then placed in the collector body beneath the absorber plate, *isolated from the fluid tubes.*

Table 11.3 summarizes the technical specifications and performance comparison of three flat plate solar collector configurations. The integration of PEG/$Mg(OH)_2$ phase change material in Case 2 significantly improved thermal efficiency and extended hot water availability. While the addition of an aluminum heat sink in Case 3 enhanced heat uniformity, it slightly reduced efficiency due to faster energy release.

- **Case 2** showed the highest improvement due to *latent heat storage* and *delayed cooling* of the collector.
- **Case 3** improved temperature uniformity but released stored heat too quickly, slightly lowering average efficiency.

11.5.1.15 Hot Water Duration

- **Case 2** maintained water temperature >38°C until *20:00 h*, significantly longer than the reference case.
- **Case 3** sustained temperatures until *19:20 h*.

11.5.1.16 Energy Storage Analysis

- **Charging mode**: PCM absorbed ~1.65 kWh (1.10 kWh as latent heat),
- **Discharging mode**: Released ~1.45 kWh during off-sunlight hours.

TABLE 11.3
Thermal Efficiency of Different Configurations

Configuration	Peak Thermal Efficiency (%)
Case 1 (Baseline)	53.6
Case 2 (With PCM)	65.4
Case 3 (PCM + Heat Sink)	53.1

11.5.1.17 Educational Insights

- PCM-based TES can significantly improve the operational window of FPSCs.
- Heat sinks improve thermal distribution but must be matched with the PCM characteristics.
- Field testing under real environmental conditions provides valuable performance data for scalable solar technologies.

11.5.2 Evacuated Tube Collector (ETC)

Evacuated Tube Collectors (ETCs) are a type of solar thermal collector widely used for medium- to high-temperature solar heating applications. They are designed to overcome the limitations of flat plate collectors, particularly in colder climates or during low irradiance conditions. ETCs minimize heat losses through vacuum insulation, offering superior thermal efficiency and enhanced year-round performance. There are different types of ETCs, as shown in Table 11.4.

11.5.2.1 Construction and Design

An ETC consists of multiple parallel rows of *double-layered glass tubes*, each tube functioning as an independent solar absorber (Figure 11.5). The construction includes:

- **Outer glass tube**: Made of highly transparent borosilicate glass, it allows sunlight to pass with minimal reflection and protects the inner components.
- **Inner glass tube**: Coated with a *selective absorptive surface* (e.g., aluminum nitride or copper oxide), it absorbs maximum solar radiation while minimizing infrared re-radiation.
- **Vacuum layer**: The space between the inner and outer tubes is evacuated to create a *vacuum insulation* barrier, drastically reducing convective and conductive heat losses.
- **Heat pipe or U-pipe**: In advanced ETCs, a *heat pipe* or *U-shaped copper tube* is inserted into the inner tube to transfer heat efficiently to the circulating fluid in the manifold.
- **Header or Manifold**: This collects and distributes the heated fluid from all tubes to the thermal storage system or load.

TABLE 11.4
Types of Evacuated Tube Collectors

Type	Working Mechanism	Typical Use
Direct Flow ETC	Heat transfer fluid circulates directly through absorber tubes	Low to medium temperature heating
Heat Pipe ETC	Uses a sealed heat pipe with phase change fluid	Domestic and commercial hot water
U-Pipe ETC	U-shaped copper pipe carries fluid and returns to header	Industrial and high-pressure systems

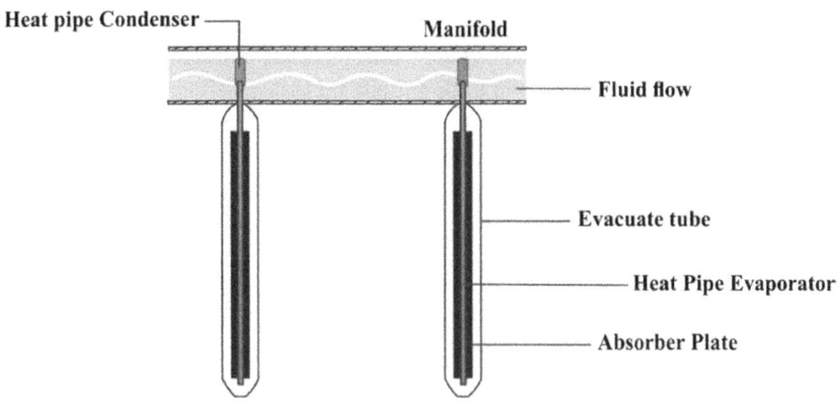

FIGURE 11.5 Cross-sectional view of a typical evacuated tube collector showing the outer and inner tubes, vacuum gap, heat pipe, and manifold [3].

11.5.2.2 Working Principle

The working of an ETC can be described in the following steps:

- *Solar radiation* passes through the transparent outer glass tube and strikes the selective coating on the inner absorber tube.
- The coating *absorbs solar energy* and converts it into heat.
- The *vacuum insulation* between the two tubes prevents heat loss to the environment.
- Heat is transferred from the inner tube to the working fluid either:
  - *Directly*, if fluid flows inside the tube (*direct flow ETC*), or
  - *Indirectly*, using a *heat pipe or U-pipe* embedded inside the tube (*indirect flow ETC*).
- The heated fluid moves to the *manifold* and then to a *thermal storage tank* or *end-use equipment* (e.g., water heater, radiator).

11.5.2.3 Performance Features

- **High efficiency**: ETCs can reach thermal efficiencies of *60–80%*, even in cold or cloudy weather.
- **Low heat loss**: The vacuum layer eliminates convective and conductive heat losses.
- **Better performance at low irradiance**: ETC performs well in winter and in low-sunlight conditions.
- **Modularity**: Tubes function independently; damaged tubes can be replaced without shutting down the entire system.
- **Operational range**: ETCs can efficiently operate in temperatures up to *200°C* or more, making them suitable for *solar absorption cooling, process heating*, and *pre-heating boiler feed water.*

11.5.2.4 Advantages

- Superior insulation and performance over flat plate collectors
- Effective in cold, windy, and high-altitude regions
- Easy maintenance due to modular tube design
- Can be combined with thermal energy storage and PCM for extended operation

11.5.2.5 Limitations

- Fragility of glass tubes; prone to breakage under impact
- Higher initial cost than flat plate collectors
- Slight risk of overheating if the system is not properly controlled
- Orientation and tilt angle must be optimized for maximum solar gain

11.5.2.6 Applications

- Domestic and commercial *hot water systems*
- *Space heating* in residential or institutional buildings
- *Solar-assisted air conditioning* and refrigeration
- *Industrial process heat* (dyeing, food processing, sterilization)
- *Preheating for boilers* in hybrid solar-fossil fuel systems

---

Case Study: Performance Enhancement of Evacuated Tube Solar Collector Using Water-Based Titanium Oxide Nanofluid [5]

BACKGROUND AND MOTIVATION

Solar collectors, particularly Evacuated Tube Collectors (ETCs), are widely used for harnessing solar thermal energy. However, conventional heat transfer fluids such as water often limit thermal performance due to their low thermal conductivity (~0.6 W/m·K). One of the viable approaches to improve heat extraction from the collector surface is by using nanofluids—base fluids with suspended nanoparticles of high thermal conductivity materials.

This case study examines the work of Mahendran et al., who investigated the use of water-based titanium dioxide (TiO_2) nanofluid as a heat transfer medium in an ETC system under tropical solar conditions.

EXPERIMENTAL SETUP

The experiments were conducted at the Pekan Campus of Universiti Malaysia Pahang (Latitude: 3°32′ N, Longitude: 103°25′ E), which experiences an average solar insolation exceeding 900 W/m^2, peaking around 1200 W/m^2.

- Collector Type: Evacuated Tube Solar Collector
- Nanofluid: Water + TiO_2 nanoparticles (30–50 nm)
- Concentration: 0.3% by volume
- Flow Rate: Fixed at 2.7 liters/min for both water and nanofluid
- Comparison Base Fluid: Water

KEY OBSERVATIONS AND RESULTS

1. Thermal Conductivity Enhancement
   The use of TiO_2 nanoparticles in water resulted in an increase in thermal conductivity, which directly contributed to improved heat transfer rates. It was observed that thermal conductivity increased with both temperature and nanoparticle concentration, making the system more responsive to diurnal variations in solar radiation.
2. Efficiency Gain
   - Efficiency using water only: ~0.58
   - Efficiency using TiO_2 nanofluid: ~0.73
   - Net efficiency improvement: 16.7%
3. Solar Responsiveness
   The nanofluid's performance showed better adaptability to the solar radiation curve, which typically starts low in the morning, peaks at around 2:00 PM, and declines thereafter.
4. Flow Rate Consideration
   Increasing water flow rate alone was found to be inefficient, as it led to diminished energy capture due to reduced residence time. Nanofluid use offered a more energy-efficient solution by enhancing thermal properties without changing flow conditions.

IMPLICATIONS

This case study highlights the significant potential of nanofluids—particularly TiO_2/water suspensions—in enhancing the efficiency of evacuated tube solar collectors. A 16.7% improvement in collector efficiency was realized under real-time equatorial solar conditions, affirming the role of nanofluid thermal conductivity enhancement as a passive method of improving collector performance. Moreover, the study underscores that even at a relatively low concentration (0.3%), nanofluids can deliver notable gains without substantial changes to system design or operation.

EDUCATIONAL TAKEAWAY

The integration of nanofluids in solar thermal systems provides an economical and efficient alternative to mechanical modifications such as increasing flow rate or collector area. This case study serves as evidence that nanofluids, when properly formulated and optimized, can play a transformative role in solar energy harvesting and thermal management.

11.5.3 Parabolic Trough Collector (PTC)

Among all solar concentrating systems, the Parabolic Trough Collector (PTC) stands as the most commercially mature and widely deployed technology globally. PTCs are a type of line-focusing solar collector, designed to track the sun along one axis and focus direct solar irradiance onto a linear receiver located at the focal line of a parabola-shaped reflector. The ability of PTCs to operate at medium to high temperatures, combined with their scalability and cost-effectiveness, has made them central to large-scale solar thermal power plants and various industrial process heat applications.

11.5.3.1 Working Principle and Configuration

The PTC operates based on the principles of geometric concentration and thermal energy conversion. A parabolic-shaped reflective mirror is used to collect and focus sunlight onto a linear absorber tube, also known as the receiver, which is positioned precisely at the focus of the parabolic reflector.

11.5.3.2 Key Components

The following are the key components (Figure 11.6):

- **Parabolic reflector**: A mirror that tracks the sun and reflects incident rays to the focal line.
- **Receiver/absorber tube**: Typically made of a metal tube with a selective coating that maximizes solar absorption and minimizes emissivity. It is enclosed within a glass envelope to reduce convective heat losses and maintain a vacuum insulation.
- **Heat transfer fluid (HTF)**: Flows through the absorber tube, extracting the concentrated thermal energy.

11.5.3.3 Operating Mechanism

1. Sunlight strikes the curved mirror surface.
2. The mirror reflects the rays to a focal line where the receiver tube is located.
3. The absorber tube heats up due to the concentrated solar energy.

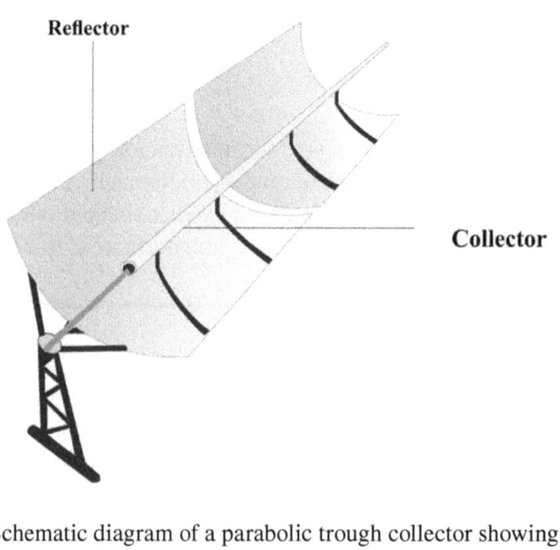

FIGURE 11.6 Schematic diagram of a parabolic trough collector showing key components and heat transfer pathway [3].

4. A heat transfer fluid (e.g., thermal oil, water, or gas) flows inside the absorber tube, absorbing this heat.
5. The heated fluid is transported to a heat exchanger, thermal energy storage unit, or directly to a system for electricity generation, desalination, or industrial use.

11.5.3.4 Temperature Range and Applications

Parabolic trough collectors typically operate in the temperature range of 50°C to 400°C. This makes them ideal for:

- Electricity generation via Rankine cycles.
- Solar cooling using absorption refrigeration systems.
- Industrial process heat (e.g., textile, chemical, and food industries).
- Desalination through multi-effect distillation (MED) or multi-stage flash (MSF) technologies.
- District heating in colder climate.

11.5.3.5 Thermal Performance and Efficiency

PTCs have a relatively high optical and thermal efficiency owing to their concentrating capability and vacuum-insulated receivers. The thermal efficiency of these systems typically ranges between 65% and 75%, depending on operating conditions, solar irradiance, collector design, and heat transfer fluid properties.

However, one of the challenges in further improving PTC performance is enhancing the heat transfer capability of the working fluid. Conventional fluids (like water, air, or thermal oil) often have limited thermal conductivity, which restricts the rate of heat extraction from the absorber surface.

Case Study: Thermal Performance of an Evacuated Tube Collector Using Water-Based Nanofluids

Source: Based on the work by Yashita Arora et al. (2020) [6]

Objective: To assess the effect of *Al_2O_3/water* and *TiO_2/water* nanofluids on the *thermal performance and efficiency* of an evacuated tube solar collector under real-time outdoor conditions.

EXPERIMENTAL SETUP

The experimental investigations were conducted on a *solar water heating system* comprising the following key components (Figure 11.7):

- **Evacuated Tube Collector (ETC)**
  - Composed of *20 parallel glass tubes* (borosilicate glass, double-walled), each containing a *U-shaped copper tube* for fluid circulation.
  - The tubes were connected to a *manifold header*, which delivered heated fluid to a storage tank.

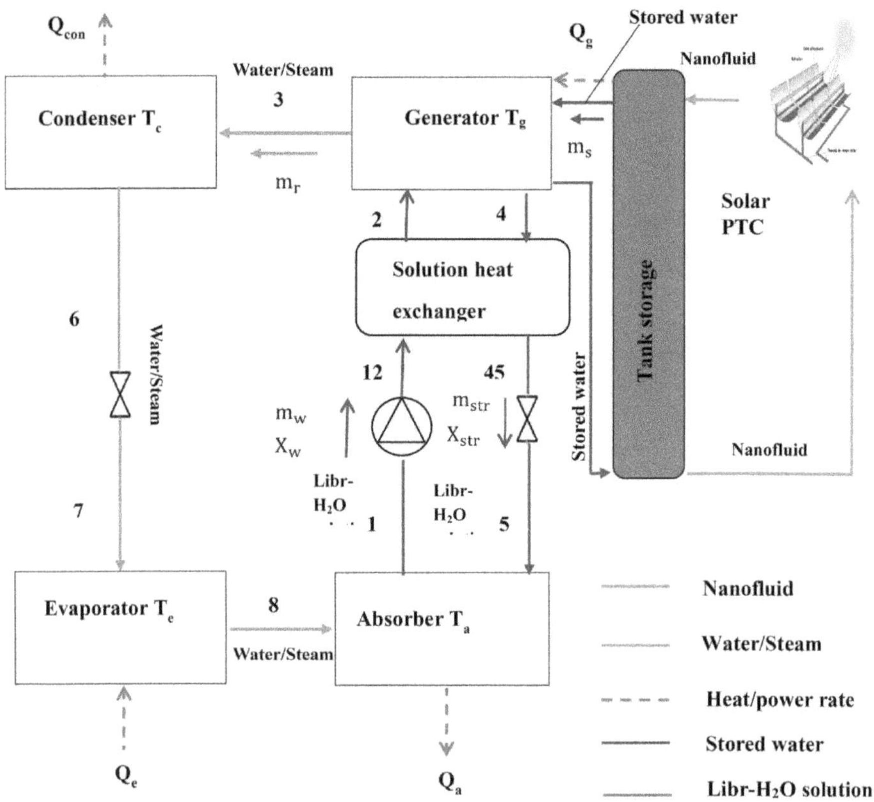

FIGURE 11.7 Schematic of the test rig for nanofluid-assisted evacuated tube collector performance analysis [6].

- **Nanofluid Preparation**
  - *Al_2O_3 and TiO_2 nanoparticles* were dispersed in distilled water using an *ultrasonic homogenizer* to ensure uniform suspension and reduce agglomeration.
  - Nanofluid concentrations were carefully controlled at *0.01 wt.%* and *0.03 wt.%* for comparison with pure water.
- **Instrumentation**
  - Thermocouples were used to record *inlet and outlet temperatures.*
  - A *rotameter* monitored the flow rate of the working fluid, fixed at *2 L/min*.
  - A *solarimeter* measured the *incident solar radiation*, while ambient temperature and wind speed were also recorded to study their influence on heat loss.
- **Storage Tank**
  - A *25-liter capacity insulated water tank* stored the heated nanofluid after passing through the collector.

TEST CONDITIONS

- Location: Outdoor testing under natural sunlight.
- Time Duration: Measurements were taken between *9:00 AM and 3:00 PM*.
- Solar Radiation: Varied from *200 W/m² to 950 W/m²* throughout the day.
- Ambient Conditions: Representative of a typical sunny day.

KEY FINDINGS

- **Improved Collector Efficiency**
  - Al_2O_3/water nanofluid at 0.03 wt.% showed the highest efficiency enhancement—up to *12.2% higher* than water.
  - TiO_2/water nanofluid also outperformed water, but slightly less than Al_2O_3.
- **Enhanced Thermal Conductivity**
  - Nanofluids exhibited *superior heat transfer properties*, reducing the thermal resistance between the collector surface and working fluid.
  - This led to *quicker rise in outlet temperature* and improved heat gain.
- **Stability and Flow**
  - Nanofluids remained stable during short-duration testing.
  - No significant sedimentation was observed, indicating the suitability of these nanofluids for short-term ETC applications.

11.6 NANOFLUID INTEGRATION IN SOLAR COLLECTORS

Thermophysical properties of traditional fluids that are used as the operational fluid, like water, ethylene glycol, and oil, are inherently constrained and therefore limit the efficiency of conventional solar thermal collectors. The thermal conductivities and specific heat capacities of these fluids are not very high, and this limits the capability of heat dissipation of the solar energy captured to the heat transfer medium, which is then used. This drawback is especially acute when it is necessary to remove high heat flux or in the conditions of transient or low-power sunlight. Consequently, with excellent conditions of sunlight exposure, the fraction of absorbed energy loses a considerable part because of inefficiencies of the target thermal transfer, which decreases the collector performance and the performance of the system.

As a way of defeating such inefficiencies, there are nanofluids that can be effectively utilized as a means of replacing such effects. The addition of highly conducting nanoparticles to conventional base fluids will dramatically increase their thermal conductivity and their heat transfer coefficient. This advancement allows quick uptake and transfer of heat from the absorber plate into the circulating liquid. Further, nanofluids have better energy storing capability, increased convective favors better, and reduced thickness of the boundary layer. Such characteristics enhance an increase in temperature throughout the collector, a quicker thermal reaction to varying irradiation, and better energy consumption. Research findings have shown that efficiency increases by 10–30% in different collector arrangements when changes are made to optimized nanofluid formulations.

Besides, the increased need for better performance is created when solar thermal systems are designed to use them in different complex applications like solar-assisted air conditioning, industrial process heating, and hybrid photovoltaic-thermal (PV/T) systems. In this type of process, stability and performance of the system with respect to variable environmental conditions will require constant and quick heat transfer. The advantage of nanofluids is that the thermophysical properties can be customized to the specific needs of the application, having accuracy in controlling thermal properties and extraction of energy. The incorporation also helps in achieving compact designs of high-efficiency collector design because it lessens the size and price of heat exchange materials. This makes nanofluids a method of performance enhancement and an essential advancement of the future-generation solar thermal technologies with regard to the goals of efficiency, minimal footprint, and the extended scope of applicability in sustainable energy systems.

11.6.1 Design Considerations for Incorporating Nanofluids in Collectors

The proper inclusion of nanofluids into solar thermal collector requires keen consideration of various parameters of their design. Compared to common heat transfer fluids, nanofluids have different thermophysical properties and higher thermal conductivity, altered viscosity, and possible stability concerns. These properties have a major effect on the thermal, hydraulic, and operational capabilities of the solar collectors. Thus, it is important to have a means of systematically selecting nanofluids, preparing them, and integrating them into the system to maximize their potential in solar energy processes.

11.6.1.1 Selection of Nanoparticles and Base Fluid

Another issue in nanofluid design is the proper choice of nanoparticles, as well as that of base fluid. Some of the more popular nanoparticles are metal oxides (e.g., Al_2O_3, CuO, TiO_2), metals (e.g., Ag, Au), and carbon-based materials (e.g., carbon nanotubes, graphene), which are dispersed into water, ethylene glycol, or oil, depending upon the range of operating temperatures. The thermal conductivity, specific heat capacity, and optical properties of the fluid depend on the nanoparticle of choice. As an illustration, CuO nanoparticles are more efficient in terms of improving thermal conductivity yet might provoke higher density and viscosity of fluids and necessitate greater pumping energy. The size, shape, and concentration of nanoparticles are also critical factors: the smaller the size (10–50 nm), the greater is its surface area available to transfer heat, but a larger amount of time could sediment or agglomerate.

11.6.1.2 Stability and Dispersion Techniques

Stability is vital to maintain uniform thermophysical properties throughout operation. Nanoparticles can settle at the bottom of the reservoir or block collector channels, in the case of poor dispersion, decreasing the effectiveness of heat transfer, and can damage the system components. Long-term stability of dispersion is commonly achieved by, for example, ultrasonication, addition of surfactant, or chemical modification of nanoparticle surfaces. Besides, in a practical system, it is possible to avoid settling by the use of mechanical stirring or recirculation of a nanofluid. To prevent corrosion or fouling, it is necessary to ensure that there is chemical compatibility between the nanoparticles, surfactants, and collector materials.

11.6.1.3 Concentration Optimization

The concentration of the nanoparticles should be optimized such that the greater the thermal performance, the smaller the hydraulic and operational difficulties associated with higher concentration of nanoparticles. Although concentration is expected on the whole to improve both thermal conductivity and heat transfer coefficient, it will also raise the viscosity, density, and the possibility of particle agglomeration. This results in an increased pressure drop, elevated pumping power, and even internal surface erosion. As such, the sweet spots are usually presented as 0.1 to 1.0% by volume, depending on the material of nanoparticles, the system design, as well as the desired performance parameters. To establish the optimum concentration under certain conditions of collector positions, experimental verification is conducted.

11.6.1.4 Thermal and Hydraulic Design Parameters

The introduction of nanofluids alters the heat transfer and flow behavior within the collector. As such, the system must be re-evaluated for parameters like:

- Heat transfer coefficient (h_{nf}).
- Reynolds and Nusselt numbers.
- Pressure drop and pumping power requirements.
- Flow regime (laminar vs turbulent).

System components such as pipes, manifolds, and absorber channels should be optimized for smooth flow and minimal thermal resistance. Microchannel or serpentine flow arrangements may further enhance heat transfer when used with nanofluids. Computational fluid dynamics (CFD) simulations can be a valuable tool for understanding and optimizing these interactions.

11.6.1.5 Material Compatibility and Fouling Control

The type of nanoparticles used and the surface characteristics of nanoparticles may attract abrasive or corrosive behavior of nanofluids. Interaction of metal oxide nanoparticles (e.g., with copper tubes, aluminum plates) can result in erosion, pitting, or fouling after prolonged exposure. Hence, internal surfaces of the collector should be made of material resistant to the chemicals or coated with protective linings. Substitution with inert or coated nanoparticles, silica, or graphene-oxide can also minimize reactivity and enhance lifespan of the system.

11.6.1.6 Compatibility with Collector Type and Application

Different solar collectors (e.g., flat plate, evacuated tube, parabolic trough) provide different thermal conditions and design issues. Nanofluids are more suitable in flat plate and evacuated tubes collectors since in such collectors convective heat exchange is the major heat exchanger. To work in concentrated solar collectors, nanofluids may have to endure higher temperatures of operation and heavier thermal cycling. Such conditions should be assessed before the nanofluid can be integrated based on its stability and resistance to thermal degradation. Also, direct systems are inefficient as nanofluids can be more stable in an indirect system with the help of a heat exchanger.

11.6.1.7 Instrumentation and Monitoring

Nanofluid performance must be monitored using highly accurate temperature, flow rate, pressure drop, and heat flux instrumentation. Sensors ought to be chosen depending on the transformed fluid attributes, and the data-collecting structures must have the capacity to identify tiny changes within the heat reaction or the deposition characteristics. It is suggested that periodic (e.g., weekly, monthly) sampling and analysis of nanofluid properties (e.g., particle concentration, pH, zeta potential) be provided to allow long-term reliability estimation.

11.6.1.8 Safety and Environmental Considerations

Though the use of nanofluids leads to significant thermal benefits, the environmental and health consequences of these fluids should also be discussed. Some nanoparticles (e.g., silver, zinc oxide) have toxic properties and thus they should be handled and disposed of carefully. The designs should be such that leaks or spillage are minimized, and all the used or degraded nanofluid should be treated like other hazardous wastes according to local laws. Of biocompatible/environmentally safe nanomaterials, silica or titania is desired in residential and institutional systems.

11.6.2 Influence on Collector Performance Parameters

The integration of nanofluids into solar thermal collectors significantly alters the performance characteristics of the system. Nanofluids, due to their superior thermophysical properties, influence key collector parameters such as the *heat removal factor (FR)*, *collector efficiency factor (F')*, and the overall *temperature rise and thermal response*. These parameters are critical in determining the effectiveness of solar collectors in absorbing, transferring, and utilizing solar energy under varying environmental and operational conditions. This section elaborates on how nanofluid incorporation modifies each of these parameters and contributes to performance enhancement.

11.6.2.1 Heat Removal Factor (FR)

The *heat removal factor (FR)* quantifies the effectiveness with which the working fluid removes absorbed solar energy from the collector surface. It is a measure of the actual useful energy gain compared to the maximum possible if the fluid were at collector plate temperature.

$$F_R = \frac{\dot{m}C_{P,nf}\left(T_{Out} - T_{in}\right)}{A_c\left[S - U_L\left(T_f - T_a\right)\right]}$$

where:
- $\dot{m}$ = mass flow rate of nanofluid.
- C = specific heat of nanofluid.
- T_{Out}, T_{in} = outlet and inlet fluid temperatures.
- Ac = collector area.
- S = absorbed solar radiation.
- U_L = overall heat loss coefficient.
- T_f, T_a = mean fluid and ambient temperatures.

Effect of Nanofluids on FR

- Nanofluids exhibit *enhanced thermal conductivity*, which improves heat absorption and transfer from the absorber to the fluid.
- The increased convective heat transfer coefficient h_{nf} results in better heat extraction, raising the value of F_R.
- Experimental results show F_R values increase by *15–30%* with nanofluids compared to conventional fluids under similar conditions.

11.6.2.2 Collector Efficiency Factor (F′)

The *collector efficiency factor (F')* represents the fraction of solar energy absorbed by the absorber plate that is transferred to the fluid per unit temperature difference between the plate and the fluid.

$$F' = \frac{1}{U_L\left(\frac{D}{k} + \frac{1}{h_{nf}}\right)}$$

where:
- U_L = overall heat loss coefficient.
- D = tube diameter.
- k = thermal conductivity of tube material.
- h_{nf} = convective heat transfer coefficient of nanofluid.

Effect of Nanofluids on F'

- The presence of nanoparticles increases $h_{nf,}$ decreasing thermal resistance between the absorber and the fluid.
- Higher hnfh_{nf}hnf enhances the rate of energy extraction, resulting in a *higher F'*, often improving system thermal performance by 10–20%.
- Studies have reported significant improvement in F' when using *CuO*, *Al_2O_3*, or *TiO_2* nanofluids in *flat plate* and *evacuated tube collectors.*

11.6.2.3 Temperature Rise and Thermal Response

The temperature rise of the working fluid and the system's ability to respond to solar radiation changes are critical in determining usability and energy efficiency in practical applications.

Influence of Nanofluids

- Nanofluids *reduce the thermal response time* of the system due to better energy absorption and faster distribution.
- They enable *higher fluid outlet temperatures* in shorter durations compared to traditional fluids.
- This rapid heating is crucial in applications requiring *intermittent energy supply*, such as solar cooking, batch heating, and thermal sterilization.
- In dynamic conditions (e.g., cloud movement), nanofluids help maintain *temperature stability*, reducing fluctuations in thermal output.

11.6.2.4 Overall Thermal Efficiency

The thermal efficiency η\etaη of a solar collector is defined as the ratio of useful heat gain to the incident solar energy on the collector surface:

$$\eta = \frac{\dot{m}C_{P,nf}\left(T_{Out} - T_{in}\right)}{A_c I}$$

where I is the incident solar irradiance.

Nanofluid Impact

- Nanofluids improve both *heat absorption* and *convection*, resulting in higher useful heat gain.
- Reported increases in collector efficiency range between *10% and 30%*, depending on nanofluid type, concentration, and collector design.
- Efficiency gains are more prominent at *lower flow rates*, where convective heat transfer dominates.

11.6.2.5 Pressure Drop and Pumping Power

While nanofluids improve thermal performance, they may also increase *viscosity*, leading to higher pressure drop and required pumping power. Hence, an optimal balance is necessary between:

- Thermal gains due to higher heat transfer rates.
- Hydraulic losses due to increased friction and flow resistance.

Design Strategy

- Use of *low-viscosity nanofluids* or hybrid nanofluids
- *Optimizing pipe diameters and flow rates*
- Employing *efficient pumps* to offset increased power consumption

11.7 ECONOMIC AND ENVIRONMENTAL IMPACT

The integration of nanofluids in solar thermal systems has attracted growing attention not only for its technical advantages in heat transfer and system efficiency but also for its potential *economic and environmental benefits*. This section evaluates the broader implications of nanofluid-enhanced solar collectors through a *cost–benefit perspective*, *life cycle analysis*, *carbon footprint mitigation*, and potential *return on investment (ROI)* in practical scenarios.

11.7.1 Cost–Benefit Analysis of Nanofluid-Based Systems

The application of nanofluids in solar thermal collectors leads to higher thermal performance and energy gains, but it also introduces additional costs due to *nanoparticle synthesis, stabilization, and handling.* A balanced cost–benefit analysis is essential to assess whether the performance improvement justifies the economic input.

- **Initial costs**: Include the purchase of nanoparticles (e.g., Al_2O_3, CuO, TiO_2), dispersion equipment (e.g., ultrasonic homogenizers), and modified pumping systems due to changes in fluid viscosity.
- **Operational savings**: Increased thermal conductivity and heat absorption translate to *lower energy consumption*, *shorter heating cycles*, and *reduced dependence on auxiliary heating systems.*
- **Payback period**: In systems where nanofluid use leads to efficiency improvements of 10–25%, the additional investment can be recovered within *2–4 years*, depending on the scale and usage conditions.
- **Scalability considerations**: For larger systems (e.g., industrial solar-assisted heating or solar desalination), the marginal cost per kWh saved tends to decline, making nanofluids more cost-effective at scale.

11.7.2 Life cycle Assessment (LCA)

Life cycle Assessment (LCA) evaluates the environmental and energy impacts of nanofluid-based systems throughout their entire life cycle—from nanoparticle production to system decommissioning.

- **Raw material extraction**: Nanoparticles like metal oxides require energy-intensive synthesis processes, which may offset short-term sustainability gains unless renewable-powered or green synthesis methods are adopted.
- **Manufacturing and dispersion**: Processes such as *sol-gel*, *hydrothermal*, or *chemical vapor deposition* introduce varying levels of emissions and waste depending on reagents and reaction conditions.
- **Use phase**: During operation, nanofluids improve collector efficiency and reduce the need for fossil-fuel-based supplemental energy sources—resulting in a net *energy saving* over the collector's life.
- **End-of-life disposal**: Potential environmental concerns include *nanoparticle leaching* and the *safe disposal or recycling* of nanofluid-containing fluids. Effective regulation and fluid reclamation techniques are essential to minimize ecological risks.

Overall, LCA studies indicate that while the initial carbon and resource footprint may be higher, the *energy payback time* is significantly improved due to efficiency gains in the use phase, making nanofluid systems *more sustainable over the long term.*

11.7.3 Carbon Footprint Reduction

Nanofluid-enhanced solar thermal systems contribute to a *substantial reduction in greenhouse gas (GHG) emissions* through:

- **Higher collector efficiency**: More efficient heat transfer reduces reliance on fossil fuel-powered auxiliary heating, especially in applications like *space heating, water heating, and process heat.*
- **Fuel displacement**: Systems replacing electric or diesel-based heating with nanofluid-enhanced solar collectors result in direct CO_2 savings.
- **Estimated CO_2 reduction**: Studies show that using nanofluid-enhanced collectors can lead to *5–20% more carbon offset* compared to conventional water-based systems, depending on the nanofluid type and operational profile.
- **System-level impact**: For example, in a 1 m^2 collector used for residential heating, switching from water to a nanofluid can save *20–30 kg of CO_2 annually*, translating to a considerable footprint reduction at a national or global deployment level.

11.7.4 Return on Investment in Real-World Solar Thermal Installations

From a commercial and residential perspective, the ROI of nanofluid-based solar thermal systems is a function of *capital costs*, *efficiency improvements*, *operating savings*, and *system lifespan.*

- **Commercial installations**: In industrial sectors such as food processing, chemical manufacturing, and pharmaceuticals, where continuous thermal loads exist, the return on investment is rapid due to high fuel displacement. Real-world pilot projects have demonstrated *ROI periods as low as 1.5–2 years* when using optimized nanofluids.
- **Residential sector**: For household solar water heating systems, nanofluids show ROI improvements, especially in regions with high solar insolation. Despite slightly higher upfront costs, *long-term savings on electricity or gas bills* lead to an attractive payback period.
- **Incentives and subsidies**: Governmental policies promoting renewable energy adoption, including *carbon credits*, *green building certifications*, and *subsidies for nanotechnology adoption*, can accelerate ROI and encourage widespread implementation.
- **Future trends**: As nanoparticle synthesis costs reduce and *standardized nanofluid formulations* become commercially available, the economic feasibility of nanofluid-enhanced systems is expected to improve further.

11.8 LIMITATIONS AND FUTURE PROSPECTS OF NANOFLUIDS IN SOLAR COLLECTOR APPLICATIONS

Despite the numerous performance advantages that nanofluids offer—such as enhanced thermal conductivity, improved heat transfer rates, and the potential to reduce collector size and increase energy efficiency—there are several important limitations and uncertainties that still hinder their widespread and long-term application in solar thermal systems. These challenges are mostly rooted in the complex physical behavior of nanofluids, the multitude of influencing variables, and the limited predictability of their thermophysical properties over time [7].

11.8.1 Stability and Sedimentation Challenges

One of the foremost limitations in using nanofluids in solar collectors is their *poor long-term stability.* Unlike conventional fluids that maintain a homogeneous phase, nanofluids are colloidal suspensions of solid nanoparticles in a base fluid, and their stability is highly susceptible to time, temperature, and operational conditions. Over time, the suspended nanoparticles tend to settle due to gravity, particularly when Brownian motion is insufficient to counteract agglomeration. This results in sedimentation, clogging of flow passages, and a decrease in the effective heat transfer capacity of the fluid.

While stabilization strategies such as *pH adjustment*, *surfactant addition*, and *ultrasonic dispersion* have been employed to minimize particle aggregation, these methods are not always effective over long durations. For example, studies have shown that after prolonged operation, the optical absorption capacity of nanofluids can drop significantly due to sedimentation, which directly reduces the amount of solar radiation converted into thermal energy. Furthermore, as aggregation progresses, the nanofluid becomes more viscous, thus increasing the pumping power required in forced circulation systems or reducing the flow rate in passive (thermosyphon) systems—both outcomes leading to reduced thermal efficiency [8].

11.8.2 Aggregation and Viscosity Effects in Evaporative Systems

Another concern arises in *evaporative or phase-change systems* such as heat pipes, where nanofluids have shown *less consistent performance.* The repeated phase change processes—particularly evaporation and condensation—cause increased particle clustering within the fluid. This aggregation impairs evaporation rates by creating thermal resistance layers around the particles or along the fluid interface. As a result, even though nanofluids typically offer better thermal transport during single-phase operation, their performance in two-phase systems may deteriorate due to increased thermal resistance and reduced heat transfer surface activity. This limitation is particularly problematic for solar thermal systems utilizing *heat pipe-based evacuated tube collectors*, where effective evaporation is critical to system efficiency [9].

11.8.3 Hysteresis Phenomenon and Temperature Sensitivity

A notable and less predictable limitation of nanofluids is the *hysteresis behavior*—a phenomenon where the physical and rheological properties of the fluid do not revert to their original state after thermal cycling. Observations show that when nanofluids are heated to certain critical temperatures (often between 60°C and 70°C for water-based suspensions), irreversible changes occur in viscosity and microstructure.

These changes degrade the nanofluid's heat transfer performance upon cooling. This is particularly concerning for *concentrated solar collectors*, which routinely operate at elevated temperatures. As the nanofluid loses its original thermal conductivity and viscosity profile due to heating, it compromises the predictability and repeatability of thermal performance, posing an engineering challenge for sustained operation [10].

11.8.4 Predictability of Heat Transfer Performance (Nusselt Number)

In conventional heat transfer analysis, the *Nusselt number* is a key dimensionless quantity used to estimate convective heat transfer. However, the prediction of Nusselt numbers for nanofluids remains inconsistent and highly sensitive to the fluid formulation. Studies indicate that the Nusselt number in nanofluid systems is not universally correlated with classical parameters like Reynolds and Prandtl numbers alone. Instead, it also depends on a wide array of nanofluid-specific variables—such as nanoparticle size, shape, material, concentration, and dispersion quality.

This variability introduces a significant challenge in creating *generalized heat transfer correlations*. Each nanofluid system may require a custom correlation developed under very specific experimental conditions. This limitation severely restricts the ease of adoption and scalability of nanofluids, as engineers cannot apply a one-size-fits-all approach to performance estimation. Moreover, the impact of hysteresis and stability degradation on long-term Nusselt number values remains largely unexplored [11].

11.8.5 Complexity in Predicting Rheological Properties

The accurate prediction of nanofluid *rheological properties*, particularly *viscosity*, is still a major research bottleneck. Numerous attempts have been made to model nanofluid viscosity using modified versions of classical equations—such as Einstein's viscosity model for dilute suspensions and Krieger–Dougherty models that incorporate particle volume fraction and packing effects. These models, however, often fall short in capturing the actual behavior of nanofluids due to the multitude of "active variables" that influence rheology.

These active variables include *nanoparticle volume fraction*, *base fluid type*, *particle size and shape*, *agglomeration tendency*, *surface charge (zeta potential)*, *presence of nanolayers*, pH, and temperature. The complex interactions between these factors make it extremely difficult to isolate the contribution of each to overall viscosity. Furthermore, experimental results often contradict each other—for example, some studies report increasing viscosity with decreasing particle size, while others observe the opposite. This inconsistency makes it challenging to build predictive rheological models, leading to uncertainties in system design, pumping requirements, and heat transfer predictions [12].

11.8.6 Challenges in Modeling Thermal Conductivity

Just as viscosity modeling is problematic, predicting *thermal conductivity* in nanofluids also presents considerable challenges. Classical models such as *Maxwell's effective medium theory* have been widely modified to include parameters such as *interfacial thermal resistance*, *particle clustering*, and *Brownian motion effects*. However, none of the proposed models consistently align with experimental data across the full range of nanofluid configurations. Researchers like Yu and Choi and others have introduced models that consider particle aggregation and interfacial layers, but even these show deviations when compared with real-world data.

The core issue is again the large number of active variables and their nonlinear interactions, which are difficult to incorporate into a unified model. As thermal conductivity plays a direct role in determining the collector's heat transfer rate, these uncertainties in modeling severely limit the confidence with which nanofluid-based solar systems can be designed or optimized.

11.8.7 Impact on Collector Flow and Thermal Modeling

Numerical simulations are widely used to predict the thermal and hydraulic performance of solar collectors. However, the *accuracy of these models is heavily dependent on the input thermophysical properties* of the working fluid. In the case of nanofluids, the inability to accurately predict viscosity, thermal conductivity, and specific heat under dynamic conditions (especially at varying temperatures and flow rates) leads to significant errors in thermal modeling. In practical terms, this may result in incorrect sizing of collector area, underestimation of pumping requirements, or failure to predict long-term degradation effects.

Moreover, due to the time-dependent nature of nanofluid stability and potential changes in fluid composition, collector models that assume constant fluid properties may become invalid over time. This mismatch between assumed and actual properties could lead to system underperformance, failure to meet thermal demands, or reduced operational lifespan of the components.

11.8.8 Future Perspective

The use of nanofluids in solar collectors holds great promise, particularly in enhancing thermal efficiency and enabling more compact and responsive energy systems. However, their practical deployment is currently constrained by significant technical challenges. These include poor long-term stability, unpredictable rheological behavior, thermal hysteresis, and difficulties in modeling key parameters like Nusselt number, viscosity, and thermal conductivity.

To overcome these barriers, future research must focus on:

- Developing *standardized preparation methods* that ensure long-term nanofluid stability.
- Creating *robust, multi-variable models* that can predict thermophysical properties under dynamic conditions.
- Conducting *long-duration performance tests* to understand degradation and hysteresis effects.
- Investigating *biocompatible and less reactive nanoparticles* to minimize corrosion and environmental impact.

11.9 SUMMARY

The chapter comprehensively discusses the role of nanofluids in advancing the efficiency and application scope of solar thermal collectors. It outlines the operational principles of various collector types and explains how the incorporation of nanoparticles into base fluids can significantly enhance heat transfer properties. Design parameters critical to successful nanofluid integration are reviewed, along with experimental data showcasing performance gains in terms of efficiency and thermal responsiveness. The limitations associated with nanofluid use, including stability concerns and unpredictable rheological behavior, are critically analyzed. Finally, the chapter evaluates the cost-effectiveness and environmental benefits of these advanced fluids, affirming their potential in sustainable energy systems. Through detailed technical insights and real-world case examples, the chapter supports the continued research and application of nanofluid-based solar thermal technologies.

REFERENCES

1. Jaiswal, K. K., Chowdhury, C. R., Yadav, D., Verma, R., Dutta, S., Jaiswal, K. S., Sangmesh, B., & Karuppasamy, K. S. K. (2022). Renewable and sustainable clean energy development and impact on social, economic, and environmental health. *Energy Nexus*, *7*. https://doi.org/10.1016/j.nexus.2022.100118.
2. Mostafizur, R. M., Rasul, M. G., & Nabi, M. N. (2021). Energy and exergy analyses of a flat plate solar collector using various nanofluids: An analytical approach. *Energies*, *14*(14), 4305. https://doi.org/10.3390/en14144305.
3. Wole-osho, I., Okonkwo, E. C., Abbasoglu, S. et al. (2020). Nanofluids in solar thermal collectors: Review and limitations. *International Journal of Thermophys*, *41*, 157. https://doi.org/10.1007/s10765-020-02737-1.
4. Algarni, S. (2023). Evaluation and optimization of the performance and efficiency of a hybrid flat plate solar collector integrated with phase change material and heat sink. *Case Studies in Thermal Engineering*, *45*, 102892.
5. Mahendran, M., Lee, G. C., Sharma, K. V., Shahrani, A., & Bakar, R. A. (2012). Performance of evacuated tube solar collector using water-based titanium oxide nanofluid. *Journal of Mechanical Engineering and Sciences*, *3*, 301–310.
6. Rejeb, O., Ghenai, C., & Bettayeb, M. (2020). Modeling and simulation analysis of solar absorption chiller driven by nanofluid-based parabolic trough collectors (PTC) under hot climatic conditions. *Case Studies in Thermal Engineering*, *19*, 100624.
7. Malika, M., & Sonawane, S. S. (2021). Low-frequency ultrasound assisted synthesis of an aqueous aluminium hydroxide decorated graphitic carbon nitride nanowires based hybrid nanofluid for the photocatalytic H2 production from Methylene blue dye. *Sustainable Energy Technologies and Assessments*, *44*, 100979.
8. Sonawane, S. S., & Sharifpur, M. (Eds.). (2023). *Nanofluid applications for advanced thermal solutions*. Elsevier. https://doi.org/10.1016/C2022-0-00868-0.
9. Thakur, P., Potoroko, I., & Sonawane, S. S. (2022). Numeric and experimental investigations of performance improvement using nanofluids in car radiators. In *Applications of nanofluids in chemical and bio-medical process industry* (pp. 133–162). Elsevier.
10. Malika, M., Bhad, R., & Sonawane, S. S. (2021). ANSYS simulation study of a low volume fraction CuO–ZnO/water hybrid nanofluid in a shell and tube heat exchanger. *Journal of the Indian Chemical Society*, *98*(11), 100200.

11. Thakur, P., Sonawane, S. S., Sonawane, S. H., & Bhanvase, B. A. (2020). Nanofluids-based delivery system, encapsulation of nanoparticles for stability to make stable nanofluids. In *Encapsulation of active molecules and their delivery system* (pp. 141–152). Elsevier.
12. Bethi, B., Sonawane, S. H., Bhanvase, B. A., & Sonawane, S. S. (2021). Textile industry wastewater treatment by cavitation combined with Fenton and ceramic nanofiltration membrane. *Chemical Engineering and Processing-Process Intensification, 168*, 108540.

12 Study of Emulsion Nanofluid Membrane

12.1 INTRODUCTION

12.1.1 Definition of Emulsion Nanofluid Membranes

Emulsion nanofluid membranes are a new and multidisciplinary membrane technology combining synergistically the unique functionalities of emulsions, nanofluids, and membrane systems. These are hybrid membranes that aim at improving the separation process in terms of performance, especially in complex multiphase systems, speeding up mass transfer and selectivity, and also increasing interaction at the interface of separation processes.

Emulsion nanofluid membranes, at their essence, are a dispersed emulsion system which more often than not is a two-immiscible-liquid assembly, that is, either water-in-oil (W/O) or oil-in-water (O/W) to which nanoparticles are added, either to the continuous or to a dispersed phase, or to the oil/water interface itself. Such a combination leads to a high-performance membrane phase due to the multi-functional properties that cannot be obtained using conventional membrane materials.

12.1.2 Key Components and Their Roles in the Membrane

12.1.2.1 Emulsions: The Structural Matrix

At the core of emulsion nanofluid membranes lies the emulsion, which serves as the structural matrix and functional transport medium. An emulsion is a heterogeneous, thermodynamically unstable system composed of two immiscible liquids—typically oil and water—where one liquid is finely dispersed as droplets within the other. These dispersed droplets become unstable so such surface-active agents are introduced to stabilize them, called emulsifiers or surfactants. These molecules reduce the interfacial tension between the immiscible phases and make coalescence of droplets impossible.

The presence of droplets with high surface area to volume ratios allows the enhanced interfacial boundary on which mass transfer, chemical reactions, and other separation processes occur efficiently. This makes emulsions particularly attractive for applications involving liquid–liquid extraction, drug delivery, ion exchange, and pollutant removal.

12.1.3 Types of Emulsions Used in Membrane Systems

The type of emulsion chosen depends on the desired separation mechanism, the solubility of the target compound, and the phase in which transport or extraction is

DOI: 10.1201/9781003628071-12

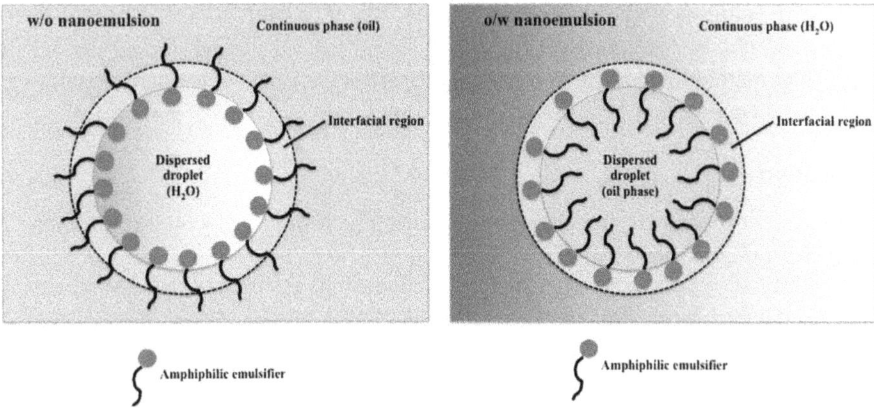

FIGURE 12.1 Illustration of a regular (oil-in-water, O/W) and reverse (water-in-oil, W/O) nanoemulsion droplet stabilized by amphiphilic emulsifiers [1].

to occur. The most commonly used emulsion systems in nanofluid membranes are as follows:

12.1.3.1 Oil-in-Water (O/W) Emulsion

In an O/W emulsion, oil droplets are dispersed within a continuous water phase (Figure 12.1). This configuration is widely used when:

- The target solute is more soluble in the oil phase.
- The external (continuous) environment is aqueous, such as in wastewater treatment or bio-separations.

Formation

- An oil phase containing active components or carriers is emulsified into water using hydrophilic surfactants (e.g., Tween 80, sodium dodecyl sulfate).
- The emulsifier molecules align at the oil–water interface with their hydrophobic tails in the oil and hydrophilic heads in water, stabilizing the droplets.

Role in Membrane

- O/W emulsions act as liquid membranes where hydrophobic pollutants or drugs in the aqueous phase can diffuse into the oil droplets.
- They allow controlled transport of lipophilic solutes and can be functionalized using oil-soluble nanoparticles.

12.1.3.2 Water-in-Oil (W/O) Emulsion

In a W/O emulsion, water droplets are dispersed in a continuous oil phase. This type is suitable when:

- The solute is hydrophilic and needs to be extracted from an organic or hydrophobic feed phase.
- The membrane design requires encapsulation of aqueous reagents or internal stripping agents.

Formation

- Water droplets are emulsified into oil using lipophilic surfactants (e.g., Span 80, sorbitan monooleate).
- Emulsifier molecules orient with hydrophilic heads inward (toward water) and hydrophobic tails outward (into oil).

Role in Membrane

- In emulsion liquid membrane (ELM) systems, W/O emulsions serve as the membrane phase.
- They encapsulate an internal aqueous phase that can react with or absorb the solute after it diffuses through the oil.

Left: In the W/O nanoemulsion, water droplets are dispersed in a continuous oil phase, with emulsifier molecules oriented such that their hydrophilic (polar) heads point inward toward the water core, while their hydrophobic (non-polar) tails extend into the surrounding oil.

Right: In the O/W nanoemulsion, oil droplets are dispersed in a continuous aqueous phase, with emulsifiers aligning their non-polar tails into the oil core and polar heads facing outward toward the surrounding water.

12.1.3.3 Multiple Emulsions: W/O/W and O/W/O

Multiple emulsions are more sophisticated structures used when multi-phase transport or controlled release is required. These systems consist of droplets within droplets, providing sequential barriers for selective transport.

i. **Water-in-Oil-in-Water (W/O/W)**
   - Structure: Tiny water droplets (internal phase) are encapsulated inside larger oil droplets, which are then dispersed in a continuous water phase (Figure 12.2).
   - **Formation**
     - First, prepare a W/O emulsion using a lipophilic surfactant.
     - Then, emulsify this W/O emulsion into water using a hydrophilic surfactant, with low-shear mixing to avoid rupturing the inner droplets.
   - **Applications**
     - Suitable for controlled release of hydrophilic solutes.
     - Common in drug delivery, enzyme encapsulation, and multi-stage extraction.

ii. **Oil-in-Water-in-Oil (O/W/O)**
   - Structure: Oil droplets are first dispersed in water, and these are then further emulsified in an outer oil phase.

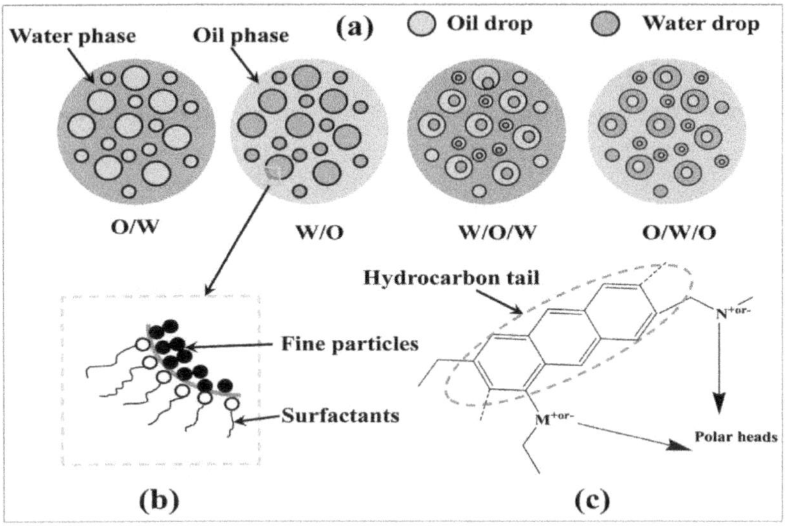

FIGURE 12.2 (a) Types of emulsions and their (b) interfacial characteristics (c) surface activity [2].

- **Formation**
  1. Prepare an O/W emulsion using a hydrophilic surfactant.
  2. Then disperse this emulsion into oil using a lipophilic surfactant.
- **Applications**
  - Used in reversible extraction processes.
  - Can protect sensitive oil-based solutes until they reach the target environment.

Emulsion nanofluid membranes leverage different emulsion types based on the nature of solute separation and target applications. In oil-in-water (O/W) emulsions, oil droplets are dispersed in a continuous phase of water, hence suitable for extracting hydrophobic solutes present in the water medium and are commonly seen in wastewater treatment and pharmaceutical separations (Table 12.1). Conversely, water-in-oil (W/O) emulsions, consisting of water dispersed in continuous oil, gave the possibility of extracting hydrophilic solutes out of organic phases and of internal phase delivery systems. Greater complexity includes water-in-oil-in-water (w/o/w) emulsions, which have water-in-oil droplets that are themselves dispersed in an outer water phase, which makes possible multi-step separations and release of sensitive compounds, for example, in pharmaceuticals and biotechnology. In the same way, oil droplets in water-in-oil (O/entering somehow not in the way it sounds/O) emulsions embrace oil dissolved in water in an oil compartment, in order to provide shelter and controlled delivery of oil-soluble materials in cosmetics, agrochemicals, and drugs. Membrane performance is improved by using the correct type of emulsion such that membrane properties match the polarity, the release profile, and the required extraction condition of the solute. Table 12.2 gives a list and description of some of the more common industrial sectors in which such emulsions may be utilized and the reasons they are used.

TABLE 12.1
Summary of Roles and Mechanisms

Emulsion Type	Continuous Phase	Dispersed Phase	Application in Membrane
O/W	Water	Oil	Extraction of hydrophobic solutes from aqueous phase
W/O	Oil	Water	Extraction of hydrophilic solutes from organic phase; internal phase delivery
W/O/W	Water	Water-in-oil droplets	Multi-step extraction; controlled release systems
O/W/O	Oil	Oil-in-water droplets	Encapsulation of sensitive oil-soluble agents

TABLE 12.2
Industrial Applications of Emulsion Types and Their Functional Roles

Industry Sector	Emulsion Types Commonly Employed	Functional Applications and Description
Petroleum	Water-in-Oil (W/O), Oil-in-Water (O/W)	During extended oil extraction, the nature of produced fluids transitions from W/O to O/W emulsions. These emulsions significantly affect the handling, separation, and processing stages in petroleum operations.
Coal-Based Chemicals	O/W, W/O, Sludge Emulsions	Emulsions emerge during coal liquefaction and gasification processes. The resulting wastewater contains stable, complex mixtures resistant to conventional degradation.
Metalworking	O/W, W/O	Emulsified fluids such as cutting oils, hydraulic oils, and coolants are used to enhance lubrication, reduce heat, and prolong tool life in machining operations.
Food Industry	O/W, W/O, Multilayer Emulsions, Pickering Emulsions	Emulsions are vital in products like milk, creams, beverages, sauces, desserts, and spreads. Pickering emulsions stabilized by proteins or particles are utilized for controlled release, lipid protection, and texture improvement.
Cosmetics & Personal Care	O/W, W/O, Multilayer Emulsions, Pickering Emulsions	Cosmetic formulations utilize emulsions to blend hydrophilic and hydrophobic components. Pickering emulsions with solid particles enhance product stability, texture, fragrance release, and color delivery in creams, lotions, and makeup.

12.1.4 Membranes: The Selective Barrier

Within the context of separation science, the notion of a "membrane" usually implies a solid, semi-permeable barrier that can be used to separate substances in terms of some physical or chemical characteristics. But when we speak about *emulsion nanofluid membranes*, we are not talking about a rigid structure but rather a dynamic, fluid membrane that is emulsified. Here, the membrane is made out of a stabilized emulsion, most commonly one involving an internal dispersed phase (either organic or aqueous), a continuous phase (oil or water), and surface-active agents (surfactants or emulsifiers) that stabilize the interface. This structure is added with nanoparticles to make *nanofluidic emulsions*, which provide the overall system with expanded stability, selectivity, and functional attributes.

Emulsion nanofluid membranes show a *re-configurable, self-healing, and highly selective medium of mass transfer*, unlike solid membranes, which are susceptible to selectivity, clogging, irreversible fouling, and also limited by size. These membranes may either be formed as free-standing emulsions in a batch system or may be immobilized, that is, onto or in a porous support medium such as an inorganic substrate (e.g., silica, alumina), or the pores of polymeric membranes (e.g., polyvinylidene fluoride). The process of immobilization gives more mechanical strength and can be used again in a continuous process. The potential changes since the possibility of immobilizing these fluid membranes on solid surfaces have brought new opportunities in industrial-scale processes, in-between the possibility of laboratory feasibility and implementation in the field.

The primary *role of these membranes* is to:

1. *Facilitate selective mass transfer* of desired chemical species from a feed phase to an internal receiving (stripping) phase.
2. *Act as a selective barrier* that prevents the passage of unwanted or interfering species.
3. *Maintain compartmentalization* of the emulsion structure to ensure that the internal phase does not leak or collapse during operation.

The mass transport through emulsion nanofluid membranes is typically governed by mechanisms such as:

- *Simple diffusion* of solutes across the membrane phase.
- *Interfacial reaction*, where the solute reacts with a carrier or extractant present in the membrane or internal phase.
- *Carrier-mediated transport*, wherein a specific carrier molecule (either dissolved or surface-bound on nanoparticles) selectively binds to the solute and shuttles it across the membrane.

Because of their high interfacial area and dynamic nature, emulsion nanofluid membranes exhibit *superior transport rates* compared to solid-state membranes. This makes them especially suitable for short-contact-time applications such as pollutant extraction, pharmaceutical separations, and selective ion recovery.

12.1.5 Functional Configuration: The Emulsion Liquid Membrane (ELM) System

One of the most advanced and widely studied configurations of these membranes is the *Emulsion Liquid Membrane (ELM)* system. The ELM represents an innovative integration of emulsion technology with membrane science and is especially useful in applications requiring high selectivity and fast kinetics.

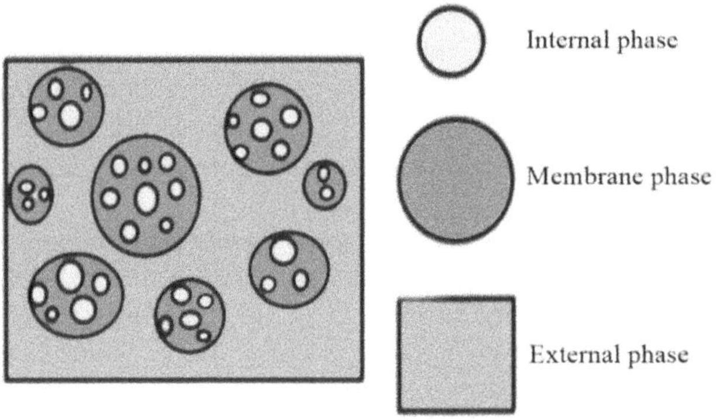

FIGURE 12.3 Structure of oil-in-water-in-oil (o/w/o) emulsion membrane system [3].

As shown in Figure 12.3, the ELM system typically comprises three key phases:

1. **External feed phase (aqueous)**—This contains the target solute to be extracted.
2. **Membrane phase (organic)**—A non-aqueous, immiscible phase containing emulsifiers/surfactants and functional nanoparticles. This phase forms the membrane around the internal droplets.
3. **Internal stripping phase (aqueous)**—A solution encapsulated within the emulsion droplets that reacts with or sequesters the transported solute.

In most applications, the membrane takes the form of a *water-in-oil (W/O) emulsion*, where the internal aqueous stripping phase is dispersed as droplets in the organic membrane solvent. This W/O emulsion is then dispersed into the feed solution (external phase), forming a *double emulsion system* (e.g., W/O dispersed in W), effectively functioning as a mobile membrane system.

12.1.5.1 Transport Mechanism in ELMs

The working mechanism of ELM involves a sequential process:

1. **Diffusion**: The target solute in the feed diffuses across the bulk aqueous phase to the surface of an emulsion droplet.
2. **Partitioning and membrane entry**: The solute partitions from the external aqueous phase into the organic membrane phase, driven by a concentration gradient or facilitated by nanoparticle-assisted binding.
3. **Migration through membrane**: The solute traverses the organic membrane, possibly aided by nanoparticle-carriers or electrochemical gradients.
4. **Stripping reaction**: On reaching the internal aqueous phase, the solute reacts with a stripping agent (e.g., acid, base, or chelating compound), rendering it non-permeable and locking it inside the droplet.
5. **Separation and reusability**: The emulsion is subsequently separated, and depending on stability and performance, can be reused.

This *cyclical and regenerative process* allows for extremely efficient single-step extraction with minimal solvent consumption.

12.2 LIMITATIONS OF EMULSION LIQUID MEMBRANES (ELMS)

Although Emulsion Liquid Membranes (ELMs) have some enviable advantages in separation science, these systems are limited by some of the critical issues. The major weaknesses are those associated with emulsion stability, membrane rupture, scale-up ability of operation, and reproducible levels of extraction efficiency.

The following is a detailed breakdown of these challenges:

12.2.1 Emulsion Instability and Phase Separation

Although Emulsion Liquid Membrane (ELM) systems offer numerous advantages over traditional separation techniques, their practical application is limited by the critical issue of emulsion stability [4]. ELMs are also thermodynamically unstable in nature, and hence tend to separate and coalesce as droplets with time. This instability is more or less blamed on membrane rupture, causing leakage of the internal phase and loss of extractant. It is possible that de-emulsification (or destabilization of the emulsion) can be caused by a number of mechanisms:

i. Coalescence of internal droplets into larger ones
ii. Unification of all emulsion globules of the feed stage
iii. Breaking of the membrane film so as to reveal direct contact between the external and internal aqueous phases
iv. Swelling or shrinking caused by water diffusion across the membrane in W/O/W systems

Additional factors contribute to acceleration of emulsion breakdown due to shear forces, thermal variations, and insufficient concentration of the surfactant that negatively contribute to the effectiveness and specificity of the ELM procedure. Therefore, to achieve stability of the emulsion over the long term, the integrity of the droplets and optimum membrane composition are to be followed.

12.2.2 Membrane Swelling, Rupture, and Leakage

Swelling involves the growth of the volume of an emulsion during extraction as a result of the entry of feed phase components into the globules of emulsion particles. It is a phenomenon that might lead to the breaking of an emulsion and loss of extraction efficiency. The viscosity of the water-in-oil (w/o) emulsion is also changed by the swelling, which makes demulsification of the emulsion difficult, thereby causing internal phase dilution. As a result, solute levels decrease in the inner droplets, and the ultimate rheological performance of the emulsion becomes worse.

Two primary types of swelling are recognized:

Osmotic swelling: The flow of water into the internal phase due to the osmotic pressure difference between the internal and the external aqueous phases, causing the solute concentration to be diluted. Some of the influencing factors are/contain the surfactant concentration, pH, type of extractant, viscosity, and ionic strength.

Entrainment swelling: It is due to mechanical forces such as the size of droplets, salinity, acidity, and the rate of agitation, which allows water to be transported and breaks emulsions.

12.2.2.1 Coalescence

Coalescence is a step where smaller emulsion droplets are recombined into bigger ones, which compromises emulsion integrity. This process typically involves:

a. Initial droplet interaction.
b. Drainage of the thin film between droplets.
c. Membrane film rupture.
d. Droplet fusion.

Interfacial tension, contact time of the droplets, and external shear forces play a role that determines the extent of coalescence. The breakup of droplets and the coalescing process are related to a fine balance in the system of emulsification, which has not been fully probed in viscous fluid behavior.

12.2.2.2 Leakage

Leakage is a process that happens when the emulsion membrane breaks, creating the passage of the internal phase or extracted solutes back to the feed phase. This interferes with the concentration gradient and lessens the power source of the mass transfer and yielding the overall mass extraction. Leakage is influenced by:

- Surfactant type and concentration.
- Viscosity and composition of the internal and external phases.
- Membrane properties (e.g., strength, elasticity).

One of the techniques that is normally used to determine membrane leakage entails measuring the change in pH experienced by the external phase. The extent of leakage (ξ) is determined as a percentage of initial internal phase volume escaped into the external phase using the initial and final volume comparisons as computed using the following equations.

$$Extent\,of\,Leakage = \frac{\begin{array}{c}percentage\,of\,internal\,phase\,volume\,that\\ escapes\,into\,the\,external\,phase\end{array}}{initial\,volume\,of\,the\,internal\,phase}\times 100 \qquad (12.1)$$

Case Study 12.1: Emulsion Liquid Membrane-Based Extraction of Copper and Cyanide Using Lix 7950 [5]

BACKGROUND AND MOTIVATION

Treatment of gold, electroplating, and organic synthesis are some industrial processes that produce large volumes of waste cyanide solutions with heavy metals like copper. Such effluents are hazardous to the environment since both of these metals and cyanide are toxic. The conventional treatment schemes, such as activated carbon adsorption and ion-exchange, are limited to low selectivity, high cost assessment, or lack of efficiency.

The Emulsion Liquid Membrane (ELM) regime and technology come in as a solution that gives simultaneous extraction and stripping multi-stage process in one stage, high mass transfer area, and low usage of chemicals. This case study looks at the application of a guanidine-based carrier, LIX 7950, in ELMs to selectively remove copper and cyanide from alkaline wastewaters.

EXPERIMENTAL OVERVIEW

The research on the optimization of ELM stability and extraction performance under different formulation conditions was undertaken. The membrane phase consisted of the following:

- Carrier: LIX 7950 (Trialkylguanidine extractant applied in the solvent extraction, specifically recovery of copper and cyanide metals in alkaline cyanide solutions)
- Surfactant: Span 80
- Solvent: Kerosene (with modifiers like dodecanol and paraffin)
- Internal Phase: 0.05 M KOH
- External Feed: Dilute copper cyanide solution ([Cu] = 19.3 mg/L, [CN] = 23.6 mg/L, pH = 9.5)

A systematic study to establish the effect of different formulation and process parameters on the ELM Stability (measured in terms of leakage and swelling) and extraction efficiency (measured in terms of copper and cyanide removal):

i. **Mixing time and speed**
   - *Optimal mixing time* was found to be *15 min at 4500 RPM.*
   - Shorter times (e.g., 5 min) led to unstable large droplets and >40% leakage.
   - Excessive mixing (>25 min) formed distorted or thin membranes prone to rupture.

ii. **KOH concentration in internal phase**
   - *0.05 mol/L KOH* ensured >90% removal of Cu and CN with minimal swelling.
   - High KOH (>0.5 mol/L) increased osmotic imbalance, causing leakage.

iii. **Effect of extractant (LIX 7950)**
   - *6% v/v* was optimal, balancing between transport efficiency and viscosity.
   - Higher concentrations did not enhance removal and could hinder dispersion.

iv. **Effect of surfactant (Span 80)**
   - Low Span 80 (3% v/v): Poor emulsion stability and only 55% extraction.
   - Optimal (4–6% v/v): >90% Cu and CN removal with leakage <2%.
   - High (>7% v/v): Increased viscosity, reducing extraction kinetics.

Figure 12.4 illustrates how varying concentrations of Span 80 (a surfactant) in the membrane phase influence:

- Leakage (%)—unintended release of internal phase into the feed.
- Swelling (%)—volumetric expansion of ELMs upon contact with feed.
- Copper and Cyanide Removal (%)—efficiency of ELM in extracting both species.

KEY INSIGHTS

- At low Span 80 concentration (≤3% v/v): Leakage exceeded 20%, and Cu/CN removal efficiency was poor (≈55%). This indicates insufficient stabilization of the emulsion.

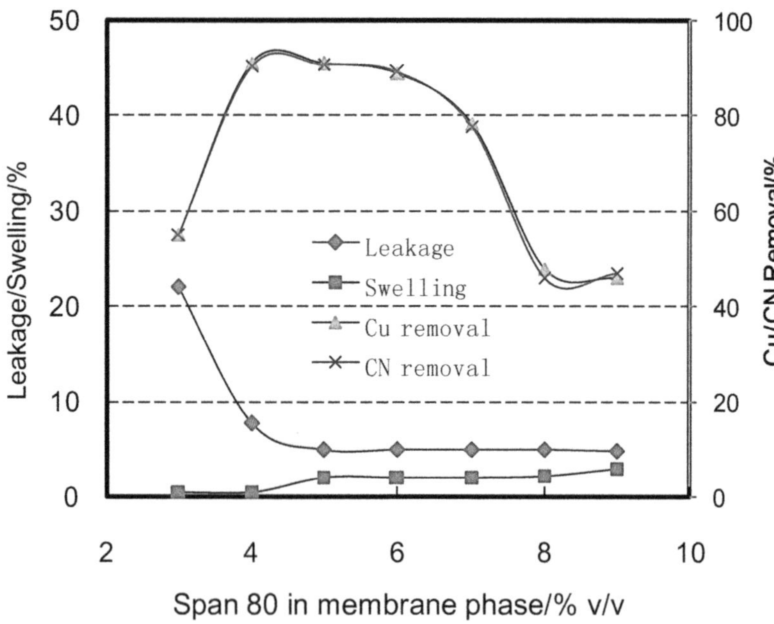

FIGURE 12.4 Effect of Span 80 concentration on ELM leakage, swelling, and removal efficiencies for copper and cyanide. Conditions: 4500 RPM emulsification for 15 min; [Cu] = 19.3 mg/L, $[CN^-]$ = 23.6 mg/L, pH 9.5 [5].

- Optimal Span 80 range (4–6% v/v): Achieved the highest Cu and CN removal (~91%) with minimal leakage (<2%) and low swelling.
- Above 6% v/v: Removal efficiency decreased due to higher viscosity and mass transfer resistance, even though stability remained.

IMPORTANCE

Span 80 ensures ELM stability by lowering interfacial tension. However, excess surfactant may create viscous membranes, limiting the diffusion of copper/cyanide ions. Thus, a narrow optimal window exists for surfactant concentration, crucial for designing efficient ELM systems.

PRACTICAL RELEVANCE

This study exemplifies how fine-tuning emulsion formulation parameters such as surfactant concentration and phase ratios can significantly impact both separation performance and emulsion stability. By achieving >90% removal efficiency with low swelling and leakage, the system presents a viable route for environmentally safe remediation of metal-cyanide-laden effluents.

12.2.3 Challenges in Practical Implementation of Emulsion Liquid Membranes (ELMs)

As much as Emulsion Liquid Membranes (ELMs) have high extraction efficiency and can be used in selective separation, there are various shortcomings associated with the practical use of this technology. The major limitation is the difficulty in reusability, as emulsion degradation leads to the loss of membrane-phase materials—such as extractants and surfactants—during operation, resulting in a complex and, in many cases, economically unviable recovery and regeneration process. Moreover, the fact that the ELM systems have low capacity for scaling down favors the laboratory scale operation compared to the industrial application of the process; the uniformity in the droplet size distribution and emulsion consistency during long run in bulk equipment has been a cause of concern. It also greatly relies on the surfactants, the latter stabilize the emulsion, but their over-utilization may interfere with the selectivity of the extraction, as it raises cost and even has some environmental issues. Moreover, although ELMs have a large interfacial area, internal mass transfer restrictions in terms of diffusion resistances, high viscosity rates, membrane thickness, and interfacial kinetics may considerably decrease the transport efficiencies, impairing their overall performance in continuous productions.

12.3 INTEGRATION OF NANOFLUIDS INTO EMULSION LIQUID MEMBRANES: MECHANISMS, BENEFITS, AND INDUSTRIAL PROMISE

The Emulsion Liquid Membrane (ELM) process is a well-developed separation technology that has the distinct feature of allowing the coupling of extraction and stripping in a one-step and one-time process. This dual functionality achieves high separation efficiency, small-scale process design, and is specifically beneficial when dealing with trace-level solute recovery. However, physicochemical instabilities, which include membrane rupture, swelling, and leaking of inner phases, coupled with the limitation of membrane-recyclability, use of large loads of surfactants in stabilizing emulsions, and lack of reproducible performance during the scale-up process, inherently limit conventional ELM systems.

Because of the disadvantages mentioned earlier, the addition of nanofluids to ELM matrices has recently become one of the disruptive approaches to mitigate these drawbacks. Nanofluids—Thermodynamically stable colloidal suspensions of engineered nanoparticles dispersed in some form of base liquid medium (including water, hydrocarbons, alcohols, or ionic liquids). When placed in the ELM system, nanoparticles are subjected to both the interfacial and bulk phase interaction on the nanoscale to microscale, whether the particles are placed in the organic membrane layer, the aqueous phase within the system or the oil–water interface. These contacts are known to fine-tune the very important membrane properties, for example, interfacial tension, rheology, diffusivity, and colloidal stability, and help confer more structural stability, transport efficiency, and functional flexibility to the ELM.

12.3.1 Overcoming ELM Instabilities through Nanofluid Integration

Among the greatest setbacks of the conventional Emulsion Liquid Membranes (ELMs) are their instability and inability to easily undergo controlled demulsification after extraction. The instability of the emulsions can also take various implications, such as the droplet coalescence, the osmotic swelling, the creaming, the Ostwald ripening, together with the flocculation, all of which cause inconvenience to the structural integrity and separation efficacy of the ELM. The stability in this segment will be characterized by the membrane remaining with its physicochemical architecture and so as not to rupture under high shear, long operation hours, or physicochemical disturbances.

To overcome these limitations, a new type of Emulsion Nanofluid membranes (ENMs) has been derived. These are a further adaptation of traditional ELMs in order to incorporate nanometer-sized solid particles into the membrane structure to provide greatly improved stability at the interfaces, as well as reinforcing structural benefits and increased extraction characteristics.

Because of their ultrafine dimensions (mostly less than 100 nm), nanoparticles possess large area-to-volume ratios and surface energy, which endows them with the capacity to adsorb firmly to oil–water interfaces. This is a type of adsorption leading to the formation of pickering emulsions, wherein nanoparticle monolayers or multilayers assemble to constitute a rigid, steric shell around the emulsion droplets. This interfacial layer has a great inhibition of coalescence and stabilizes the droplets of different sizes—sub-micron to several microns under both static and dynamic conditions.

More so, the position of such nanoparticles at the interface is thermodynamically controlled and reacting under minimum free energy principle, which confirms the fact that once the particles are adsorbed then they get trapped irreversibly at the oil–water interface. This interfacial jamming leads to an energy barrier against droplet deformation, collapse, or coalescence and so promotes the emulsion stability, even under conditions of turbulent flow.

12.3.2 Procedure for the Preparation of Emulsion Nanofluid Membranes (ENMs)

The development of ENMs involves a series of carefully controlled steps that ensure the homogeneous dispersion of nanoparticles, emulsion stability, and functional performance for pollutant extraction. The schematic process flow is illustrated in Figure 12.5, depicting the conceptual understanding of a *water-in-oil-in-water (W/O/W)* system utilized for solute extraction.

Phase Definition

- **External phase (feed solution):** Aqueous solution containing target pollutants, such as heavy metals or pharmaceutical residues.
- **Membrane phase:** An organic phase comprising the carrier fluid, extractant, dispersed nanoparticles (nanofluid), and optionally, a stabilizing surfactant.
- **Internal phase (stripping agent):** An aqueous solution containing a stripping or complexing agent to selectively receive the extracted solute from the membrane.

12.3.2.1 Step-by-Step Protocol

12.3.2.1.1 Component Selection

The foundation of ENM synthesis lies in the judicious selection of components tailored for the target application. This includes:

- **Carrier fluid:** Typically, a hydrophobic organic solvent such as kerosene, hexane, or toluene.
- **Extractant:** A chemical agent (e.g., D2EHPA, LIX 84, Aliquat 336) that selectively binds with the solute of interest.
- **Nanoparticles:** Engineered nanomaterials (e.g., Fe_3O_4, SiO_2, TiO_2, graphene oxide) are chosen based on their interfacial activity, compatibility with the membrane phase, and functionality (e.g., magnetic, antimicrobial, catalytic).
- **Surfactant (optional):** Span 80, Tween 80, or CTAB may be added in small amounts if needed to reduce interfacial tension and assist emulsification.

12.3.2.1.2 Nanoparticle Dispersion Preparation

To ensure a uniform distribution of nanoparticles within the carrier fluid, the nanofluid is prepared by dispersing the nanoparticles using:

- **Ultrasonication:** Utilizes high-frequency acoustic energy to break agglomerates and achieve nanoscale dispersion.
- **Mechanical stirring or vortexing:** Ensures preliminary homogenization.

- **Surface modification:** In cases where particle wettability needs to be adjusted (e.g., hydrophobization of metal oxides), functional groups (–OH, –COOH, $–NH_2$) may be grafted onto nanoparticle surfaces.

The resulting *nanofluid* must be colloidally stable, avoiding sedimentation or agglomeration during the emulsification process.

12.3.2.1.3 Emulsification (W/O Phase Formation)

The nanofluid is emulsified with the internal aqueous stripping phase to form a *water-in-oil (W/O)* emulsion. The objective is to encapsulate fine aqueous droplets (containing stripping agent) within a continuous organic phase.

- **Techniques Employed**
  - High-pressure homogenization
  - Rotor-stator systems
  - Probe sonication

Process parameters such as *rotational speed*, *temperature*, and *energy input* are optimized to ensure monodisperse droplet formation and consistent nanoparticle distribution.

12.3.2.1.4 Formation of W/O/W Emulsion (Double Emulsion)

The W/O emulsion is subsequently dispersed into the *external aqueous feed solution*, resulting in a *W/O/W structure*, as shown in Figure 12.5. In this configuration:

- The internal aqueous phase is encapsulated within oil droplets.
- These oil droplets are, in turn, dispersed in the external aqueous medium.

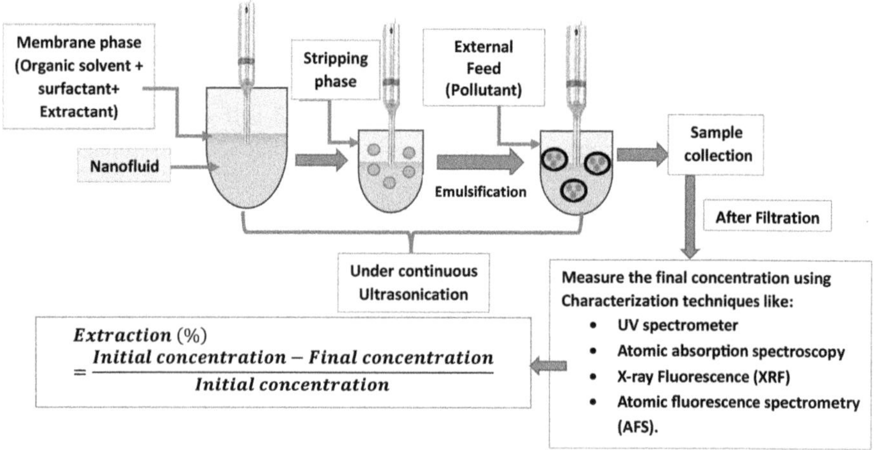

FIGURE 12.5 Methodology for pollutant extraction using ENM-based separation systems.

Nanoparticles residing at the oil–water interface provide a robust barrier, stabilizing the double emulsion and ensuring controlled solute transport.

12.3.2.1.5 Stabilization

To enhance long-term stability:

- *Surfactants or co-surfactants* may be introduced in trace amounts to reinforce nanoparticle adsorption and reduce interfacial tension.
- *pH and ionic strength* are adjusted to optimize zeta potential and prevent aggregation.
- Stabilization is monitored under static and dynamic conditions to assess resistance to creaming, sedimentation, and shear-induced rupture.

12.3.2.1.6 Characterization of ENMs

Accurate characterization is critical to validate the performance and reproducibility of the synthesized ENM. Techniques include:

- **Dynamic light scattering (DLS):** Measures average droplet size and polydispersity index (PDI).
- **Scanning electron microscopy (SEM):** Visualizes the spatial distribution of nanoparticles at the emulsion interface.
- **Rheological analysis:** Determines the viscoelastic properties and flow behavior, important for scaling and process control.
- **Microscopy (optoscope):** Direct visualization of droplets under various magnifications

This section explores the various mechanisms through which nanofluid integration enhances ELM performance, offering insights into their practical applications and future potential.

12.3.2.1.7 Implementation, Optimization, and Performance Evaluation of ENM Systems

The formulated Emulsion Nanofluid Membrane (ENM) is then used in a particular separation or purification process, in case successful characterization has been attained. There are two typical ways: (i) to directly apply the ENM as a liquid phase extraction medium or (ii) to impregnate the emulsion onto a porous support structure to take the form of a Supported Liquid Membrane (SLM). Under the direct mode, the ENM is dispersed to come in close physical contact with the emulsion droplets in a mixing vessel holding the aqueous feed solution, thus duly exposing the pollutant-ridden exterior phase.

The ENM is blended homogeneously with the feed solution with moderate agitation (~10 minutes) in order to favor mass exchange, that is, as shown in Step 3 of Figure 12.5. After the extraction, emulsion feed is transferred to a separating funnel to attain phase separation by gravity (Step 4 of Figure 12.5), where aqueous raffinate phase (treated) and an emulsion-rich phase separate under the influence of the difference in the density of the phases.

The performance of the extraction process would be assessed by taking periodic ~ 5 mL aliquots of the raffinate at various particular times of contact (at 5-minute intervals) and measuring their residual pollutant content either by UV-Visible spectroscopy or by X-ray fluorescence (XRF) (Step 5 of Figure 12.5). Such measurements give time-resolved access to transport kinetics and efficiency of the ENM system.

The extraction efficiency (%) is quantified using the following equation:

$$Extraction\,Efficiency(\%) = \frac{C_{initial} - C_{final}}{C_{initial}} \tag{12.2}$$

$C_{initial}$ and C_{final} are the pollutant concentrations in the feed phase before and after extraction, respectively.

Furthermore, to streamline performance factors such as the concentration of the nanoparticles, the type and concentration of the surfactant, the conditions of emulsification, the main variables are investigated methodically with the view to optimizing their performance. This type of iterative tuning makes sure that the ENM has high selectivity of extraction, mechanical stability, and process endurance of various pollutants of differing nature and concentration levels.

12.4 EXTRACTION MECHANISM IN EMULSION NANOFLUID MEMBRANES (ENMS)

The Emulsion Nanofluid Membranes (ENMs) are developed rapidly across Emulsion Liquid Membranes (ELMs) that are supposed to maximize the efficiency of extraction, stability, and selectivity. In ENMs, nanoparticles are incorporated into the organic phase of a water-in-oil-in-water (W/O/W) emulsion system at a point of high stability so as to enhance the performance of solute transport capabilities and break through the natural drawbacks of classical ELMs, that is, insufficient emission times and emulsion instability (due to surfactant reliance and low reusability).

The main "extraction" role of ENMs is a desired process in which pollutants (e.g., metal ions, dyes, pharmaceuticals) are selected by an external aqueous feed to get into an internal aqueous stripping phase, which is separated by a nanofluid-stabilized oil membrane. This membrane can be viewed to be both a barrier and a facilitator in the sense that it selectively determines which solutes can pass through its matrix. The functionality advantage of a nanoparticle integration includes enhanced mass transfer, interfacial stabilization, and chemical reactivity on a molecular level, which essentially alters the transport dynamics of the system. The general extraction process involves five consecutive stages, namely partitioning, diffusion, chemical interaction, interfacial stripping, and desorption or regeneration. All the stages depend severely on the number, localization, and activity of the nanoparticles captured throughout the membrane phase.

12.4.1 Partitioning: Solute Migration from Feed to Membrane Phase

The first step in pollutant extraction within ENMs is partitioning—the transfer of solute molecules from the aqueous feed solution into the hydrophobic emulsion membrane phase. This process is governed by the distribution coefficient (K_d) of the solute, which reflects its relative solubility between the aqueous and organic phases.

$$K_d = \frac{C_{Membrane}}{C_{Feed}} \tag{12.3}$$

Here, $C_{membrane}$ and C_{feed} are the solute concentrations in the membrane and feed phases, respectively.

12.4.2 Partitioning in ENMs Is Enhanced Through

- **Nanoparticle-Laden Interfaces**: Nanoparticles (e.g., Fe_3O_4, SiO_2, TiO_2) located at the oil–water interface can increase the local affinity for certain solutes due to their surface chemistry.
- **Carrier-Mediated Solubilization**: The membrane may also contain extractants that chemically bind with the solute, altering its solubility profile and driving it into the membrane.
- **Pickering Stabilization Effect**: Nanoparticles form a physical shell at the oil–water boundary, promoting consistent interfacial surface area for sustained partitioning.

This step is crucial, as it sets the stage for transmembrane transport by building the concentration gradient needed to drive diffusion through the membrane.

12.4.2.1 Diffusion through the Nanofluid Membrane Matrix

Following partitioning, solute molecules must traverse the oil membrane phase, which is enriched with dispersed nanoparticles. In conventional ELMs, this diffusion is relatively straightforward. However, in ENMs, the process is more complex due to the microstructural interactions between solutes and nanoparticles.

12.4.2.1.1 Nanostructure-Induced Tortuosity

Nanoparticles dispersed within the membrane create a tortuous diffusion pathway, increasing the path length for solute migration. This geometric constraint can reduce the diffusion rate but also:

- Increases contact time with functional sites.
- Enhances selectivity by promoting preferential solute–nanoparticle interactions.

12.4.2.1.2 Surface-Driven Transport

Functionalized nanoparticles can temporarily adsorb solutes, facilitating a form of surface hopping diffusion. Rather than traveling purely via Brownian motion, solutes may:

- Bind to nanoparticle surfaces.
- Desorb and rebind multiple times en route to the inner interface.

This dynamic process increases selectivity while reducing membrane fouling and back-diffusion.

12.4.2.1.3 Influence of Nanoparticle Type and Concentration

- Hydrophilic nanoparticles (e.g., SiO_2) tend to concentrate near the inner aqueous interface, accelerating delivery into the stripping phase.
- Hydrophobic nanoparticles (e.g., functionalized carbon nanotubes) stay in the organic core, modulating solute retention time.

Optimization of nanoparticle concentration and dispersion stability is thus essential to tune the diffusion kinetics within the ENM structure.

12.4.2.2 Chemical Interactions within the Membrane Phase

The presence of reactive or functionalized nanoparticles introduces new chemical interaction pathways that fundamentally alter the membrane's behavior. Unlike traditional ELMs that rely solely on passive solute transport, ENMs enable in situ chemical reactions, including the following:

12.4.2.2.1 Complexation

Functionalized nanoparticles (e.g., Fe_3O_4 with carboxyl or amine groups) can form metal–ligand complexes, stabilizing specific ions such as Pb^{2+}, Cr^{6+}, or Cu^{2+}. This ensures:

- Enhanced selectivity for target ions.
- Suppression of competitive transport.

12.4.2.2.2 Ion Exchange

Some nanoparticles (e.g., zeolite-coated TiO_2 or layered double hydroxides) facilitate ion-exchange reactions:

- Releasing membrane-bound ions (e.g., Na^+, H^+).
- Capturing target ions from the feed phase.

12.4.2.2.3 Redox Reactions

In redox-sensitive systems, nanoparticles such as nZVI (nano zero-valent iron) can reduce or oxidize solutes during membrane transit, altering their chemical form for easier stripping or precipitation. These mechanisms enable ENMs to act not just as passive barriers but as chemically active interfaces, enabling higher selectivity and reaction-controlled transport.

12.4.2.3 Interfacial Transport into the Stripping Phase

Once the solute reaches the internal interface of the emulsion droplet, it must transition from the membrane phase into the aqueous stripping phase to complete the extraction process.

This interfacial step is driven by:

- Chemical potential difference between the membrane and the stripping phase.
- Stripping reagent reactivity, typically selected to bind, precipitate, or displace the solute.

12.4.2.3.1 Enhanced Interfacial Delivery

Nanoparticles near the internal interface (due to phase affinity or interfacial energy minimization) can:

- Concentrate solutes locally.
- Present reactive groups to aid desorption.

For instance, Fe_3O_4 nanoparticles functionalized with EDTA may bind Pb^{2+} during membrane transit and release it into an EDTA-stripping phase via ligand exchange.

12.4.2.3.2 Kinetic Facilitation

In ENMs, interfacial delivery is faster than in traditional ELMs due to:

- Reduced surfactant interference.
- Higher stability under agitation.
- Lower risk of droplet coalescence.

Thus, nanoparticles contribute to both mass transfer acceleration and interface stabilization, ensuring uninterrupted solute flow.

12.4.2.4 Desorption and Reusability of ENMs

A key innovation in ENMs is the potential for controlled regeneration and multiple use cycles—critical limitations in traditional ELMs. After extraction, solutes may be:

- Recovered from the stripping phase (via pH adjustment or precipitation).
- Separated from nanoparticles (if binding is reversible).
- Reactivated by simple washing or demulsification.

12.5 COMPARATIVE ANALYSIS OF EMULSION LIQUID MEMBRANE (ELM) AND EMULSION NANOFLUID MEMBRANE (ENM) SYSTEMS

12.5.1 Extraction of Pharmaceuticals

Emulsion Liquid Membrane (ELM) and Emulsion Nanofluid Membrane (ENM) are two of the new generation separation technologies that will extricate and extract the contaminants in the aqueous environment with high efficiency and specifically the trace pharmaceutical pollutants like norfloxacin and tetracycline. Nevertheless, their compositions, maturation mechanisms, and enhancement performance strategies differ greatly.

In the conventional ELM formulation, as shown in the study by Muthusaravanan et al., the investigators used water-in-oil emulsion in which a biosurfactant (soapnut saponin) was chosen to emulsify the emulsion mixture. Isoamyl alcohol is used as a solvent, poly-butyl succinimide as a stabilizer, and sodium hydroxide as the internal stripping phase in the membrane phase. The main role performed by this system is to shift the solute (norfloxacin) across the membrane into the internal phase (in and out through diffusion) of the external aqueous phase. It is demonstrated in the paper that optimization by using Artificial Neural Network (ANN) and Response Surface Methodology (RSM) can make an extraction efficiency of up to 91.27%, which demonstrates that a natural surfactant-based ELM system can be both effective and green. Nevertheless, ELM stability is a major obstacle, where stability is usually influenced by the surfactant level, rate of emulsification, and rupture of membranes [6].

Alternatively, the ENM system that was used in the removal of tetracycline instead incorporates nanofluids, which are normally nanoparticles suspended in the membrane phase, in addition to the increased exchange of mass and structural stability that is attained. Contacting Fe_3O_4 or TiO_2 nanoparticles within the membrane is used in this case as a method of enhancing the interfacial surface area, as well as ensuring greater control over the extraction dynamics. These nanoparticles not only ensure enhanced permeability and selectivity but also have extra features like magnetic separation and antimicrobial properties. The use of nanoparticles allows cutting down the amount of surfactants needed to stabilize the emulsion, both structurally and thermally, and improve its self-healing and antifouling properties. Consequently, ENMs exhibit stronger extraction kinetics, are more tolerant to the variations in feed concentration, and may possibly have a longer working life than the traditional ELMs [7].

Although both systems are based on the same principles of operating with the use of emulsion-based interphases, ENMs have multifaceted benefits because of the integration of nanotechnology. Nevertheless, ENMs also bring new complications into play, including nanoparticle agglomeration, increased formulation complexity, and increased cost implications. Also, the mathematical or computational methods used to model the mass transfer and interfacial processes involved in ENMs are more complex in nature owing to extra degrees of freedom that the nanomaterials equip.

In summary, ELMs are an efficient, non-hazardous green solution to pollutant elimination; however, ENMs await at their high-tech corner by adopting nanofluids to support the membranes in terms of functionality, selectivity, and durability. The final decision is always on which kind of solvent to use, ELM or ENM, which is dependent on the characteristics of the contaminant, efficiency that is required, stability of the system, and cost effectiveness.

12.5.2 Wastewater Treatment: Methylene Blue Dye Removal (Table 12.3)

Key Insights

- **Stability** is the biggest advantage of ENM over ELM. Nanoparticles (Al_2O_3) in ENM provide structural support to the emulsion membrane, allowing for longer operational periods and resistance to phase separation.

TABLE 12.3
Comparative Study: ELM vs. ENM for Methylene Blue Removal

Criteria	Emulsion Liquid Membrane (ELM)	Emulsion Nanofluid Membrane (ENM)
Reference	Gadade et al. 2023 [8]	Malika et al. 2023 [9]
Composition	Surfactant (e.g., Monumal 80 or Span 80) + organic diluent (kerosene, n-heptane, etc.) + stripping phase (NaOH)	Surfactant (Span 80) + kerosene + *Al_2O_3 nanofluid* + stripping phase (NaOH or H_2SO_4)
Mechanism	Extraction through W/O/W emulsion, where dye diffuses from the external to internal phase via the membrane	Same mechanism but enhanced by *nanoparticles* that increase stability and interfacial area
Extraction Efficiency	~99.57% with kerosene as diluent and 0.05 M NaOH in 12 minutes	99% with kerosene + Al_2O_3 nanofluid and 0.1 M NaOH in 15 minutes
Emulsion Stability	Thermodynamically unstable over time; susceptible to swelling, coalescence, or phase separation	Significantly *improved stability* (up to 10 days) due to the inclusion of nanoparticles, which prevent coalescence
Optimization Parameters	Surfactant concentration (5 wt%), contact time, agitation speed (400 rpm), feed/strip ratio	Surfactant concentration (0.03 vol%), ultrasonication (15 W, 15 min), stripping agent type, and concentration
Droplet Size Control	Globule size controlled by agitation (typically 0.1–2 mm), unstable at high stirring speeds	*Ultrasound-assisted* emulsion formation produced *nano-sized droplets* (~17.3 nm), enhancing surface area and extraction rate
Preparation Method	Mechanical stirring at 2500 rpm, no ultrasonication	Combined magnetic stirring and *ultrasonication*, improving emulsion uniformity and droplet dispersion
Limitation	Sensitive to surfactant concentration and mechanical mixing conditions; rapid destabilization can occur	Requires specialized equipment (ultrasonicator); higher initial complexity but offers *better long-term performance*

- **Extraction efficiency** in both systems is very high (~99%), but ENM achieves this with a lower surfactant concentration and enhanced process control.
- **Ultrasonication** plays a vital role in ENM by producing smaller, more uniform droplets, increasing interfacial area and mass transfer rate.
- **Scalability**: ENM is considered more promising for industrial applications due to its higher stability and reduced dependency on precise mechanical mixing conditions.

12.6 CHALLENGES AND LIMITATIONS

12.6.1 Scalability Issues

Despite their laboratory success, ENMs face challenges when transitioning to pilot or industrial scale. Maintaining monodispersity of emulsions, consistent nanoparticle dispersion, and interfacial stability under continuous flow or high-shear conditions remains difficult. Process control becomes complex due to the sensitivity of the emulsion system to mixing rates, nanoparticle concentration, and fluid dynamics [10].

12.6.2 Long-Term Stability and Fouling

While nanoparticles enhance droplet stability, ENMs are still prone to long-term instability due to coalescence, Ostwald ripening, or nanoparticle desorption. In continuous operations, fouling caused by the accumulation of solutes or microbial biofilms at the oil–water interface can reduce membrane performance and hinder reuse.

12.6.3 Nanoparticle Leaching and Environmental Safety

Leaching of nanoparticles into the treated effluent poses significant environmental and health risks. The toxicity of metal oxides or engineered nanomaterials (e.g., TiO_2, Fe_3O_4) requires strict post-treatment monitoring and nanoparticle recovery strategies to prevent ecological contamination and ensure compliance with environmental safety standards.

12.7 FUTURE PROSPECTS

12.7.1 Smart Responsive Emulsion Membranes

The development of stimuli-responsive ENMs—capable of responding to temperature, pH, light, or magnetic fields—could enable tunable selectivity, dynamic self-healing, and programmable extraction cycles. These "smart" membranes may allow on-demand release, autonomous phase separation, or in situ regeneration, greatly enhancing operational flexibility [11].

12.7.2 Green Nanomaterials for Sustainable Membranes

Future ENM systems are expected to utilize biodegradable or biosafe nanoparticles (e.g., cellulose nanocrystals, graphene derivatives, or protein-functionalized silica) to minimize environmental risks while preserving functional performance. This shift toward eco-friendly formulations will be critical in applications such as food, cosmetics, and biomedicine [12].

12.7.3 AI-Assisted Design and Predictive Maintenance

Artificial Intelligence (AI) and Machine Learning (ML) can revolutionize ENM optimization by enabling predictive modeling of emulsification behavior, nanoparticle-surfactant interactions, and extraction kinetics. Real-time data analytics can also facilitate predictive maintenance, detect membrane failure precursors, and suggest optimal operating conditions, improving process reliability and energy efficiency [13, 14].

12.8 SUMMARY

Emulsion Nanofluid Membranes (ENMs) offer a powerful platform that synergizes the dynamic transport properties of emulsions with the stability and functional versatility of nanofluids. Their W/O/W structure enables simultaneous extraction and stripping in a compact format, while embedded nanoparticles enhance interfacial stability, mass transfer rates, and functional selectivity. Compared to traditional Emulsion Liquid Membranes (ELMs), ENMs address key limitations such as emulsion rupture, swelling, and surfactant dependency.

Despite these advantages, challenges such as scalability, nanoparticle leaching, and long-term emulsion integrity must be addressed through advanced formulation techniques and systemic process integration. Emerging developments in green nanomaterials and intelligent membrane systems point to a promising future for ENMs in fields ranging from industrial wastewater treatment to pharmaceutical purification. Their ability to couple separation, reactivity, and adaptability in a single phase-rich interface positions ENMs as a transformative technology in next-generation separation science.

REFERENCES

1. Kupikowska-Stobba, B., & Kasprzak, M. (2021). Fabrication of nanoparticles for bone regeneration: New insight into applications of nanoemulsion technology. *Journal of Materials Chemistry B*, *9*(26), 5221–5244.
2. Tian, Y., Zhou, J., He, C., He, L., Li, X., & Sui, H. (2022). The formation, stabilization and separation of oil–water emulsions: A review. *Processes*, *10*(4), 738. https://doi.org/10.3390/pr10040738.
3. Sadeghi, S., Zargarzadeh, L. (2025). Gibbsian surface thermodynamic analysis of emulsion liquid membranes. *Scientific Reports*, *15*, 6401. https://doi.org/10.1038/s41598-025-90774-x.
4. Hussein, M. A., Mohammed, A. A., & Atiya, M. A. (2019). Application of emulsion and Pickering emulsion liquid membrane technique for wastewater treatment: An overview. *Environmental Science and Pollution Research*, *26*(36), 36184–36204.
5. Lu, D., Chang, Y., Wang, W., Xie, F., Asselin, E., & Dreisinger, D. (2015). Copper and cyanide extraction with emulsion liquid membrane with LIX 7950 as the mobile carrier: Part 1, emulsion stability. *Metals*, *5*(4), 2034–2047. https://doi.org/10.3390/met5042034.
6. Muthusaravanan, S., Priyadharshini, S. V., Sivarajasekar, N., Subashini, R., Sivamani, S., Dharaskar, S., & Dhakal, N. (2019). Optimization and extraction of pharmaceutical micro-pollutant-norfloxacin using green emulsion liquid membranes. *Desalination and Water Treatment*, *156*, 238–244.
7. Kohli, H. P., Gupta, S., & Chakraborty, M. (2019). Stability and performance study of emulsion nanofluid membrane: A combined approach of adsorption and extraction of ethylparaben. *Colloids and Surfaces A: Physicochemical and Engineering Aspects*, *579*, 123675.

8. Gadade, P. R., Sardare, M. D., & Chavan, A. R. (2013). Studies of extraction of methylene blue from synthetic waste water using liquid emulsion membrane technology. *The Canadian Journal of Chemical Engineering*, *91*(1), 84–89.
9. Malika, M., Pargaonkar, A., & Sonawane, S. S. (2023). Application of emulsion nanofluid membrane for the removal of methylene blue dye: Stability study. *Chemical Papers*, *77*(7), 3967–3977.
10. Malika, M., & Sonawane, S. S. (2020). Effect of nanoparticle mixed ratio on stability and thermo-physical properties of CuO-ZnO/water-based hybrid nanofluid. *Journal of the Indian Chemical Society*, *97*(3), 414–419.
11. Sonawane, S., Thakur, P., & Paul, R. (2020). Study on thermal property enhancement of MWCNT based polypropylene (PP) nanocomposites. *Materials Today: Proceedings*, *27*, 550–555.
12. Malika, M. M., & Sonawane, S. S. (2019). Review on application of nanofluid/nano particle as water disinfectant. *Journal of Indian Association for Environmental Management (JIAEM)*, *39*(1–4), 21–24.
13. Malika, M., Bhad, R., & Sonawane, S. S. (2021). ANSYS simulation study of a low volume fraction CuO–ZnO/water hybrid nanofluid in a shell and tube heat exchanger. *Journal of the Indian Chemical Society*, *98*(11), 100200.
14. Thakur, P., Sonawane, S., Potoroko, I., & Sonawane, S. H. (2021). Recent advances in ultrasound-assisted synthesis of nano-emulsions and their industrial applications. *Current Pharmaceutical Biotechnology*, *22*(13), 1748–1758.

13 RSM and ANN Studies for the Intensification of Nanofluid Application

13.1 THE NEED FOR OPTIMIZATION AND PREDICTIVE TOOLS IN NANOFLUID APPLICATIONS

The exponential increase of thermal and energy systems with high efficiency has triggered a dramatic progress in studying the area of heat transfer fluids, which has resulted in the area of nanofluids as a very promising category of materials. Nanofluids are artificially made using nanoparticles, regularly under 100 nanometers, that are suspended in conventional base fluids, including water, ethylene glycol, or oil. They are also able to produce unique thermal properties, such as increase in thermal conductivity, high convective coefficients of heat transfer, and adjustable rheology, among others, and this feature has positioned them as potential game-changers in a broad scope of operations that include power industries, automobiles, electronics cooling, chemical systems, as well as renewable energy systems. Nevertheless, even though quite evident, the real application of nanofluids is restricted by the fact that their operation in different operating conditions is partitioned in many cases with difficulty.

Among the first things to be considered before engaging in the use of optimization and predictive technologies like RMS and ANN in the improvement of thermal properties are the major thermal characteristics of nanofluid that determine their effectiveness. Such properties are density, specific heat, thermal conductivity, surface tension, and viscosity. All these parameters are important in establishing the adequacy of heat transfer, stability, and flow characteristics of nanofluids in real-life situations. Measurement of these properties with enough precision is the starting point experimentally of any modeling or simulation effort. Common instruments for assessing these properties are illustrated in Figure 13.1.

Besides, the behavior of nanofluids depends on many connected parameters, including the concentration of the nanoparticles, their shape, size, properties of the base fluid, flow regime, and temperature gradients. The variables do not work in isolation, but instead, there is a complex and non-linear interaction between them in ways that render experimental studies by themselves inadequate to the attainment of a complete understanding, or to an optimum application. It is here that optimization and predictive modeling tools become vital. Conventional empirical models cannot be applied that well and precisely to reveal the entire behavior of nanofluid systems. Thus, complex calculation methods are needed to develop, examine, and optimize nanofluid-based processes.

DOI: 10.1201/9781003628071-13

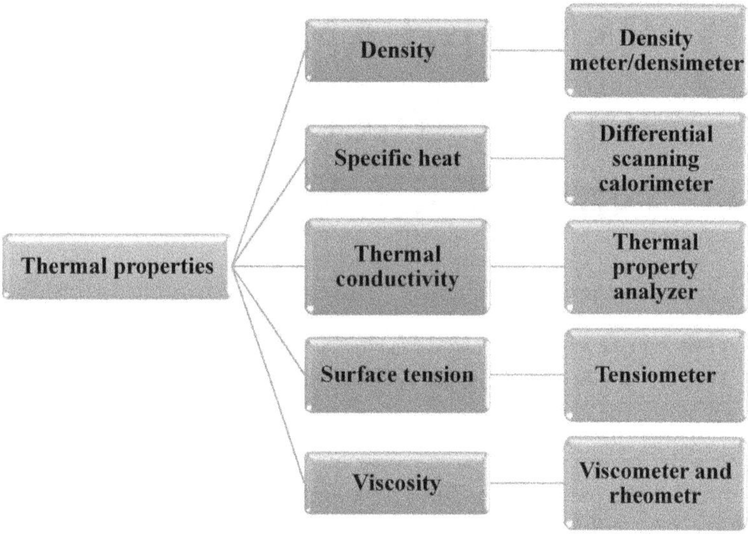

FIGURE 13.1 Thermal properties of nanofluids and instruments used for their measurement [1].

Response Surface Methodology (RSM) and *Artificial Neural Networks (ANNs)* are among the best and common authentic tools used as far as this is concerned. RSM is a statistical method that is used to formulate empirical models via designed experiments, thereby allowing the examination of whether multiple input variables and interactions are encouraged or curtailed in regard to one or various output responses. The given approach makes it possible to quite effectively find the best operating conditions with the minimum number of experiments carried out, and the maximum amount of information generated. At the same time, ANNs provide a machine-learning way to model nonlinear systems in a data-driven approach. They also have the ability to detect intricate patterns in big datasets and can make wider predictions on diverse operating conditions despite the inadequacy of conventional theoretical models.

The combination of RSM and ANN in nanofluid studies provides not only a better forecasting potential but also gives way to process intensification, that is, the improvement of process performance, energy efficiency, and system reliability. This, in industry terms, means the introduction of more efficient cooling systems, better optimization of energy utilization, better products, and less effect on the environment. The intelligent control methods, real-time monitoring of the system, and the digital twins, facilitated by the use of these tools, are an essential part of the transition to smart manufacturing and *Industry 4.0.*

The given chapter addresses the interactivity between the methodologies of RSM and ANN, and their use in nanofluids. It gives a critical analysis of the applications of these tools in modeling, predicting, and optimizing the behavior of nanofluids in diverse industrial settings. Their applications in thermo-systems, as renewable energy tools, in process engineering, in electronic cooling, and in biomedical technologies are discussed. Through knowledge and practice of these superior methodologies, engineers and researchers can open up new levels of efficiency, design of the system, and establish a path to the wider industrial adaptation of nanofluid technologies.

13.2 ROLE OF RSM AND ANN IN PROCESS INTENSIFICATION

Since the usage of nanofluids expands in sophisticated thermal systems, efforts to achieve greater efficiency of the process, enhanced process control, and design precision become more important. Process intensification refers to the methodological optimization of process performance with respect to innovative approaches that save expended energy, material use, and increase functional output. Such nanofluid intensification in reality is not simply a question of choosing an appropriate base fluid or nanoparticle: this needs to be highly optimized through multiple interacting parameters within realistic feasibility constraints. This difficulty has pointed to the further attraction to computational and statistical hypotheses that can transcend traditional experimentation.

Response Surface Methodology (RSM) is essential in the implementation of a controlled experiment whose informational content is high. Instead of testing the effect of individual variables, it is the case that RSM adds together the overall effects of many of the variables at the same time, revealing trends and correlations that can be missed. The approach produces a response surface model, which can be exploited to traverse the parameter space in the most efficient manner, thus realizing system outputs of interest—be it a higher thermal conductivity or a lower viscosity of the system. Combining design of experiments (DoE) ideas with regression-based equations enables RSM to enable engineers to make informed change decisions that would enhance process stability and operation accuracy.

In addition, the Artificial Neural Networks (ANN) understand a data-aligned modeling appraisal that has the potential to define complex and non-linear dependencies in nanofluid systems. ANNs, unlike statistical models, need not be based upon predetermined equations; ANNs learn based on experimental or operating data and can refine the internal structure to improve predictive capability over time. These flexibility properties render them particularly useful when dealing with a complex or dynamic system, with varying or non-linear interactions in the inputs. As an example, multi-step responsive control strategies and predictive maintenance measures may be trained by ANN models describing real-time behavior of nanofluids with different loads.

Nanofluid-based process design, which involves the integration of RSM and ANN, can be seen as a big step forward in terms of engineering approach. RSM is along with constructed optimization and understandable translation, and ANN is preferable with nonlinear processing and large data. Collectively, they constitute a hybrid framework that is not only able to improve the accuracy of predictions but is also automatable, scalable, and multi-objective optimizable. The synergy is increasingly used in industrial applications in heat exchangers, refrigeration systems, machining operations, and solar energy harvesting units, where the ability to model fluid behavior accurately is directly related to the performance of the device.

Essentially, the implementation of RSM and ANN allows shifting the paradigm of passive selection of fluids to active process design. These devices enable scientists/engineers to develop a more intelligent and flexible system that can react to the operation requirement with a higher level of efficiency, and it is a key milestone in the development of nanofluid technologies.

13.3 FUNDAMENTALS OF RSM AND ANN

13.3.1 Response Surface Methodology (RSM)

13.3.1.1 Introduction to RSM

In applied sciences and engineering, it is necessary to know how many input variables are related to single- or multiple-output response in order to develop efficient and optimized systems. To help us model and analyze such intricate relationships as they occur, other statistical and mathematical techniques, outlined to formulate, is *Response Surface Methodology (RSM)*, specifically applied in case of optimizing a process or product. RSM was developed in the 1950s and is a combination of *Design of Experiments (DoE)* principles, regression analysis, and optimization theory with the aim of investigating how various variables could influence the performance of a desired outcome.

13.3.1.2 What Are DOE's

13.3.1.2.1 Design of Experiments (DoE): A Fundamental Framework for Scientific Investigation

Design of Experiments (DoE) is a statistical and systematic methodology that enables experimentation to be performed so that valid and objective conclusions can be drawn about the impact of various factors on one or more responses. Instead of manipulating one input variable each time, which is ineffective and frequently fails to simulate the influence of interaction effects, DoE enables a researcher to examine the concurrent effect of multiple variables and their interactions with very few experimental runs.

The DoE has its origins in the earlier groundbreaking contributions made by Sir Ronald A. Fisher in the early 20th century, who codified statistical ideas that were later to revolutionize how experiments are designed and analyzed. DoE can also give a methodical approach to investigate the influence of important input parameters in engineering, manufacturing, biotechnology, and nanotechnology, namely nanofluid systems, and how input parameters can be optimized.

13.3.1.2.2 Key Principles of DoE

At its core, DoE is built on several key principles:

1. **Factorial structure**: Instead of isolating one variable at a time, factorial experiments test all combinations of variables (called *factors*) at different levels, providing a full picture of how the inputs affect the outputs.
2. **Randomization**: Experimental runs are randomized to eliminate bias caused by uncontrolled factors such as environmental changes or equipment fluctuations.

3. **Replication**: Repeating experimental runs helps assess the variability (or noise) inherent in the system and increases the reliability of conclusions.
4. **Blocking**: Used to account for known sources of variability (such as time of day or equipment type), blocking helps isolate the effects of the factors of interest.

13.3.1.2.3 Why DoE Is Crucial in Nanofluid Applications

Nanofluid systems are governed by complex relationships among thermal conductivity, particle concentration, flow rate, temperature, and viscosity. Traditional experimentation would require exhaustive testing of all combinations, which is impractical and resource-intensive. DoE addresses this by:

- *Reducing the number of experiments* without compromising the quality of insights.
- *Revealing interaction effects* between variables that cannot be detected through one-variable-at-a-time methods.
- *Providing a mathematical model* that can be used for prediction and optimization.

For example, in a study involving nanofluids for cooling electronic devices, DoE can help determine how nanoparticle type, concentration, and base fluid interact to affect heat transfer and pressure drop.

13.3.1.2.4 Types of Experimental Designs in DoE

Several designs are used in DoE, each suited to different purposes:

- **Full factorial designs**: Explore all possible combinations of factor levels. Suitable when the number of factors is small.
- **Fractional factorial designs**: Use a subset of combinations to reduce the experimental burden. Useful in early screening phases.
- **Central Composite Design (CCD)**: Augments factorial designs with axial and center points to model curvature in the response surface. Often used in RSM.
- **Box–Behnken Design (BBD)**: Efficient quadratic design that avoids extreme values. Useful for safe or constrained systems.
- **Mixture designs**: Applied when the factors are proportions of a mixture (e.g., ratios of nanoparticles in different solvents).
- **Plackett–Burman designs**: Screening designs that help identify the most influential factors among many.

In the context of *nanofluid applications*, where factors such as nanoparticle type, volume concentration, flow velocity, and temperature significantly influence thermal and flow behavior, RSM provides a powerful toolset for systematically investigating these effects. Unlike traditional one-variable-at-a-time approaches, RSM considers interactions and curvature effects, making it ideal for processes where multiple parameters act simultaneously and nonlinearly.

13.3.1.3 Conceptual Basis and Mathematical Formulation

At the core of RSM lies the construction of a mathematical model—usually a second-degree polynomial—that approximates the true functional relationship between the input variables and the system response. This model takes the general form:

$$Y = \beta_0 + \sum_{i=1}^{k} \beta_i X_i + \sum_{i=1}^{k} \beta_{ii} X_i^2 + \sum_{i<j}^{k} \beta_{ij} X_i X_j + \varepsilon \tag{13.1}$$

where:

- β_0 is the intercept.
- β_i are the linear coefficients.
- β_{ii} are the quadratic coefficients.
- β_{ij} represent interaction terms.
- ε is the random error.

Further, Y represents the response (e.g., heat transfer coefficient), X_i and X_j are input variables (e.g., nanoparticle concentration, flow velocity), and ε captures the unexplained variability or noise in the system. The coefficients β are estimated using regression analysis. This empirical model is capable of revealing curvature and interaction effects, making it highly suitable for physical systems that exhibit non-linear behavior, as is often the case in nanofluid dynamics. This model allows not only for the prediction of responses but also for the identification of optimal operating conditions and sensitivity to individual variables.

13.3.1.4 Design of Experiments (DoE): The Foundation of RSM

The predictive and optimization capabilities of Response Surface Methodology (RSM) depend fundamentally on the quality and structure of the data collected during the experimental phase. This data must represent the behavior of the system across the range of variables being studied. *Design of Experiments (DoE)* provides the formal strategy for organizing experimental trials such that the resulting data can support accurate, reliable, and interpretable modeling.

DoE ensures that all relevant variables, and especially their *interactions and nonlinear effects*, are adequately captured without requiring exhaustive testing of every possible combination. Instead of relying on random or sequential testing, DoE employs *structured experimental matrices*, allowing researchers to extract maximum information from a minimal number of experiments. In RSM, two experimental designs dominate the practice of building second-order polynomial models: *Central Composite Design (CCD)* and *Box-Behnken Design (BBD)*. Each has unique features, advantages, and suitable contexts for application.

13.3.1.4.1 Central Composite Design (CCD)

Central Composite Design is one of the most frequently used DoE frameworks for building quadratic models in RSM. It is particularly powerful when there is a need to model *nonlinear responses* or estimate curvature in the response surface.

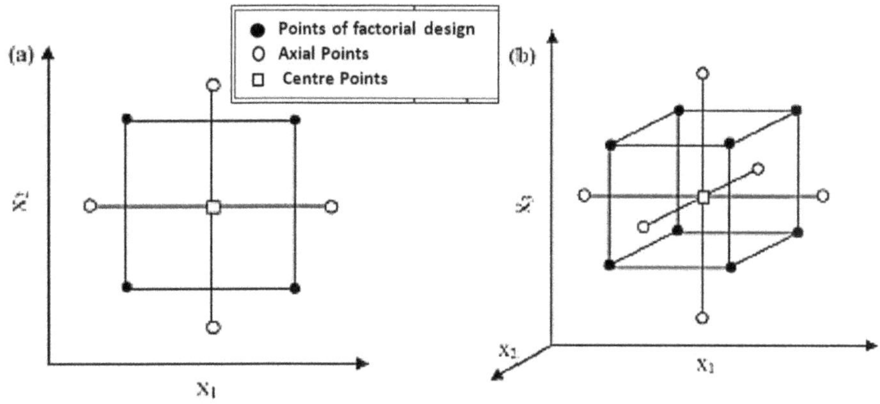

FIGURE 13.2 Geometry of a Central Composite Design (CCD) in 2D [2].

(Illustrate a 2D design space showing 4 factorial points, 4 axial points placed along each axis, and 1 center point)

A CCD consists of three components (Figure 13.2):

- **Factorial or fractional factorial points**: These are the corner points of the design space, representing all possible combinations of the high and low levels of the input factors.
- **Center points**: Placed at the midpoint of all factors, center points are replicated to estimate experimental error and test for curvature in the response surface.
- **Axial points (star points)**: These are located along each axis, at a specified distance from the center, to allow estimation of quadratic (squared) terms. The distance of these points from the center is typically denoted by the parameter α\alphaα, which determines the rotatability of the design.

13.3.1.5 When to Use CCD

- When a *nonlinear response is anticipated* and a full exploration of curvature is required.
- When *axial (extreme) values* of variables are experimentally feasible and safe.
- When one wants to build a *rotatable design*, which maintains uniform prediction error throughout the design space.
- Suitable for *model validation*, as CCD allows a comprehensive estimate of both main and interaction effects, as well as curvature.

For example, in nanofluid systems, CCD can be used to evaluate how both *nanoparticle concentration* and *inlet temperature* affect *heat transfer rate*, especially when the response does not follow a straight line.

13.3.1.5.1 Box-Behnken Design (BBD)

Box-Behnken Design is another widely adopted RSM design, notable for its efficiency and safety. Unlike CCD, it *does not include axial points* and hence avoids extreme values of the variables, which may be *inaccessible or unstable* in certain experimental setups.

In a BBD:

- Experimental points are located at the *midpoints of the edges* of the design space, along with replicated center points.
- Each factor is combined with every other factor at three levels (low, medium, and high), but not all combinations are tested.

13.3.1.6 When to Use BBD

- When the experimental region must *avoid extreme conditions*—for example, high nanoparticle concentrations that may cause agglomeration or system instability.
- When a *relatively small number of experimental runs* is desired.
- When a *spherical or ellipsoidal design space* is preferred.
- Best for systems where *intermediate values* provide more reliable data than boundary values.

BBD is often used in applications such as optimizing *coolant flow rate and particle size* in nanofluid-enhanced heat exchangers, where testing extreme concentrations or temperatures might lead to clogging or system failure.

13.3.1.7 Important Point

The choice between CCD and BBD depends on the *nature of the process*, *the safety of boundary testing*, and *available experimental resources*. For complete, high-resolution modeling of processes with strong curvature, CCD is ideal. For safer, more economical studies where extreme conditions are impractical, BBD provides an efficient alternative.

13.3.1.8 Key Stages in RSM Application

13.3.1.8.1 Selection of Critical Variables and Experimental Range

The process begins with identifying key process parameters that significantly influence the response. For nanofluid systems, these typically include nanoparticle volume fraction, particle type, base fluid, temperature, and Reynolds number. Literature review, prior experiments, or screening designs such as *Plackett-Burman* can be used to determine which variables warrant deeper exploration.

13.3.1.8.2 Choosing an Appropriate Experimental Design

RSM relies on statistically planned experiments that enable efficient exploration of the parameter space. The most widely used designs include:

- **Central Composite Design (CCD):** This design includes factorial points, axial points (to explore curvature), and center points (to estimate experimental error). It is particularly suited for developing quadratic models.

- **Box-Behnken Design (BBD):** This design requires fewer experiments and is ideal when operating conditions should avoid extreme values. It is constructed from combinations of midpoints on the edges of a multidimensional cube and center points.
- **3-level full factorial design:** Appropriate when only a few variables are involved (typically three or fewer), this design tests every possible combination of three levels (low, mid, high) of the variables.

13.3.1.8.3 Model Building and Validation

After conducting the designed experiments, data are used to fit a polynomial model using the **least squares regression** method. This model is validated through various diagnostic tests:

- **R^2 and adjusted R^2:** Measure how well the model explains the variability in the response.
- **Lack-of-fit test:** Checks whether the model adequately describes the experimental data.
- **Residual analysis:** Evaluates the distribution and independence of errors.

These steps ensure that the developed model is statistically sound and capable of making reliable predictions.

13.3.1.8.4 Visualization and Interpretation

After validation, graphical simulations are obtained via the model that displays graphical representations of the model, for example, *3D response surfaces plot*. Such plots assist in visualizing the influence of the combinations of the variables on the response and grant an intuitive equation of the behavior of the system. Figure 13.3 shows a plot of response surfaces and contours in three dimensions that serves as an illustration during the analysis of response surface methodology (RSM). The visualizations play a key role in the explanation of how the responses of a system are sensitive to changes in two independent variables—which are labeled in this case under *Factor A* and *Factor B*. These representations play an important role in optimization works in which the objective would be to determine optimum or critical operating conditions. All plots give an indication of different types of surface characteristics: (a) minimum, (b) maximum, and (c) saddle point.

13.3.1.8.4.1 Minimum Response Surface

Figure 13.3(a) shows a convex surface to present a minimum point of the response surface. The bottom part of the surface represents the minimum of the response variable, which may be regarded as an expression of the most efficient or cheapest state of a specific process. In the example, the shape of the surface and the concentric contour lines around the center point, which implies that any action beyond the center in either direction of Factor A or Factor B, is a contributor to the increase in the response. That means that the point of centrality is a real local minimum and should be optimized in case one aims at minimizing variables like viscosity, pressure drop, or thermal resistance. Practically, this could be applied to establish the minimum, that is, controllable nanoparticle concentration and flow rate, which would not exceed the thermal transport performance of the desired nanofluid cooling system.

13.3.1.8.4.2 Maximum Response Surface

Figure 13.3(b) displays a concave surface, which explains a maximum response. There, the surface is peaked in the middle of the design space, and values go down toward the edges. Such a surface is something to wish for in cases where the optimization criterion is to maximize an advantageous property like thermal conductivity, Nusselt number, or heat transfer rate. The contour plot, associated with them, again depicts concentric circles, though here the maximum is the global optimum. Optimization of maximizing enhancements in thermal performance is achieved by engineers in a bid to establish the most desirable set of variables, for example, the nanoparticle concentration being high and the flow velocity medium.

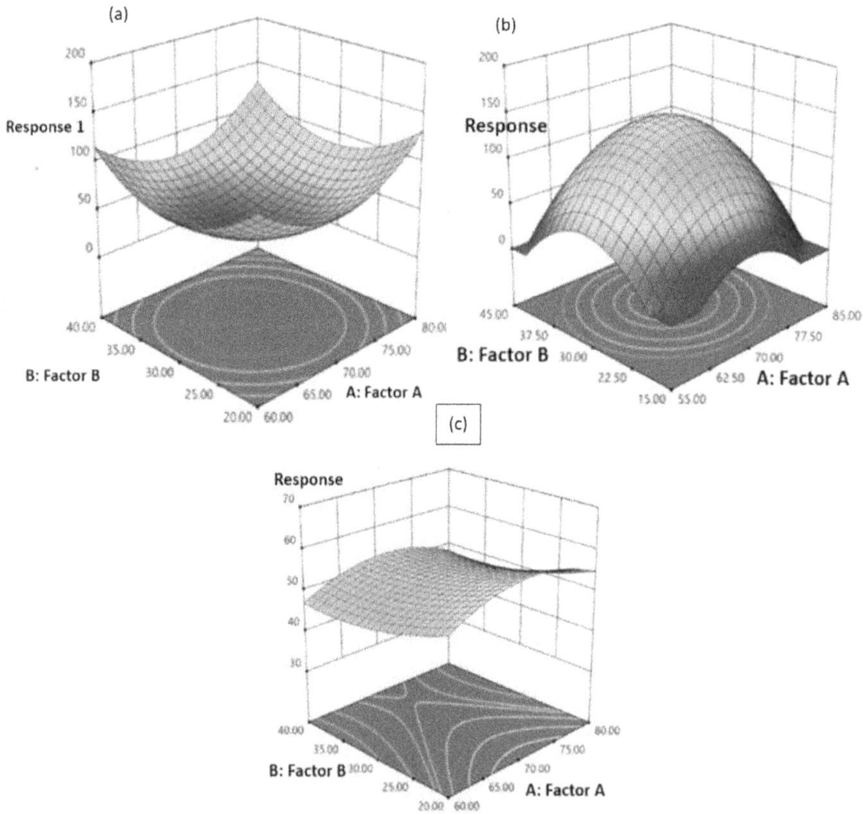

FIGURE 13.3 Interpretation of 3D response surfaces and contour plots in RSM: (a) Minimum, (b) Maximum, (c) Saddle point response surface.

13.3.1.8.4.3 Saddle Point Surface

Figure 13.3(c) represents a saddle point where the surface rises in one dimension and falls in another direction. This kind of stationary point is neither maximum nor minimum. Rather, it is a change zone of the response landscape. Saddle points are usually unstable in optimization problem and should be avoided as desired operating points. The lines of contour here come in the form of crossing curves as opposed to concentric circles, as in the case of the MARZEN denominations. In complex systems, saddle points are common when the response to a combination of variables is part of conflicting response dynamics. As an example, raising one parameter, like temperature, can increase the heat transfer rate in a nanofluid system, but at a certain level of nanoparticles loading, the system becomes unstable or agglomerates in which the increase is canceled.

These graphical tools also assist in identifying *saddle points*, *maxima*, or *minima*, and in understanding the nature of variable interactions.

13.3.1.8.5 Optimization Techniques

Optimization in RSM can be carried out using both graphical and numerical methods:

- **Graphical optimization:** Overlay plots of multiple contour maps can identify operating zones where all performance criteria are satisfied. This is useful when optimizing more than one output (e.g., maximizing heat transfer and minimizing viscosity).
- **Numerical optimization (Desirability function):** For more complex systems, a desirability function transforms each response into a scale from 0 (undesirable) to 1 (most desirable). The combination of input variables that maximizes the overall desirability index is selected as the optimal point.

13.3.1.8.6 Software Platforms for RSM

Several commercial and open-source software packages support RSM implementation. These platforms offer user-friendly interfaces for design generation, statistical analysis, visualization, and optimization.

13.3.1.8.6.1 Design-Expert

Design-Expert is one of the most powerful and user-friendly tools specifically built for RSM and DOE. It offers a library of experimental designs (CCD, BBD, etc.), automated regression analysis, and interactive 3D surface plots. The software includes desirability-based optimization features, making it ideal for both single- and multi-objective problems.

13.3.1.8.6.2 Minitab

Minitab is widely used in quality control and process engineering. Its DOE module supports factorial, response surface, and mixture designs. It offers integrated ANOVA, regression diagnostics, and contour plotting, which are critical for validating RSM models.

13.3.1.8.6.3 JMP

JMP (from SAS) provides a visual, interactive environment for data analysis and DOE. It supports advanced modeling, scripting, and dynamic plotting, ideal for researchers who want both analytical power and intuitive design.

13.3.1.8.6.4 MATLAB and Python

For custom applications, platforms like MATLAB (with the Statistics and Machine Learning Toolbox) and Python (using libraries such as pyDOE, statsmodels, and scikit-learn) allow full control over design, model building, and simulation. These tools are suited for users with programming expertise who wish to integrate RSM into larger workflows or combine it with machine learning.

13.3.2 Case Study: Optimization of Heat Transfer Rate in Magnetized Hybrid Nanofluid Flow Using RSM and ANOVA

13.3.2.1 Introduction and Background

Advancements in the exponential growth of thermal management needs in energy and aerospace systems have availed the evolution of high-performance coolants like hybrid nanofluids. Such fluids are a combination of two or more nanoparticles, including diamond and silver, to enhance thermal conductivity, specific heat, and electrical conductivity, thus being ideal for high-flux requirements. Put to practical application under systems with magnetohydrodynamic (MHD) effects and porous media to heat transfer by radiation, these fluids provide revolutionary effects in the control of heat flow across stretching surfaces.

The present case study constitutes a computational and statistical optimization of perfect heat transfer in a hybrid nanofluid consisting of diamond silver nanoparticles in water, which flows over a permeable exponentially stretching sheet. The thermal radiation, the magnetism field, the Darcy dissipation, the suction, as well as the nonlinear buoyancy are included in the investigation. Combining Response Surface Methodology (RSM) with Analysis of Variance (ANOVA), a predictive model is established to find the optimal Nusselt number, that is, the rate of convective heat transfer.

13.3.2.2 Methodology: Role of RSM in Process Optimization

Governing equations are a set of equations based on Navier–Stokes equations, energy balance equations, solved numerically through the use of a shooting-type fourth-order Runge-Kutta method. Nonlinear effects like Joule heating, Darcy drag, and radiative dissipation effects are considered in the system. After generation of a valid dataset on the basis of these simulations, Response Surface Methodology is in turn used to determine a second-order polynomial model. The key input factors considered are as follows:

- **Magnetization (M)**: 0.1 to 0.5
- **Darcy number (Da)**: 0.2 to 0.4
- **Thermal radiation (Rd)**: 0.1 to 0.3

A *Central Composite Design (CCD)* is used to structure 20 experimental runs. The outputs are *Nusselt numbers* obtained for each parameter combination.

13.3.2.2.1 Importance of ANOVA in Model Validation

To ensure statistical soundness, the RSM model is validated using *Analysis of Variance (ANOVA)*. Table 6 in the original paper summarizes the results, indicating:

- **F-values > 2,600,000** for the model and all main effects.
- **P-values < 0.0001**, confirming that all factors and their interactions are statistically significant at a 95% confidence level.
- **R^2 = 100%**, suggesting that the model perfectly fits the observed data.

ANOVA ensures that the model is not only predictive but also robust and free from overfitting or collinearity. The inclusion of interaction and square terms allows the model to capture both *nonlinear trends* and *synergistic effects* between parameters.

13.3.2.2.2 Residual Analysis for Model Reliability

Figure 13.4 shows residual plots for normality, independence, and homoscedasticity of Nusselt number model.

This figure illustrates four essential residual diagnostics:

- **Normal probability plot**: Residuals align closely to a straight line, confirming normal distribution.

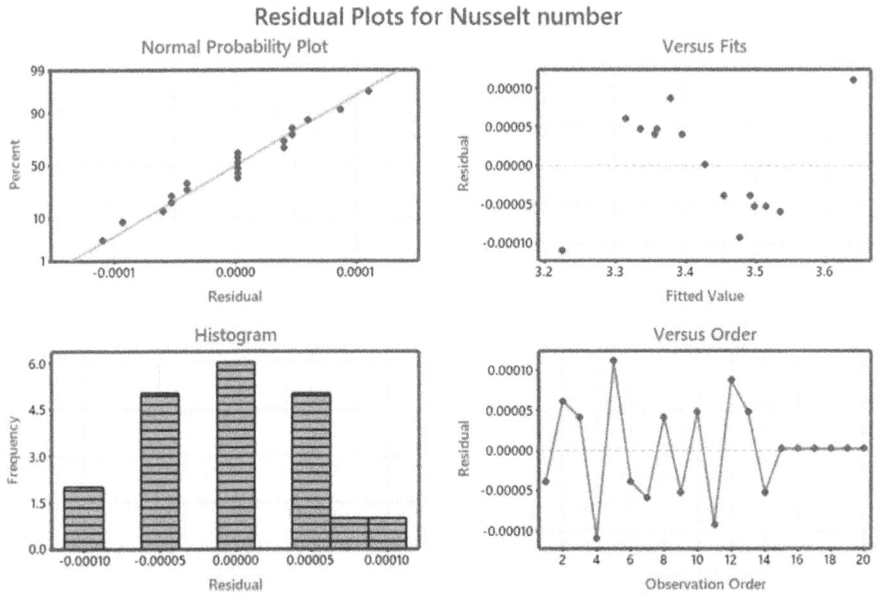

FIGURE 13.4 Residual plots for Nusselt number [3].

- **Residuals versus fitted values**: No discernible pattern suggests *homoscedasticity*, that is, equal variance across the model domain.
- **Histogram of residuals**: Displays a bell-shaped distribution, reinforcing the assumption of normality.
- **Residuals versus observation order**: Random scatter supports *independence of errors*, indicating no autocorrelation.

These diagnostic plots affirm the *statistical health of the model*, validating the predictive power of the RSM-ANOVA combination.

13.3.2.2.3 Optimized Model Equation: The Core Outcome of RSM

At the heart of any successful application of Response Surface Methodology (RSM) lies the *optimized model equation*—a statistically derived mathematical expression that encapsulates how selected process variables influence a system's output. This equation, typically quadratic in form, serves as a *predictive and interpretive tool* that connects theoretical modeling with empirical data. In the case study involving *magnetized hybrid nanofluid flow*, the optimized equation models the *Nusselt number*, which quantifies the heat transfer rate, as a function of three key parameters: *magnetization (M)*, *Darcy number (Da)*, and *thermal radiation parameter (Rd)*.

The full optimized model obtained through RSM is:

$$\begin{aligned} Nu &= 3.42700 - 0.06990\,M - 0.04869\,Da + 0.08915\,Rd \\ &+0.00146\,M^2 + 0.00071\,Da^2 - 0.00250\,Rd^2 + 0.00206(M \cdot Da) \\ &-0.00226(M \cdot Rd) - 0.00156(Da \cdot Rd) \end{aligned} \tag{13.2}$$

This equation is more than a curve-fitting result—it is a compact mathematical *surrogate model* that represents the underlying physics and interaction behavior of the system. It enables process engineers and researchers to make fast, accurate predictions of the Nusselt number for any combination of M, Da, and Rd within the tested experimental range.

13.3.2.2.4 Structure and Interpretation

The optimized model is composed of three types of terms:

a. **Linear terms (first-order effects)**
   - −0.06990M, −0.04869Da, and +0.08915Rd represent how each factor individually affects the Nusselt number. For example, the negative sign before M and Da implies that increasing either variable tends to reduce the heat transfer rate.

b. **Quadratic terms (curvature or second-order effects)**
   - $+0.00146M^2$, $+0.00071Da^2$, and $-0.00250Rd^2$ indicate nonlinearity. These terms capture the response surface's curvature and allow the identification of *optima*—points where the response reaches a maximum or minimum rather than varying linearly.

c. **Interaction terms (cross-variable effects)**
   - +0.00206(M·Da), −0.00226(M·Rd), and −0.00156(Da·Rd) represent how the combination of two variables together influences the response in a way not explainable by individual effects alone. These are critical in systems where coupled physical phenomena, such as magnetic suppression and porous resistance, affect performance synergistically.

13.3.2.2.5 *Engineering Insights*

- *Magnetization and porosity* act as resistive forces, reducing fluid velocity and convective heat transfer.
- *Thermal radiation* significantly improves heat transfer, especially in high-temperature or radiatively active systems.
- *Hybrid nanofluids* consistently outperform pure fluids and mono-nanofluids in terms of both velocity and temperature profiles.
- *Suction* enhances boundary layer stability but reduces heat transfer; *injection* introduces instabilities that can increase local turbulence.

13.3.2.2.6 *Application Scope*

This case study demonstrates the power of combining *physics-based modeling* with *statistical optimization techniques*. The RSM-based predictive model—validated via ANOVA—offers an efficient and accurate tool for designing *thermal systems* in:

- Electronics cooling.
- Nuclear reactor moderation.
- Geothermal energy extraction.
- Solar thermal systems.
- Advanced aerospace materials processing.

Such models drastically reduce the need for physical prototyping and enable *real-time design adjustments*, marking a significant step forward in *process intensification and smart engineering.*

13.4 ARTIFICIAL NEURAL NETWORKS (ANN)

13.4.1 INTRODUCTION

In the domain of thermal and fluid engineering, the modeling and predictions of system behavior often entail nonlinear multivariable issues with complex and even uncertain resulting interactions between parameters involved. This takes on an extreme form with regard to nanofluids, which are engineered suspensions of nanoparticle dispersion suspended in traditional base fluids. Such innovative fluids have superior heat conductivity, specific heat capacity, and viscosity characterizations, and therefore are very promising in the next generation heat transfer systems. Yet, their thermophysical performance is very much dependent on a set of interrelated variables, such as nanoparticle volume fraction, particle size and shape, kind of base fluid, flow regime, temperature, and geometrical setting of the flow field.

The traditional deterministic or empirical models of these kinds of systems, for example, polynomial regression, curve fitting, or analytical correlations, sometimes failed to model them adequately. Such classical methods usually involve simplifying assumptions, neglecting higher-order terms, or lacking the ability to adequately fit the underlying experimental or simulation measurements that are inherently nonlinear. Consequently, an increasing tendency to more adaptive, intelligent methods of modeling with the ability to learn based on data rather than to depend entirely on first-principles formulations has ensued.

Artificial neural networks (ANNs) represent a very remarkable solution in this regard. ANNs are an instance of artificial intelligence and were influenced by the information processing of biological neurons, and are adaptive, non-parametric, as well as nonlinear learning systems. ANNs can learn complex input–output mappings, generalize limited experience to new situations, and identify unknown functions with application to solving or enhancing performance at many tasks, such as classification, character recognition, prediction, control, waveform compression, and pattern recognition. ANNs are useful in many areas, but the main components of ANNs have been developed separately: hardwired interactions of connected processing components, massively parallel processing, and recognition of complex mappings. That is why they can perfectly fit to predict the performance metrics of nanofluid systems and optimize them when the traditional physics-based models are not enough.

Universal function approximators feature is one of the characteristic traits of ANNs. The Universal Approximation Theorem assures that a large neural network, with temperature raised to a high enough value, can represent any continuous function with any desired accuracy. Such ability enables ANNs to discover the physical relationships embedded in the experiments or computational fluid dynamics (CFD) data, without enforcing any explicit statements (or any assumptions) about the model structure. To nanofluids, this is specifically beneficial since the phenomenon of agglomeration of the particles, thermophoresis, Brownian motion, and intensity of resistance at an interface tend to combine in sophisticated manners, and this is rather tough to evaluate analytically.

Moreover, nanofluid flows usually experience non-Newtonian fluid behavior, flow instabilities, and sensitivity to boundary conditions (particularly when conditions are turbulent or transitional). ANNs give a strong framework that allows characterization of these phenomena by multi-layered models where non-linear activation functions and non-static learning algorithms are required. Consequently, they are also able to simulate not merely steady-state responses, such as Nusselt number or pressure drop but also transient effects, such as thermal stratification, pulsating flows, or temperature dependence of viscosity.

The deployment of ANNs in nanofluids research cuts across engineering systems such as microchannel heat sinks, solar thermal collectors and enhanced refrigeration cycles, and biomedical cooling devices. ANN models have also been able to shorten vast computational time, enhance the accuracy of predictions, and allow real-time decision-making in both of these systems when ANN is integrated into control algorithms. In addition, ANNs are used as surrogate models when optimization methods are also used in the process (genetic algorithms (GA), particle swarm optimization

(PSO), or multi-objective systems). In these cases, ANNs enable the exploration of design space and finding the best operating conditions to be achieved much faster than with other techniques.

13.4.2 Structure and Learning Algorithms of Artificial Neural Networks (Using MATLAB "nnstart")

Artificial Neural Networks (ANNs) are mathematical algorithms that give a brain-like computational model based on layers of processing units, or neurons, arranged into an input layer, one or more hidden layers, and an output layer. The form and internal topology of an ANN will determine how the ANN can learn patterns and relations in data. Every single neuron in the network is supplied with input, which is transformed with the help of an activation regularity and delivered to neurons in the next layer. Training the network such that it does this is the main aim of the network.

13.4.2.1 Step 1

A typical feedforward neural network (the most typical ANN structure), data pass one way (as illustrated in Figure 13.5), without iteration or feedback. The input layer has a neuron for each corresponding feature in the data. The invisible layers, which can be one or more layers of user-defined number of neurons, apply nonlinear transformations to the input. The result is finally obtained by the output layer and may consist of a continuous value (regression tasks) or a label of a class (classification

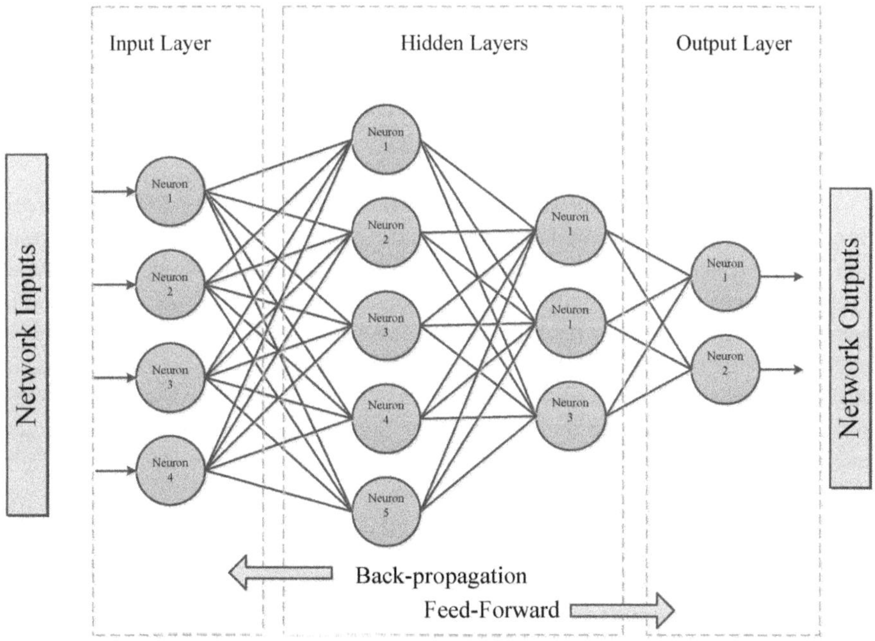

FIGURE 13.5 Architecture of a typical feedforward and neural network [4].

tasks). It is also important to specify activation functions that are used in the hidden layers (e.g., sigmoid, hyperbolic tangent, or ReLU) to add the non-linearity to the model and be able to learn non-trivial functions.

Using the Neural Network Toolbox GUI (referred to as nnstart) in MATLAB, it is possible to build and train an ANN. The tool, which is an interactive interface, offers a visual structure of design, training, validation, and the simulation of neural networks without manual code writing.

Initializing the users' access to nnstart interface requires entering the name of the interface, that is nntool, in the MATLAB command window. Using the GUI, it is possible to import input and target data that can be in the MATLAB workspace as either matrices or tables. Once the data is imported, a user forms a network by defining significant structural characteristics like the number of hidden layers, the number of neurons in each layer. The type of network defaulted in nnstart is the feedforward backpropagation network, and this is executed using the feed-forwardnet function.

After determining the network structure, a user can choose a training algorithm, that is, the algorithm through which the network learns the data. The most commonly used training algorithms in nnstart include:

Levenberg–Marquardt (trainlm): It is an improvement of the Gauss–Newton algorithm and proves extremely effective when training over small- to medium-sized datasets. It exhibits rapid convergence, and it is applicable to the problems of function approximation.

Bayesian Regularization (trainbr): This is an addition to the objective estimator done by including regularization functions, thus avoiding overfitting. It can be especially helpful in cases when the set of data is noisy or when the network is complex.

Scaled Conjugate Gradient (trainscg): An algorithm that is memory efficient, and when working with larger datasets, it works well when you have less than optimal computational resources.

Training is initiated by forward propagation whereby data to be trained is forced through the network layer-by-layer, and fitted outputs are estimated. Such estimations are contrasted against real target values via a loss function, which in the case of a regression problem, is usually the mean squared error (MSE). Backpropagation is then used by the network, where the error is sent back to the network of the previous layers by virtue of the chain rule of calculus, unless there are no previous layers. In this process loss gradient of each weight of the network is calculated, and weights are moved toward the direction that reduces loss. Such an iterative optimization is then carried out for a number of epochs as defined or till convergence terms (e.g., slight change in validation performance) are achieved.

Automatic division of data into training, validation, and test segments can also be performed by using the option of automatic data division, which is available in the form of a directory named:nnstart. Weapons are optimized using the training set, generalization during the training on the validation set, and the unbiased evaluation of the final model using the test set. MATLAB gives visualization in real time during training, represented as a performance plot, regression plot, and training state plot, to allow visualization of convergence as well as possible problems with overfitting or underfitting.

After training, the trained network can be evaluated on fresh data, exported to the workspace, or saved to be used later. It may also be applied to controls or simulations. The trained ANN works as a model that takes new sets of input values and gives out predictions with the acquired weights and biases.

13.4.3 Feedforward and Feedback Mechanisms

13.4.3.1 Feedforward Neural Networks

Feedforward networks represent the most widely used ANN architecture. The data in this style flow in one direction only, and this is the input layer, through the hidden layers to the output layer, with no cycle or looping. The inputs of layers are processed, and the results of one layer are fed into the next layer. The non-linear activation functions used by the hidden layers are sigmoid, tanh, or ReLU, and they allow the network to discover non-relational relationships. Feedforward networks are especially useful when one would be dealing with the issue of a stationary relation between input and output in the regression and classification problems.

The main strengths of feedforward networks are their ease, stability, and quick turnaround training rates. Their architecture is devoid of dynamic states and recursive loops, which simplify their analysis, training, and interpretation. It also eliminates the chances of instability in learning due to a lack of feedback.

13.4.3.2 Backpropagation (Feedback Learning Mechanism)

Backpropagation is not a network architecture but a learning algorithm used to train both feedforward and feedback neural networks. It is based on the chain rule of calculus and allows the network to minimize the error between predicted and actual outputs by iteratively updating the internal weights. The training process consists of two phases:

1. **Forward propagation**: Input data are propagated through the network layer by layer to produce an output.
2. **Backward propagation**: The error is calculated (typically using Mean Squared Error) and propagated backward through the network to compute gradients. These gradients are used to update the weights using an optimization method such as gradient descent, Levenberg–Marquardt, or Adam.

Thus, the network structure is feedforward, but the training process uses backward propagation of error to adjust weights.

13.4.3.3 Which Is Preferred?

Feedforward neural networks trained through backpropagation are popular in most engineering applications, and they are considered to be computationally efficient, accurate, and easy to implement in data modeling systems, system identification, and system optimization. Time-series and dynamic systems with the previous outputs affecting future behavior would be more appropriate to feed-back (or recurrent) networks. Nevertheless, they may be more computationally demanding to train and more likely to get stuck in convergence problems.

13.4.3.4 Simple Example: ANN Modeling for Nanofluid Systems

In the context of nanofluid modeling, feedforward neural networks trained with backpropagation algorithms have shown exceptional capability in capturing the complex thermophysical behavior of nanofluids. These systems involve nonlinear relationships between variables such as nanoparticle volume fraction, flow velocity, temperature, thermal conductivity, and viscosity. Experimental or simulation datasets can be used to train the ANN, allowing it to predict important outputs like Nusselt number, pressure drop, or effective thermal conductivity.

Using MATLAB's nnstart, one can import nanofluid datasets and configure a feedforward network by specifying:

- Input features (e.g., φ, Re, T).
- Output targets (e.g., Nu, μ, k_{eff}).
- Network topology (number of hidden neurons and layers).
- Training algorithm (e.g., Levenberg–Marquardt for high-speed convergence).

The ANN model is then trained using a training/validation/testing split (e.g., 70/15/15) to ensure that the model generalizes well. After training, the model's performance is evaluated using metrics like Mean Squared Error (MSE), Root Mean Square Error (RMSE), Mean Absolute Percentage Error (MAPE), and the coefficient of determination (R^2), and the trained network can be exported and used for rapid prediction and design optimization of nanofluid thermal systems.

In such applications, feedforward neural networks remain the preferred structure due to the static input–output relationship and the need for high computational efficiency, particularly when embedded in real-time simulation or optimization loops.

13.4.3.4.1 Step 2: Training, Validation, and Testing Datasets

Once the neural network structure is defined, the next essential step in the modeling workflow is the segmentation of the dataset into training, validation, and testing subsets. This data partitioning strategy is critical to ensure the generalization capability of the artificial neural network and to prevent overfitting—a condition where the model performs well on the training data but poorly on unseen data.

The full dataset, which consists of paired input and target values, is typically divided into three parts.

13.4.3.4.1.1 Training Dataset

The training set is the largest portion of the dataset and is used to adjust the weights and biases of the neural network during the learning phase. This subset is fed into the network in multiple iterations (epochs), and the learning algorithm—typically backpropagation with gradient-based optimization—minimizes the difference between the network's predictions and the true target values. The network uses this set to discover the mapping from inputs to outputs and to internalize the data structure. Typically, 60–70% of the available data is allocated to the training set.

13.4.3.4.1.2 Validation Dataset

The validation set is used to monitor the model's performance during training, but it does not participate in weight updates. Its role is to assess how well the network generalizes to new data, helping to identify issues such as overfitting or underfitting. When the network begins to perform significantly better on the training set than on the validation set, it is a sign of overfitting. In such cases, early stopping can be applied, halting the training process at the point where validation error starts to rise. The validation set is also useful for hyperparameter tuning, such as selecting the number of hidden neurons or adjusting the learning rate. Typically, 15–20% of the data is reserved for validation.

13.4.3.4.1.3 Testing Dataset

The testing set provides a final, unbiased evaluation of the trained network's performance. It is completely independent of both training and validation data and is used only after the training process is completed. The testing phase helps assess the true predictive power and generalization ability of the network under real-world conditions. Performance metrics such as Mean Squared Error (MSE), Root Mean Square Error (RMSE), Mean Absolute Percentage Error (MAPE), and the coefficient of determination (R^2) are calculated on the test set to determine the model's accuracy and reliability. Usually, 10–20% of the data is allocated to testing.

In MATLAB's neural network environment (via nnstart or programmatic methods), this division is automatically performed unless manually specified. The default split is 70% training, 15% validation, and 15% testing, though users can customize these ratios based on dataset size and application requirements. Proper data division is fundamental to the robustness and credibility of any neural network model. It ensures that the network not only learns the training data but also generalizes well to previously unseen conditions, which is crucial for the deployment of ANN models in real-time prediction and decision-making systems.

13.4.4 How ANN Differs from RSM and Why It Is Needed in Nanofluid Modeling

Although both Artificial Neural Networks (ANN) and Response Surface Methodology (RSM) are very effective instruments of modeling, prediction, and optimization tools, they are highly different regarding the basic structure, assumptions, and possibilities of their implementation. A more well-known and statistically based method is the classical method known as RSM, which constructs explicit mathematical models, typically as a set of polynomial equations to estimate the relationship between independent input variables and a response variable. ANN, which is machine-learning approach to data mining that is based on how biological neural processes adapt to information, is self-learning by adjusting from internal weighting within a network design with no preconceived mathematical representation.

The main differences are connected with the approach toward modeling. RSM relies on a user-specified second-order polynomial regression model on the basis that the connection between inputs and the output is assumed to be continuous, distinctively differentiable, and that higher-order contributions (change) of quadratic terms and their interactions are noticeable and sufficient. Although effective in systems that turned out with relatively smooth and moderately nonlinear behavior, it is not so accurate in systems in which the response surface is very nonlinear, which involve discontinuities, or features higher-order interaction effects, which is typical of nanofluid systems. In contrast, ANN is a non-parametric, data-driven technique. It does not presuppose any relationship form between inputs and outputs. As opposed to that, it creates internal representations based on raw data learning, thus being intrinsically meant to work with high-dimensional, nonlinear, and noisy datasets.

Another critical difference is in model interpretability versus flexibility. RSM provides clear-cut equations, which are simple to understand and also viewable as contour and surface graphs. These properties make it particularly handy in line with identifying main effects, interaction terms, and optimal points at which to operate systems whose physical behavior is easily understood or at least close enough to be linear. ANN models, however, act as black-box predictors. They give high predictive accuracy but fail to produce a closed-form equation, and interpreting the inner structure of the trained networks is not a straightforward task. However, the trade-off is tolerable in most of the engineering applications where prediction is favored over interpretability, for example, in a complex heat transfer system involving nanofluids.

Computationally, RSM has fewer data points, and it is not computationally demanding. Thus, it can be applied where the acquisition of data is very costly or a limited number of experimental runs can be run. ANN, in its turn, needs more data to be trained and avoid overfitting. Much more data, however, are becoming available over time by using computational fluid dynamics (CFD) simulations, conducting experiments, and collecting data through the use of sensors, so the data-intensive nature of ANN is not the current constraint but a mere opportunity to utilize its learning capabilities.

These differences are even more eminent in such a distinctive area of nanofluid applicability. The behavior of nanofluids is affected by the particle–particle interactions, nonlinear enhancement of the thermal conductivity, regimes of flow changes, and viscosity variations because of temperature. These characteristics often defy simple polynomial approximations. As an example, in the addition of nanoparticles, although thermal conductivity could be enhanced at concentrations above a certain value, the enhancement will be detrimental to the flow performance because of the increased viscosity and the agglomeration tendency, which have non-monotonic trends and can hardly be characterized with quadratic-related functions. ANN models in these instances can parameterize such complex nonlinear dependencies where no manual specification of the interaction terms or transformations is needed.

In addition, ANN is mainly beneficial in real-time prediction and control systems. Control algorithms embedded with ANN models may be appropriate in cases of desired adaptable or dynamic thermal control or coolant fluid flow in nanofluid-based heat exchangers, owing to the flexibility and dynamic nature of ANN models and not their RSM counterparts, whose nature is non-spatial and equation-based.

13.4.5 Advantages in Non-Linear Modeling and Prediction

ANNs have significant advantages as far as modeling and forecasting of systems with non-linear multivariable relationships are concerned. Classical mathematical modeling methods, for example, linear regression or polynomial fit, assume a fixed functional form and are also restricted to models allowing only very simple, non-complex dependencies. On the contrary, ANNs do not need previous knowledge of the underlying mathematical structure. Rather, they are trained directly on the data to perform non-trivial input–output mappings, and these properties uniquely make them effective tools in dealing with highly nonlinearity systems.

One of the most significant advantages of ANNs is their ability to serve as universal function approximators. The Universal Approximation Theorem states that a single-hidden-layer artificial neural network with a finite number of hidden neurons has the capability of approximating any continuous function with any degree of accuracy, given proper data and training. This theoretical background enables ANNs to simulate a broad variety of systems whose analytical forms are way too complex or simply nonexistent.

One more area where ANNs excel is the capability of capturing relationships among many variables that are not necessarily intuitively associated, and using them to make predictions that are linearly inseparable. The behavior of individual input variables on the response of nonlinear systems tends to have a complicated multi-dimensional surface relationship in the sense that the resultant behavior of a single input may vary depending on input values of other inputs. The processing layout of ANN architectures based on a stacking of layers and nonlinear activation functions is in principle capable of learning and representing these kinds of relationships. This renders them very applicable in applications entailing pattern recognition, parameter estimation, inverse modeling, and dynamic identification of systems.

The second important strength is that ANN modeling is data-driven. In contrast to traditional models that demand a substantial theoretical appreciation of a system under investigation, ANNs rely only on the accessibility of quality as well as representative data. After the training process, the ANN is completely capable of learning how the system works and making decisions that predict system behavior very fast, in some cases even in real time, which is most priceless in practical use cases like process controls, diagnostics, and adaptive optimization. Also, ANNs can efficiently operate with noisy or incompletely received or partially corrupted data, as they have a tendency to generalize the pattern even in cases where some flaws appear in the set during the training.

ANNs have adaptive learning behavior in predictive modeling, which implies that they are able to update and perfect their internal parameters as additional data are presented to them. This is what qualifies them to be used in systems whose environmental conditions or operating variables vary with time and so must be reflected by the model. ANN models can likewise deal with minor-scale issues in smaller feature spaces or upgrade to manage vast, ultra-dimensional issues without adjusting the fundamental algorithmic framework on account of their scalability.

Lastly, ANNs are compatible with various optimization and hybrid modeling frameworks. They are readily combined with genetic algorithms (GA), particle swarm optimization (PSO), fuzzy models, or finite element models to improve prediction, control, and decision-making. This flexibility predisposes them as a modeling option of choice in multi-disciplinary engineering fields where non-linear behaviors prevail.

Case Study: Prediction of Relative Viscosity of Al_2O_3–Engine Oil Nanofluid Using an Optimized Artificial Neural Network (Hemmat Esfe et al. 2021)

BACKGROUND AND OBJECTIVE

The accurate prediction of nanofluid viscosity is a critical challenge in thermal-fluid engineering, particularly when dealing with complex oil-based nanofluids such as Al_2O_3 dispersed in 10W40 engine oil. The relative viscosity (μ_{nf}) of such nanofluids depends on several nonlinear variables, including nanoparticle volume fraction (ϕ), temperature (T), and shear rate (γ). Experimental measurements of μ_{nf}\ across a range of operating conditions are both time-consuming and costly, making the development of predictive models a practical necessity.

This case study demonstrates how a feedforward Artificial Neural Network (ANN) was employed to model and predict the relative viscosity of Al_2O_3/engine oil nanofluid with high accuracy. The study explored 400 different ANN architectures to determine the optimal structure for maximum predictive accuracy, aiming to reduce experimental overhead while maintaining physical fidelity.

MODELING APPROACH AND DATA HANDLING

The ANN was developed using a multilayer perceptron (MLP) architecture trained via the Levenberg–Marquardt algorithm (LMA) (Figure 13.6), a robust optimization technique known for fast convergence in nonlinear regression problems. The model considered three input variables:

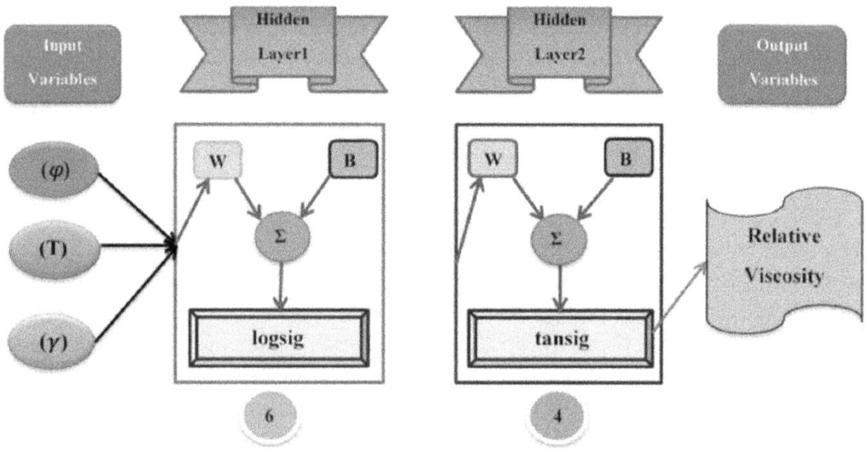

FIGURE 13.6 Best ANN structure selected to predict the Al_2O_3/10W40 nanofluid [5].

- ϕ (nanoparticle volume fraction): 0.25% to 2%
- Temperature (T): 5°C to 55°C
- Shear rate (γ): experimental range unspecified but systematically varied

The output of the network was the relative viscosity (μ_{nf}) of the nanofluid.

A dataset of 134 experimental samples was divided into three parts:

- 70% for training
- 15% for validation
- 15% for testing

This standard split ensured model generalization and robustness against overfitting.

13.4.5.1 ANN Structure and Optimization

To determine the optimal network structure, the authors systematically evaluated 400 different network configurations, varying the number of neurons in the hidden layers and the combination of transfer functions. The optimal ANN architecture was found to be:

- Two hidden layers.
- Six neurons in the first layer.
- Four neurons in the second layer.
- Transfer functions: *logsig* in the first hidden layer, *tansig* in the second, and *purelin* in the output layer.

13.4.5.2 Performance Evaluation

As shown in Figure 13.7 performance of the ANN was assessed using:

- Regression coefficient (R): 0.995838.
- Mean Squared Error (MSE): 4.14469×10^{-8}
- Margin of Deviation (MOD): between –2% and +2% for the entire dataset.

These metrics indicate a very high level of predictive accuracy, with most predicted values aligning closely with the experimental data.

13.4.5.3 Comparison with Empirical Correlations

To benchmark ANN performance, the authors also developed an empirical correlation using Response Surface Methodology (RSM) based on cubic polynomial regression. While the correlation achieved reasonable accuracy ($R^2 \approx 0.967$), it failed to match the ANN in both precision and generalization. Figure 13.7 clearly demonstrates that ANN predictions adhered more closely to the experimental values across all temperature and volume fraction ranges.

Thus, ANN outperformed the RSM-based model, especially under complex input conditions where higher-order nonlinearities and interactions dominate the response behavior.

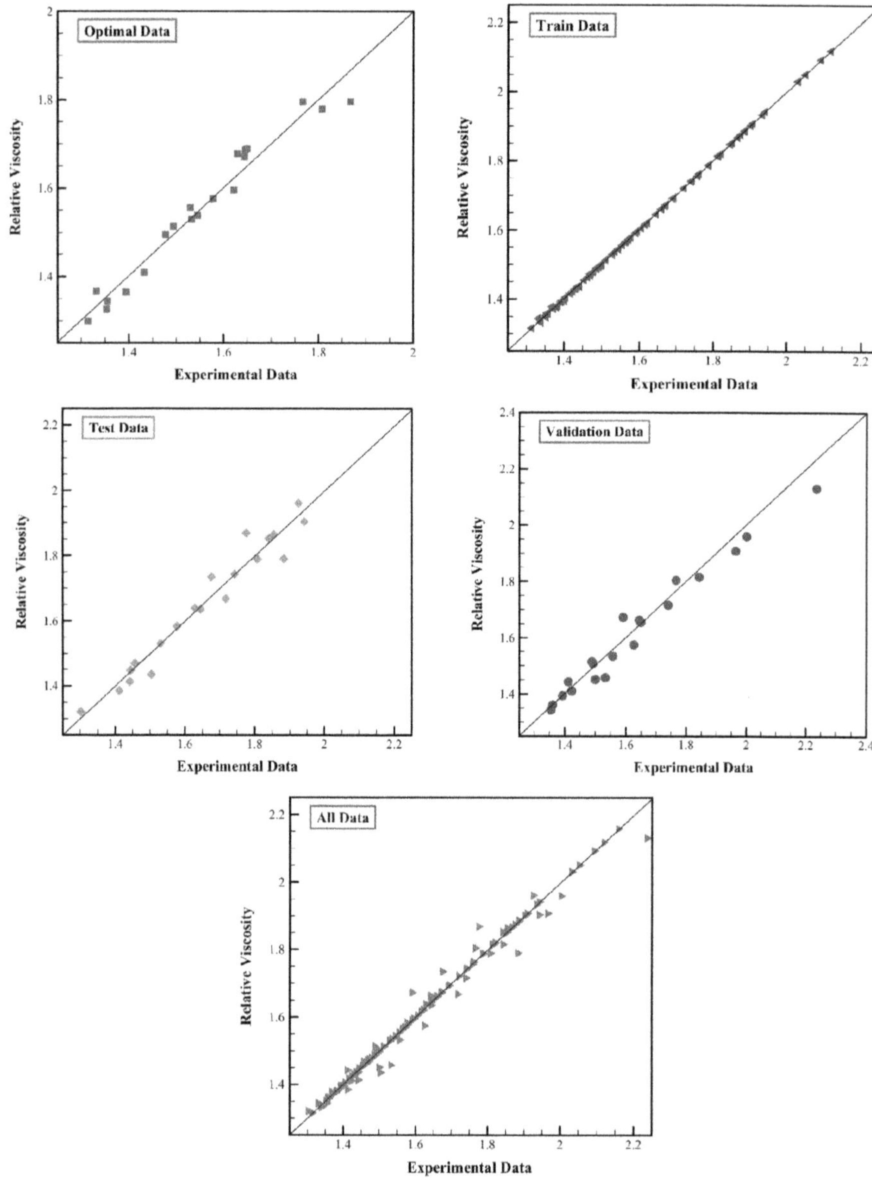

FIGURE 13.7 Comparison of experimental and ANN-predicted relative viscosity values for Al_2O_3–Engine Oil nanofluid [5].

13.4.5.4 Significance of ANN in Nanofluid Prediction

This study highlights several reasons why ANNs are superior for modeling complex nanofluid systems:

1. **Flexibility in learning nonlinear patterns**: Unlike empirical correlations, which assume fixed mathematical forms, ANNs can learn arbitrary nonlinear mappings directly from data.
2. **High accuracy and generalization**: The optimized ANN achieved extremely low error margins with strong regression coefficients across training, validation, and test sets.
3. **Cost and time efficiency**: Once trained, the ANN can predict viscosity instantly, eliminating the need for costly and repetitive experimental campaigns.
4. **Scalability**: The ANN framework can be extended to predict other thermophysical properties, such as thermal conductivity, specific heat, or density, using the same methodological framework.

13.5 HYBRID MODELING AND OPTIMIZATION APPROACHES

In complex engineering systems, particularly those governed by highly nonlinear, multivariable interactions, single modeling approaches may exhibit limitations in accuracy, interpretability, or computational efficiency. To overcome these limitations, *hybrid modeling strategies*—particularly the coupling of *Response Surface Methodology (RSM)* with *Artificial Neural Networks (ANN)*—have emerged as powerful tools for enhanced prediction, optimization, and design.

RSM is a well-established statistical technique that constructs polynomial regression models based on a designed set of experiments. It provides clear interpretability, allows for sensitivity and interaction analysis, and offers a structured route for response optimization. However, RSM assumes the response surface can be adequately represented by low-order polynomials (usually quadratic), which may not hold in systems exhibiting strong nonlinearity, discontinuities, or asymptotic behavior. This is where the coupling with ANN becomes advantageous.

Artificial Neural Networks are data-driven models capable of capturing arbitrarily complex nonlinear mappings between inputs and outputs. Unlike RSM, ANN does not require a predefined functional form. However, ANNs are often criticized for being "black box" models, offering limited physical insight and requiring large datasets for effective training. Moreover, they can be computationally expensive to train and prone to overfitting if not properly validated.

By combining RSM and ANN in a *coupled or hybrid framework*, it is possible to exploit the complementary strengths of both methods:

- RSM provides an *initial model structure* and can be used to *generate synthetic training data* over a well-designed experimental space.
- ANN is then trained on either the RSM-generated dataset or combined experimental/simulated datasets, allowing it to *learn higher-order nonlinearities and residual patterns* not captured by the polynomial model.

- Alternatively, RSM can be applied to *post-process and interpret* ANN predictions, offering insights into variable sensitivities and optimization trajectories.

This coupling is implemented in various configurations. One common strategy is to first apply RSM to model the primary response and extract its trend, and then use ANN to model the residuals or prediction errors as a function of the same inputs. The final prediction is obtained by summing the RSM output and the ANN-corrected residual, thus achieving improved accuracy. Another approach uses ANN to map input parameters to hidden feature spaces, which are then analyzed using RSM to extract interpretable patterns and optimize decision variables.

Such *RSM–ANN frameworks* have been shown to yield superior predictive accuracy compared to standalone RSM or ANN models. They are especially effective in situations where experimental datasets are limited but computational or surrogate models are available to extend the training domain. In parametric optimization studies, this hybrid approach enables *global exploration via RSM* and *local refinement via ANN*, creating an efficient, high-resolution predictive system.

Case Study 1: Optimizing Viscosity Predictions of GNP-Alumina Hybrid Nanofluids: A Comparative Analysis of RSM and ANN Modeling Techniques (Borode et al. 2024)

BACKGROUND

In the realm of hybrid nanofluids, the accurate prediction of rheological properties such as viscosity is essential for optimizing thermal systems and fluid transport processes. GNP (Graphene Nanoplatelets)–Alumina nanofluids have attracted attention for their superior thermal conductivity and stability. However, modeling their viscosity behavior, especially under varying mixing ratios and temperatures, poses significant challenges due to the non-linear and complex interactions between fluid components [6].

To address this, the study compared *Response Surface Methodology (RSM)*, a statistical modeling tool, with *Artificial Intelligence (AI)* approaches—specifically *Artificial Neural Networks (ANN)* and *Adaptive Neuro-Fuzzy Inference System (ANFIS)*—for predicting both absolute viscosity (μ_{nf}) and relative viscosity (μ_{rel}).

EXPERIMENTAL SETUP OVERVIEW

- **Base Fluid**: Ethylene glycol
- **Nanoparticles**: GNP and Alumina
- **Mixing Ratios Tested**: 0, 0.33, 1, 3
- **Temperatures**: 15°C to 55°C
- **Measurement Tool**: Digital viscometer
- **Data Analysis**: MATLAB R2020a (ANN and ANFIS), Design Expert 13 (RSM)

TABLE 13.1
Model Optimization Highlights

Method	Model Basis	Optimization Parameters	Best Fit Metrics
RSM	Quadratic polynomial regression	Source model (linear, 2FI, quadratic, etc.)	$R^2 = 0.9986$ for μ_{nf}; 0.8835 for μ_{rel}
ANN	Two-layer feed-forward neural net	Neuron count, activation functions	$R^2 = 0.9997$ for μ_{nf} with 4 neurons; 0.9817 for μ_{rel} with 5 neurons
ANFIS	Fuzzy logic + neural learning	Membership functions (type and count)	$R^2 = 0.9999$ for μ_{nf}; 0.9872 for μ_{rel}

Table 13.1 summarizes the performance of RSM, ANN, and ANFIS models in predicting the absolute viscosity (μ_{nf}) and relative viscosity (μ_{rel}) of GNP-Alumina hybrid nanofluids under varying conditions. While RSM offered reliable results through polynomial regression, its predictive strength was limited compared to AI-based approaches. ANN improved accuracy by optimizing neuron architecture, demonstrating strong correlation values and lower error metrics. However, the ANFIS model—combining fuzzy logic and neural networks—emerged as the most robust, achieving the highest R^2 values and lowest error indices. This comparison clearly highlights ANFIS as the superior tool for modeling nonlinear fluid behavior.

KEY FINDINGS

- *RSM* effectively modeled viscosity using a *quadratic model*, capturing significant relationships between temperature and mixing ratio. It provided relatively accurate predictions with minimal complexity.
- *ANN* offered a highly accurate, data-driven alternative, especially effective when the optimal neuron architecture was selected (4 neurons for μ_{nf}5 neurons for μ_{rel}). However, training stability depended on neuron configuration and data division.
- *ANFIS*, integrating the advantages of both ANN and fuzzy logic, *outperformed* RSM and ANN in both predictive accuracy and generalization. It achieved the *lowest error metrics across all categories*, including RMSE, MAE, MAPE, and RSS.

13.5.1 Conclusion: Which Was Better? ANN or RSM

While *RSM* remains a robust, interpretable statistical tool ideal for experimental design and sensitivity analysis, *ANN* and *ANFIS* clearly outperformed *RSM* in terms of *prediction accuracy* for both absolute viscosity (μ_{nf}) and relative viscosity (μ_{rel}).

Between *ANN and ANFIS*, *ANFIS* was superior, as it effectively captured non-linearities and combined rule-based learning with neural adaptation. This synergy enabled the *most precise and reliable viscosity predictions*, with R^2 *values as high as 0.9999.*

13.5.2 Significance for Process Industries

The integration of *RSM and AI models* like *ANN and ANFIS* allows for:

- Reduction in experimental workload.
- Accurate property forecasting for design input.
- Efficient optimization of hybrid nanofluids across various operating ranges.

This case study reinforces the role of *AI-assisted modeling techniques* as essential tools for advancing the application of *hybrid nanofluids* in *thermal systems, electronics cooling, and chemical process intensification.*

Case Study 2: Multi-Model Optimization and Prediction of ZnO Nanofluid Thermal Performance in Automotive Radiators (Kaladgi et al. 2021)

BACKGROUND AND MOTIVATION

Automobile engine cooling systems demand improved thermal efficiency to reduce fuel consumption and emissions. Nanofluids, due to their superior thermal properties, offer a potential upgrade over conventional coolants. In this case, *zinc oxide (ZnO)–water nanofluids* were evaluated for enhancing the thermal performance of a car radiator [7].

The novelty of the study lies in combining multiple modeling and optimization approaches: *Taguchi method*, *Grey Relational Analysis (GRA)*, *Response Surface Methodology (RSM)*, and *Artificial Neural Network (ANN)* modeling to ensure robust optimization and accurate predictive modeling (Table 13.2).

EXPERIMENTAL SETUP SUMMARY

- **Base Fluid**: Distilled Water
- **Nanoparticles**: ZnO (zinc oxide)
- **Volume Concentration**: 0.025–0.1%
- **Flow Rate Range**: 3 to 7 L/min
- **Inlet Temperature Range**: 50 to 70°C
- **Parameter of Interest**: Thermal Performance Factor (TPF)
- **Design Tools**:
  - Taguchi L9 orthogonal array for experimentation
  - Grey Relational Analysis (GRA) for multi-objective optimization
  - RSM for regression-based prediction
  - ANN for non-linear predictive modeling

TABLE 13.2
Modeling and optimization framework

Method	Purpose	Key Outcome
Taguchi Method	Design of Experiments (DoE)	Efficient experimental setup using L9 array
GRA	Multi-response optimization	Simultaneous enhancement of heat transfer and flow behavior
RSM	Regression modeling and optimization	Developed a second-order model for TPF ($R^2 = 0.9881$)
ANN	Data-driven predictive modeling	Achieved $R^2 = 0.9962$ with lowest RMSE = 0.0085

KEY FINDINGS

- The *optimal condition* for maximum thermal performance (TPF) was achieved at:
  - **Flow Rate**: 7 L/min.
  - **Nanoparticle Concentration**: 0.1%.
  - **Inlet Temperature**: 70°C.
- *RSM* provided a statistically significant quadratic model and facilitated surface plots to visualize the interactions between variables. The model had high accuracy with $R^2 = 0.9881$.
- *ANN* captured the complex nonlinear relationship among input variables more effectively. It outperformed RSM with a higher *R^2 of 0.9962*, demonstrating *better generalization* and *lower prediction errors* (RMSE, MAE, MAPE).

13.5.3 Conclusion: Which Was Better, RSM or ANN?

While *RSM* provided a solid statistical model and was essential for process visualization and early-stage optimization, *ANN clearly outperformed it* in prediction accuracy, particularly for nonlinear and multi-parametric dependencies. ANN's adaptability to complex interactions makes it the better modeling tool in data-rich environments.

However, the *combined approach*—using Taguchi-GRA for experiment design and parameter selection, RSM for regression analysis, and ANN for predictive learning—provides a *comprehensive and complementary optimization strategy* [8–10].

13.5.4 Industrial Implication

This integrated modeling and optimization framework offers a powerful roadmap for:

- Designing high-efficiency nanofluid-based heat exchangers.
- Optimizing thermal systems in automobiles.
- Reducing energy losses and emissions.

The approach demonstrates how *hybrid intelligence* (statistical + AI) enhances decision-making in *thermal system design* [11–13].

13.6 SUMMARY

This chapter provides a comprehensive exploration of advanced modeling and optimization strategies for nanofluid applications, focusing on different nanofluids in various industrial sectors. It emphasizes the growing need for intelligent tools to manage the nonlinear and multivariable behavior of nanofluids. The chapter introduces RSM as a statistical tool that helps understand variable interactions and optimize thermal performance through design of experiments (DoE), especially using Central Composite Design (CCD) and Box-Behnken Design (BBD).

Artificial Neural Networks (ANN), conversely, are presented as machine learning tools that excel in capturing complex nonlinear patterns without needing predefined equations. Through MATLAB's nnstart, users can create feedforward neural networks for predicting thermal behaviors like viscosity, Nusselt number, and heat transfer rates. ANN's superior performance is demonstrated in multiple case studies, such as modeling the viscosity of Al_2O_3–engine oil nanofluids and optimizing hybrid nanofluids under MHD conditions.

The chapter culminates in showcasing hybrid RSM–ANN approaches that combine the clarity of statistical regression with the flexibility of AI-based learning. This hybrid framework proves particularly effective for industrial applications where real-time control, efficiency, and reduced experimentation are critical. In conclusion, while RSM aids in interpretability and initial model building, ANN outperforms in complex prediction tasks, and their integration represents a powerful pathway for future process engineering.

REFERENCES

1. Yasmin, H., Giwa, S. O., Noor, S., & Aybar, H. Ş. (2023). Reproduction of nanofluid synthesis, thermal properties and experiments in engineering: A research paradigm shift. *Energies*, *16*(3), 1145. https://doi.org/10.3390/en16031145.
2. Kidane, S. W. (2021). Application of response surface methodology in food process modeling and optimization. In *Response surface methodology in engineering science*. IntechOpen.
3. Baithalu, R., Mishra, S. R., & Panda, S. (2025). A statistically predictive heat transfer rate model in a radiative magnetized hybrid nanofluid flow with dissipation over a sheet elongating exponentially: Combined RSM & ANOVA. *Partial Differential Equations in Applied Mathematics*, *13*, 101031.
4. Abdolrasol, M. G. M., Hussain, S. M. S., Ustun, T. S., Sarker, M. R., Hannan, M. A., Mohamed, R., Ali, J. A., Mekhilef, S., & Milad, A. (2021). Artificial neural networks based optimization techniques: A review. *Electronics*, *10*(21), 2689. https://doi.org/10.3390/electronics10212689.
5. Hemmat Esfe, M., & Toghraie, D. (2021). An optimal feed-forward artificial neural network model and a new empirical correlation for prediction of the relative viscosity of Al2O3-engine oil nanofluid. *Scientific Reports*, *11*, 17072. https://doi.org/10.1038/s41598-021-96594-z.

6. Borode, A., & Olubambi, P. (2024). Optimisation of artificial intelligence models and response surface methodology for predicting viscosity and relative viscosity of GNP-alumina hybrid nanofluid: Incorporating the effects of mixing ratio and temperature. *The Journal of Supercomputing*, *80*(4), 4841–4869.
7. Kaladgi, A. R., Afzal, A., Manokar, A. M., Thakur, D., Agbulut, U., Alshahrani, S., & Subbiah, R. (2021). Integrated Taguchi-GRA-RSM optimization and ANN modelling of thermal performance of zinc oxide nanofluids in an automobile radiator. *Case Studies in Thermal Engineering*, *26*, 101068.
8. Malika, M., & Sonawane, S. S. (2024). Ecological optimization and LCA of TiO2-SiC/water hybrid nanofluid in a shell and tube heat exchanger by ANN. *Proceedings of the Institution of Mechanical Engineers, Part E: Journal of Process Mechanical Engineering*, *238*(1), 45–55.
9. Malika, M., & Sonawane, S. S. (2021). Application of RSM and ANN for the prediction and optimization of thermal conductivity ratio of water based Fe2O3 coated SiC hybrid nanofluid. *International Communications in Heat and Mass Transfer*, *126*, 105354.
10. Malika, M., & Sonawane, S. S. (2021). Statistical modelling for the Ultrasonic photodegradation of Rhodamine B dye using aqueous based Bi-metal doped TiO2 supported montmorillonite hybrid nanofluid via RSM. *Sustainable Energy Technologies and Assessments*, *44*, 100980.
11. Thakur, P., Sonawane, S. S., Sonawane, S. H., & Bhanvase, B. A. (2020). Nanofluids-based delivery system, encapsulation of nanoparticles for stability to make stable nanofluids. In *Encapsulation of active molecules and their delivery system* (pp. 141–152). Elsevier.
12. Bethi, B., Sonawane, S. H., Bhanvase, B. A., & Sonawane, S. S. (2021). Textile industry wastewater treatment by cavitation combined with Fenton and ceramic nanofiltration membrane. *Chemical Engineering and Processing-Process Intensification*, *168*, 108540.
13. Sonawane, A. V., Rikame, S., Sonawane, S. H., Gaikwad, M., Bhanvase, B., Sonawane, S. S., . . . & Gaikwad, R. (2024). A review of microbial fuel cell and its diversification in the development of green energy technology. *Chemosphere*, *350*, 141127.

14 Future Prospects of Nanofluid Applications for Energy Conservation

14.1 INTRODUCTION

14.1.1 Importance of Energy Conservation in Addressing Climate Change and Resource Depletion

At the advanced systems level, energy conservation is not just a means of cutting utility bills or enhancing process efficiencies, but it is a critical development of the global strategy of mitigating anthropogenic global warming and checking the unsustainable exhaustion of natural resources that exists. Thermodynamically, energy conservation implies reducing exergy destruction in or by the industrial, residential, and transportation processes, and optimization of the usable work output of a particular energy input. This is particularly urgent due to the already increasing demand for primary energy worldwide, which is driven by urbanization, industrialization, and digitalization.

In the light of climate systems, a profusion of combustion of fossil fuels in heat exchange and power generation processes is a significant source of CO_2 and short-lived climate pollutants (SLCPs) in the form of black carbon and methane. The Intergovernmental Panel on Climate Change (IPCC) predicts that more than 70% of worldwide greenhouse gas emissions are a result of energy consumption and production. In that regard, energy conservation measures, and particularly, the measures aimed at improved thermal management, are central to the achievement of lower emission intensity per unit of energy service produced.

The ability to increase the prominence of the heat transfer mechanisms decreases the thermal resistance in heat exchangers, engines, cooling loops, and energy recovery systems. This is simply put into the consumption of less fuel in fossil-driven systems and greater efficiency yield in renewable systems such as solar collectors and geothermal systems. The decreased system requirement of parasitic power to losses and pumping also enhances the overall system coefficient of performance (COP), thus ensuring that energy utilized is anchored on thermodynamic efficiency frontiers.

In addition, the overexploitation of resources, especially the critical minerals, fossil fuels, and fresh water, has become unsustainable in most parts of the world. The conventional heat transfer liquids, for example, water, ethylene glycol, or artificial oils, demand huge amount and are frequently used below optimal thermal conductivity states, thus demanding more energy to reach thermal balance. Nanofluids present an avenue of uncoupling performance-volume, supplying the capacity to combine

DOI: 10.1201/9781003628071-14

small-scale and high-performance systems, accomplishing a small material and energy print.

14.1.2 Current Limitations in Conventional Energy Transfer Systems

The transfer of thermal energy is fundamental to nearly all industrial and domestic energy systems, power production, chemical production, refrigeration, and electronic cooling. The basic shortcomings of the conventional heat transfer systems limit the application and performance, energy efficiency, and flexibility, regardless of decades of optimization. The predicates are due to intrinsic features of the material (such as cost, loss, and efficiency), system-level (such as inefficiencies, grid availability, and cost), geometrical, and environmental sustainability requirements. Within the background of world energy conservation and decarbonization targets, it is essential to solve these bottlenecks as a precondition to use thermal management of the next-generation energy systems. This part analyzes critically the multi-dimensional boundlessness of the traditional heat transfer media and structures.

14.1.2.1 Low Thermal Conductivity of Base Fluids

The main drawback of traditional systems of energy transfer is low thermal conductivity of common working fluids (e.g., water (~0.6 W/mK), ethylene glycol (~0.25 W/mK), and mineral oils (~0.13 0.2 W/mK)). Such low values are an order of magnitude smaller than that of solid conductors such as metals or engineered nanomaterials. The heat transport rate within such fluids is thus naturally bound to be limited, making large temperature gradients or large surface areas necessary to effect the desired exchange rate of thermal energy. This not only raises the amount of energy required to pump the fluids and circulate the fluids but also increases the entropy generation, thereby reducing the second-law effectiveness of the system.

14.1.2.2 Geometrical and Space Constraints

Conventional systems tend to use high surface area heat exchangers (e.g., finned tubes, coiled systems, or shell-and-tube designs) to make up the shortfall in thermal conductivity. This however comes with certain design and space limitations, mainly where the space is rather small, including processors, cars, portable medical equipment, or aircraft. Volumetric efficiency is everything in these areas, and traditional fluid systems cannot deliver large thermal loads without incurring the risk of thermal bottlenecks or thermal overloads.

In addition, the trade-off between heat transfer surface area and pressure drop becomes pressing. Surface area enlargements normally increase hydraulic resistance, which requires more pumping power and makes operations expensive.

14.1.2.3 Laminar–Turbulent Transition Limitations

In numerous realistic systems, particularly flows with low Reynolds numbers, viscous fluids are in the laminar flow regime, where mixing is dismal, and convective heat transfer coefficients are inefficient. Although turbulence is used to promote a better convective heat transfer, in several cases, turbulence is to be avoided because of the pressure drop penalties and higher energy cost. There is an inherent performance compromise between the flow-induced mixing (which increases convective heat transfer) and pressure drop (which can add to parasitic losses).

In addition, the presence of the thermal boundary layers during the laminar flows prevents the full ability of the heat dissipation, especially in the system of microchannels and a heat exchanger with no plate. Failure to break down these layers by conventional fluids, except through massive energy contribution, restrains their use in low-Reynolds number-based flow conditions.

14.1.2.4 Limited Specific Heat and Thermal Diffusivity

Another important factor regarding transient thermal systems is specific heat capacity. Conventional fluids are sometimes not able to provide the specific heat or thermal diffusivity needed to absorb high-temperature transients or thermal shock. For example, water, despite its relatively high specific heat (~4.18 kJ/kg·K), cannot be used in high-frequency heating or ultra-fast cooling because of its low thermal diffusivity.

This limitation is particularly relevant in:

- Thermal management of pulsed power electronics.
- Phase-change energy storage systems.
- Cryogenic applications where thermal inertia is critical.

14.1.2.5 Scaling, Fouling, and Degradation

Conventional fluids are also prone to scaling, fouling, and chemical degradation over prolonged operation. For instance:

- Water leads to mineral scaling (e.g., $CaCO_3$, $Mg(OH)_2$) under high-temperature operation, which severely degrades thermal conductivity at interfaces.
- Glycol-based coolants can oxidize and polymerize, leading to viscosity increase and formation of sludge or gels.
- Organic thermal oils degrade into volatile or polymeric products, reducing heat transfer efficiency and increasing system maintenance needs.

These fouling effects not only reduce the effective heat transfer area but also increase the thermal resistance, degrading performance over time and necessitating periodic shutdowns for cleaning or fluid replacement.

14.1.2.6 Thermal Stability Limitations

Thermal stability is a crucial concern in high-temperature industrial applications, such as concentrated solar power (CSP), oil refineries, and power plant condensers. Conventional fluids exhibit a narrow operational thermal window:

- Water boils at 100°C (1 atm) and requires pressurization for higher temperature operation.
- Organic oils degrade above 300–350°C.
- Glycols decompose thermally and oxidatively above 200°C.

This severely limits their applicability in emerging high-temperature technologies like:

- Supercritical CO_2 Brayton cycles.
- High-flux solar receivers.
- Thermal cracking reactors.

Moreover, operating under pressurized conditions increases mechanical design complexity, safety hazards, and cost.

14.1.2.7 Environmental and Safety Concerns

Many conventional heat transfer fluids are toxic, flammable, or non-biodegradable, raising serious environmental and health risks. For example:

- Ethylene glycol is highly toxic if ingested.
- Synthetic refrigerants (e.g., HFCs, HCFCs) have high global warming potential (GWP) and are now restricted under the Kigali Amendment.
- Hydrocarbon oils pose fire and explosion hazards under leak or rupture conditions.

These challenges limit their use in sustainable and green energy applications, where environmental compatibility is paramount.

14.1.2.8 Pumping Power and Energy Losses

Due to low thermal conductivity and moderate viscosity, conventional fluids require high flow rates to achieve acceptable thermal performance. This leads to:

- Elevated pumping power requirements.
- Increased parasitic energy losses.
- Lower overall system efficiency.

In HVAC, data center cooling, and industrial process loops, the pumping system can account for 10–20% of the total energy consumption, creating a persistent drain on energy resources. Furthermore, viscosity-induced head losses become more pronounced in miniature or long-loop systems, exacerbating energy inefficiency.

14.1.2.9 Inflexibility to Dynamic or Transient Loads

Modern energy systems demand real-time adaptability to variable loads, especially in renewable-dominated grids. Conventional fluids lack dynamic reconfigurability or "smart" behavior, that is, the ability to change their thermophysical properties in response to external stimuli. Consequently, conventional thermal systems operate under static design assumptions, resulting in a significant mismatch between capacity and demand during off-peak or surge conditions.

For example, in solar thermal systems:

- Early morning or cloudy conditions may lead to fluid underheating and energy wastage.
- Peak sunshine periods may cause thermal overshoot, risking fluid degradation or system failure.

Without real-time tunability, conventional systems must over-engineer thermal capacity to handle worst-case scenarios, leading to oversizing, capital wastage, and lower operating efficiency during normal operation.

14.1.2.10 Incompatibility with Emerging Compact and Wearable Technologies

With the rise of miniaturized and flexible electronics, conventional cooling strategies have reached their physical limits. Wearable electronics, biomedical implants, compact IoT sensors, and flexible photovoltaics demand localized, passive, and efficient thermal management. Traditional fluids, due to their bulk volume, pumping requirements, and rigid flow geometry constraints, are unsuitable for such applications.

Moreover, thermal interface materials (TIMs) used in electronic assemblies typically exhibit poor coupling with conventional fluids, leading to interfacial thermal resistance and local hotspots. This incompatibility results in performance degradation, reduced device life, and reliability concerns.

14.1.2.11 Lack of Integrated Functionality

Conventional fluids are often single-function media—limited to heat transfer roles. In modern multifunctional systems, the demand is rising for working fluids that can:

- Act as thermal carriers and energy storage media.
- Exhibit antimicrobial or self-healing behavior.
- Enable real-time diagnostics via embedded sensing or colorimetric feedback.
- Provide electromagnetic shielding or triboelectric functions.

Traditional fluids fall short in enabling such integrated functionalities, forcing reliance on separate systems for heat management, sensing, and actuation—adding complexity, weight, and cost.

14.2 ROLE OF NANOFLUIDS IN THE GLOBAL ENERGY TRANSITION

14.2.1 Contribution toward Carbon Neutrality and Energy Efficiency

Because of the enhanced and tunable thermophysical properties, nanofluids have proved to be a necessary catalyst in the realization of carbon neutrality and an increase in energy efficiency in several sectors of the energy industry. High thermal conductivity, enhanced specific heat, and high capacity to raise the convective heat transfer coefficients enable nanofluids to handle the inadequacies that are inherent in the conventional heat transfer systems, where exergy destruction, high temperature difference, and entropy generation tend to be the overall contributors. The inefficiencies cause an increase in the amount of energy needed to attain the desired level of thermal management, which literally increases carbon emissions and energy waste. These problems are alleviated by nanofluids since they facilitate fast and homogeneous heat transfer, thereby allowing the thermal systems to work near thermodynamic reversibility, where driving temperature differences are low and fewer repetitions are taken per cycle.

In fossil fuel applications, including industrial boilers, thermal power plants, and combustion engines, the nanofluids lower the total heat transfer resistance, hence requiring less fuel to reach the best operation temperatures. This leads to reduced direct (Scope 1) CO_2 emissions per unit of useful thermal energy supplied. In air-conditioning units and data center cooling infrastructure, along with high-performance computing modules powered by electricity, nanofluids help relieve pressures on compressors and circulation pumps, which in turn lowers the electrical energy requirements and hence indirect (Scope 2) emissions, especially where grid-supplied electricity remains heavily carbon-intensive.

In addition to their traditional usages, nanofluids are of great importance to improve the efficiency of renewable energy systems where they are applied as high-performance working fluids in solar thermal collectors, concentrated solar power (CSP) receivers, photovoltaic-thermal (PV/T) systems, and geothermal loops. They allow a higher efficiency of conversion of thermal to electrical energy due to the possibility of both raising the heat removal factor and lowering stagnation losses, as well as the ability to enhance the temperature windows of operation. Nanofluids with customized optical and thermal characteristics in the CSP systems can selectively absorb solar energy and enhance heat storage in molten salts or thermal oils, hence providing the capability of long-term dispatchability and the ability to achieve greater capacity factors. Likewise, in geothermal systems where multiphase flow and the scaling and low conductivity media affect the performance, microscopic fluids heavily reduce the performance of the system regarding the production of heat in the subsurface and deposition-associated problems.

Furthermore, nanofluids are central to the increase in thermal energy storage (TES) technologies. Dispersing nanoparticles into phase change materials (PCMs) or molten salts, one can substantially improve the rates of melting and solidification, improve effective thermal diffusivity, and minimize thermal hysteresis and subcooling. The facilitation can make load balancing and mitigation of peak demands more efficient, as well as support increasing storage use levels, which is crucial in the process of incorporating intermittent renewable sources, such as sun and wind, into the grid. The effect of this peak shaving and energy time-shifting ability is to directly facilitate the decarbonization of power generation through a decline in the dependence on fossil-fuel-based peaking plants.

The nanofluid contribution of miniaturizing the system and material efficiency is also a relatively ignored but critical aspect. Improved heat transfer efficiency enables reduction of heat exchanger, condenser, and thermal management module sizes, therefore mitigating the amount of raw material, energy used in manufacturing, and transportation footprint the same cumulatively affecting a reduction in the embodied carbon of thermal infrastructure. Where mass and volume are severe constraints (as in airplanes, automobiles, and portable electronics), this can add immense value in the energy efficiency and flexibility of design.

Dynamic thermal control of smart and load-varying systems can also present adaptive nanofluid control. In contrast to more traditional fluids, nanofluids can be engineered to be field-responsive (e.g., magneto-, electro-, or thermo-responsive) to allow in situ property tuning to changing load conditions. Such flexibility is needed before the next generation of applications, such as building-integrated energy systems, wearable electronics, and AI-based HVAC networks, as the thermal loads will be greatly variable. The dynamic thermal response not only contributes to the efficiency of operation but also reduces the potential risk of oversizing and underutilization that occur in the majority of static thermal systems and also causes an inefficient consumption of carbon resources.

When all these myriads of advantages are put together, then the strategic nature of nanofluids in helping to cut operational and life cycle emissions of thermal systems becomes even more prominent. By fixing wastefulness in materials, parts, and systems, nanofluids have massive potential to permit carbon mitigation objectives in the frameworks that have been claimed globally, such as the Paris Convention. Moreover, with decarbonization tasks expanding to the broadening industry, building energy utilization, transport, and data-supporting infrastructure, the dependence on nanofluids to empower low-carbon warm innovativeness turns out to be increasingly significant. Nanofluids in this regard should not be viewed as a step forward but rather a shift in paradigm of managing, storing, and consuming thermal energy in a carbon-constrained world.

14.2.2 Potential to Support Clean Energy Technologies (Solar, Wind, Hydrogen, Geothermal)

It is believed that nanofluids have a significant prospect of improving the efficiency, reliability, and efficacy of the novel renewable energy systems, specifically in the thermally limited applications, for example, solar thermal conversion process, generation of hydrogen, thermo-geothermal energy acquisition, and subsidiary in the wind turbine. Nanofluids have also been seen to exhibit the best optical and thermal properties in solar energy systems, which help in achieving a high level of solar absorption, reduced re-radiation loss, and high efficiency of convective heat transfer in direct absorber solar collectors (DASCs), evacuated tube solar energy systems, and concentrating solar power (CSP) receivers' solar systems. Synthesized nanoparticles (ones consisting of metal oxides or carbon-based products, or plasmonic metals) can be customized to be spectrally selective absorbers, which will absorb the maximum amount of solar radiation to generate thermal energy, and extend the temperatures at which the heat transfer fluid can be used beyond those that the oils or water can handle.

In the production of hydrogen and in thermochemical hydrogen processes, especially thermochemical water splitting cycles (e.g., CuCl, Zn-ZnO), photothermal catalysis or photoelectrochemical (PEC) cells, nanofluids have a dual use, as particularly thermally efficient vectors of heat, and also as catalysts because of the catalytic activity of suspended nanoparticles. The advancement in photothermal reactors using nanofluid has gained a more substantial quantum yield and higher rate of evolution of hydrogen in concentrated sunlight or laser illumination.

In geothermal cooling systems, nanofluids also have excellent advantages as thermal fluids in terms of their ability to extract the thermal energy efficiently at the underground hot spot by increasing the thermal conductivity of hot brine or hot water used to circulate in the system, and also in regard to preventing the control of scale and mineral deposition formation because surface modified nanoparticle offers such advantages easily. They may give a higher heat recuperation per unit mass flow rate than other mechanisms used in Enhanced Geothermal Systems (EGS), permitting economical uses in lower temperature rates.

As related to wind energy, nanofluids are not applied directly to energy harvesting but constitute life-saving assistance where high power densities are important: thermal management of power electronics, generators, and control systems. The use of nanofluids in cooling loops or thermosyphons within a nacelle or within an inverter system helps to minimize thermal stresses on the equipment, extending the equipment life and interfacing reliability with the grid. Additionally, the capacity of nanofluids to be used in hybrid systems energy like the solar-thermal products and the hydrogen co-generation plants also increases their importance in the decarbonization of the world energy mix. Accordingly, nanofluids are being developed into functional enablers that supplement the applicability, efficacy, and adoption of various clean energy technologies and technologies to narrow the discrepancy between the efficacy of theoretical designs and reality that is emulated during the real-world functioning.

14.2.3 Integration in Smart Energy Grids and Sustainable Infrastructures

The introduction of the nanofluids into the smart energy grids and sustainable infrastructures is an important breakthrough in realizing the real-time thermal responsive capacity, decentralized energy control, and energy-use optimization in interconnected systems. With the smartening of energy grids through to intelligent, cyber-physical systems where the energy path is bi-directional, renewable penetration levels are high, and the control system digitalized, the importance of thermal management plays a pivotal role—particularly in managing trade-offs between power quality, reliability, and end-to-end energy efficiency in energy grids.

Nanofluids help address this issue at the material level, giving such key infrastructure points as transformer cooling systems, grid-interfacing inverters, battery energy storage units (BESS), and combined heat and power (CHP) modules high-performance, adaptive thermal control. Their increased capability to transport heat enables thermal transient dissipation to occur on shorter time scales and, thus, enables relatively short-duration peak temperature excursions to be reduced in sensitive electronics and helps to enable load-following and frequency response capabilities in distributed energy resources (DERs). Nanofluid-based systems have the capability to dynamically adjust heat transfer as an occupancy-based measure depending on occupancy, existing conditions surrounding the buildings, and grid commands, making them ideal candidates for demand side management, thermal load shifting, and attenuating peak load demands.

In addition, the nanofluids incorporated in the thermal energy storage systems (sensible and latent) can be used to increase grid stability to achieve time-of-use arbitrage integration of the heat/cooling and stored at low-demand or renewable-rich times, and when it is needed on the grid during peak load times. Infrastructure sustainability-wise, the concept that the HVAC can be miniaturized, the amount of refrigerant used can be lowered, and that hybrid solar thermal collectors can be coupled with nanofluids means a direct decrease of embodied carbon, greater building energy performance index (EPIs), and an increase in LEED or BREEAM scores. In advanced applications, AI-managed nanofluid circuits can talk with building management systems (BMS) as well as with supervisory control and data acquisition (SCADA) systems, enabling predictive thermal control by a machine learning model pretrained by occupancy, weather, as well as grid dynamics. The flexible heat routing of nanofluids, magneto- or electro-responsive, may transform energy-efficient data centers and industrial parks, which work as microgrids. Briefly, the potential impact of nanofluids to influence the operation, optimization, and sustainability of smart energy and infrastructure systems will not be just that but rather an active contribution to such systems.

14.3 NEXT-GENERATION NANOFLUIDS

14.3.1 Development of Biodegradable and Green Nanofluids

The *next-generation nanofluids* evolution process is inherently connected with the current world focus on green engineering, life cycle sustainability, and circular economy system. Although conventional nanofluids are thermally efficient, they are likely to contain synthetic nanoparticles and petroleum-based base fluids that pose environmental and health risks at manufacturing, deployment, and disposal stages. This encouraged the highly developed studies of biodegradable and environmentally benign nanofluids, as they employ eco-friendly synthesis routes, biogenic surfactants, and non-toxic, renewable base fluids (vegetable oils, glycerol, ethyl lactate, or bio-alcohols). Nanoparticles made of plant ashes (e.g., rice husk silica, banana-peeled-derived carbon), bio-waste-derived oxides, or naturally occurring minerals, such as montmorillonite and halloysite, which create little ecological impact and toxicity, are included in biodegradable nanofluids. The fluids have excellent thermal conductivity improvements and, in many instances, match the improvements of synthetic equivalents when developed through surface-functionalization with biopolymers, for example, chitosan, gum arabic, alginate, lignin, and can be used as both the thermally conducting and the stabilizing agent, obviating surfactants that bioaccumulate or disrupt aquatic ecosystems.

In addition, green nanofluids are becoming source-specific (single-use or closed-loop) in biomedical cooling, solar thermal systems in remote locales, food-grade process equipment, where environmental compatibility is paramount as well as thermal performance. According to Life Cycle Assessment (LCA) studies, these fully biodegradable nanofluids can have more than 70% reduction in carbon footprint with cradle-to-grave impacts in comparison with conventional ethylene glycol or mineral oil nanofluids. Significantly, studies are moving toward production of nanoparticles in situ within biodegradable base fluids in low-energy, solvent-free processes, such as microwave-assisted green chemistry or biosynthesis, where hazardous precursors are omitted, and the process waste is minimized. The innovations conform to *ISO 14040/44 standards* and new code requirements on eco-design and end-of-life recapture of the functional fluids in process and energy sectors. Therefore, it can be suggested that the use of green nanofluids is not an alternative but rather a strategic step toward the realization of sustainable thermal systems that would achieve harmonization of high-performance heat transfer with environmental responsibility.

14.3.2 Use of Functionalized Nanoparticles (Magnetic, Luminescent, Thermally Responsive)

Functionalization of nanoparticles and subsequent integration within the nanofluid complex has greatly contributed to the ability of the nanofluids to perform more than being mere sources of passive heat carriers—providing a multi-functional thermal platform that can be utilized in an adaptive-reactive, targeted, and energy-smart construct. Functionalization in order to achieve specific physicochemical characteristics, the surface of the nanoparticles or their structure can be modified—this is called functionalization and turns the conventional nanofluids into smart (active) materials; these can include magneto-responsiveness, luminescence, or thermal reactivity. The common magnetic nanofluids, consisting of superparamagnetic Fe_3O_4 or γ-Fe_2O_3 nanoparticles, can have a field-responsive rheology and thermal transport properties and enable out-of-contact control of flow, orientation, and heat removal via applied magnetic fields. It has been used in precision electronic cooling, localized hyperthermia in biomedical equipment, and magnetically aided solar thermal collectors, where convection induced by a field can be used to increase the efficiency of energy capture.

Doped rare earth oxide nanoparticles, for example, Y_2O_3:Eu^{3+}, $LaPO_4$:Ce^{3+}, become luminescent nanoparticles in nanofluids, which can provide bi-functionality: optical sensing and thermal to electrical conversion as they respond to thermal fluctuations by emitting wavelength-specific signals. These materials are also essential in photothermal and photocatalytic, as they are used to perform in situ diagnostics of temperature or light-heating in confined microfluidic geometries or energy-harvesting surfaces.

Meanwhile, thermally responsive nanoparticles, such as phase-change nanocapsules or thermochromic material, offer dynamic thermal buffering properties. On mixing with nanofluids, these nanoparticles experience reversible phase or structural transition (e.g., solid–liquid conversion, or re-arrangement of structure) over certain temperature ranges and/or intervals, which may provide thermal capacities and facilitate self-regulating heat exchange. Also under development are Janus particles and core-shell structures (e.g., silica-coated gold or polymer-grafted ZnO) to incorporate many functionalities side-by-side, such as magnetic, thermal, and optical capabilities, to be applied to multimodal control in dynamic heat management applications.

The same type of functional nanofluids is being adapted to closed-loop thermal networks with real-time management of heat transport affinely coupled with external control (magnetic, optical, thermal) via *AI-based control systems.* The use of functionalized nanoparticles, such as thermoregulated smart windows, adaptive radiative cooling materials, is an important turning point to programmable heat transfer materials that will address the requirements of next-generation energy systems and *Industry 5.0* architectures.

14.3.3 Hybrid Nanofluids for Tailored Thermal and Rheological Properties

In order to overcome such constraints of mono-component nanofluids, a next-generation solution in the form of hybrid nanofluids was developed: these fluids are prepared by dispersing two or more types of nanoparticles into a base fluid. Placement and combination of nanoparticles with thermophysical features judiciously are always helpful; hybrid nanofluids can lead to a synergistic improvement of heat transfer without increasing viscosity too many viscosities. In other words, by mixing and matching nanoparticles with thermophysical characteristics that complement each other (e.g., arrangements of nanoparticles with high thermal conductivity (e.g., CuO, Al_2O_3), low density (e.g., SiO_2), or certain heat intensification capacity (e.g., TiO_2). This thermodynamic programmability enabled by hybridization allows optimization toward application-specific warmth where thermal conductivity, viscosity, specific heat, and density can actually be programmed accurately according to the conditions of use; flow regime, heat flux, and temperature sensitivity.

Further, the particle heterogeneity is common in hybrid systems by virtue of utilizing spherical particles with rod-like or plate-like particles (e.g., Al_2O_3-CNT or graphene-TiO_2) to elevate the percolation temperature of the thermal conductivity and preserve Newtonian-like behavior within the moderate shear. This optimization is of special value in miniaturized heat exchangers, electronic cooling systems, and compact solar thermal devices, where both high thermal performance (balance to comprehend the balance between high thermal performance and low pumping power) and low pumping power are important. Commenting on the non-linear behavior of thermal enhancement in hybrid nanofluids, both experimental and molecular-level dynamic studies have revealed that the thermal enhancement behavior in hybrid nanofluids is non-linear, and hence there exist interparticle synergies in the form of phonon bridging, electrostatic interactions, or other micro-convection effects at the solid–liquid interface.

Moreover, through hybridization, multifunctionality can be realized whereby a nanoparticle can be used as a thermal enhancer and another can provide magnetic, catalytic, or anti-fouling properties such that a nanofluid can be used in multiphysics context, such as magneto-convective solar collector, photothermal reactor, or desalination system using membranes. Difficulties in achieving colloidal stability, avoiding agglomeration, and transparently modeling non-Newtonian fluid behavior in arbitrary flow regimes remain to be solved, but in the past few year's surfactant-free synthesis, pH-responsive stabilization, and rheological prediction and reconstruction have greatly advanced the feasibility of these proposals. As machine learning approaches are incorporated (including formulation predictions and process intensification strategies), hybrid nanofluids will allow the development of personalized high-performance thermal fluids that meet the performance requirements of low-carbon and energy-smart technologies in the future.

14.3.4 Phase-Changing and Self-Healing Nanofluids for Thermal Storage

One of the new promising areas in thermal energy storage is phase-changing nanofluids (PCNFs), a form of advanced nanofluids that combine both phase change materials (PCMs) and functional nanoparticles. Through synergetic combinations of the PCM-involved latent heat storage capacity and nanoparticle-created enhanced thermal conductivity and structural modulation, these fluids show improved energy density and thermal response. Embedded fluid PCMs experience solid–liquid phase transitions at desired temperature intervals, resulting in isothermal absorption or supply of heat. Addition of high-conductivity nanoparticles (e.g., graphene, Al_2O_3, Cu) or nanoshell-encapsulated PCM ions can simultaneously boost the charging/discharging rates and spatial thermal homogeneity, essentially negating the inherent drawback of low thermal conductivity of most conventional PCMs.

The therapeutic self-healing nanofluids also offer an added layer of rigidity and survival time to thermal systems because once subjected to external stresses, degradation, or agglomeration, they can automatically restore their working status, for example, colloidal stability or thermal conductivity. Such actions are facilitated by functional groups whose binding is stimuli-responsive or reversible supramolecular association (e.g., hydrogen bonding, π–π stacking, ionic interactions) incorporated on nanoparticle surfaces. As an example, the pH or temperature-sensitive ligands on nanofluids can dynamically re-disperse nanoparticles in case they are detected to be destabilized, eliminating any significant loss in performance after numerous working cycles. These intelligent fluids make the need for external maintenance minimal, so they would fit in the remote or autonomous energy applications such as solar thermal plants or distributed energy storage facilities.

A significant breakthrough in this field is nano-enhanced microencapsulation of PCMs that utilize smart polymers, processes of mesoporous silica, or metal-organic frameworks (MOFs), which serve as both PCM-containers and thermal bridges. The shells not only alleviate the PCM leakage and volume change in phase transitions but also enable programmable melting points, higher surface-to-volume ratio, and thermal reliability in multiple cycles. In addition, it is possible to incorporate magnetically active or light-reactive nanoparticles in these nanofluids that can be externally controlled using a magnetic or optical field, which has been shown to permit temperature locking or unlocking on demand, essential to load balancing in the smart grid, and solar intermittency compensation.

System-level-wise, the phase-changing and self-healing nanofluid can easily be incorporated into latent heat thermal energy storage (LHTES) systems, and they have the ability to accelerate charging/discharging, thermal buffer, and long-term storage. Its uses may involve solar power plants, industrial waste thermal recovery, and building-integrated thermal storage. Nevertheless, issues of thermal stability of cycling, encapsulation stability, and low-cost scalability remain problematic, currently tackled via nanoengineering interfacial regions of PCM and matrix, stabilization methods, and AI-aided development of materials. These high-performance nanofluids are becoming an important enabler of high-performance, fault-tolerant, and smart energy storage systems to form the backbone of future energy network systems.

14.4 TECHNOLOGICAL INNOVATIONS SUPPORTING NANOFLUID ADOPTION

14.4.1 3D Printing of Nanofluid-Enabled Microchannels

The adoption of nanofluids in complex heat transfer systems has been greatly advanced by the introduction of 3D printing (additive manufacturing) methods that can accommodate the thermal management design at a microscopic dimension. Three-dimensional printing techniques/high-resolution printers that have a significant microfluidic-enabled cooling channel fabrication capacity include stereolithography (SLA), direct ink writing (DIW), selective laser sintering (SLS), and two-photon polymerization (2PP), which have all provided grounds to explore microfluidic-enabled cooling structures and are uniquely threatened in the constructional completion of nanofluids and are thermally optimized with applications in mind.

Microchannels that are compatible with nanofluid and manufactured using 3D printing enable flow passages to be geometrically fine-tuned to enable nonlinear meandering, fractal networks, porous structures, or biomimetic heat engine topographies that can facially disintegrate thermal boundary layers and increase surface area to volume ratios. Such elaborate flow patterns can be used to dramatically increase the level of convective heat removal when applied to high-thermal-conductivity nanofluids (e.g., CuO/water, graphene-ethylene glycol, or hybrid nanoparticle dispersions) and make possible very high energy-density, compact, and efficient electronic-device, photonic device, and microreactor thermal management systems.

One of the main innovations here is the co-design of materials and nanofluids whereby 3D printable matrices are designed to be thermally compatible, chemically resistant, and mechanically sound in the presence of suspended nanoparticles. Newer developments can be seen in the use of nanocomposite resins impregnated with heat-conducting fillers (e.g., BN, CNTs, AlN) to form hybrid structures, which, by their synergetic environment together with nanofluid flow in channel, form a more effective heat transfer device by incorporating a thermally conductive wall. Besides, FGMs may be printed, yielding areas with different thermal conductance, hence allowing more spatially optimal heat flux paths.

The other new frontier is the integration of sensor-loaded 3-D-printed channels that could facilitate real-time monitoring of flow non-equilibrium, temperature gradients, and nanoparticle dispersion dynamics. With this data-driven philosophy and the assistance of artificial intelligence in feedback regulation, it is possible to make the flow adaptive, spot the beginning of fouling or clogging, and stabilize the thermal loads autonomously, increasing the life and the reliability of the systems.

Also, the opportunity to produce a rapid prototype and mass-customized nanofluidic heat exchange systems of electric vehicles, wearables, and biomedical implants to meet the specific geometric needs is also made possible with the use of 3D printing. This is because the capability to combine multifunctionality capabilities in one 3D-printed design, including electrical insulation, mechanical support, and dynamic cooling, has made these systems essential to future energy-efficient, space-saving technologies.

Some of the setbacks, despite the advances, include nanoparticle sedimentation in the course of curing, the influence of roughness in the flow of nanofluids, and the scalability of microchannel arrays in an industrial system of cooling. Efforts are being made to come up with post-printing surface modifications, ultrasmooth coating on the internal walls, and self-lubricating nanofluid formulations in order to overcome these drawbacks. However, the combination of 3D printing and nanofluids stands to transform the design, performance envelope of compact, high-performance thermal systems in a full range of critical energy applications.

Case Study: Enhancing Thermal Performance Using 3D-Printed Nanofluidic Heat Sinks (Dixit et al. 2022)

In a recent experimental study, the thermal performance of *3D-printed mini heat sinks* with optimized geometries of fin structures (Figure 14.1) was tested in a thermally active fluid Al. The heat sinks were produced with high-resolution stereolithography (SLA)-based 3D printing, which allowed complicated geometrical features to be created, such as cylindrical pin fins, lattice structures, and fractal microchannels, which would have been difficult to manufacture with conventional machining or prohibitively expensive.

The addition of nanofluids in such tailor-made microarchitectures proved to show considerable improvement in heat transfer coefficient (HTC) and thermal uniformity, especially at low Reynolds numbers, where laminar conventional flows are very much prone to the development of thermal boundary layer. The 3D printed fins with their large surface area-to-volume ratio, together with the improved thermal conductivity and micro-convection properties of the nanofluid, have contributed to a thermally enhanced system that showed an improvement of up to 58% in terms of thermal performance over traditional water-cooled metal heat sinks.

Besides, thermal imaging showed that the formation of hotspots and temperature gradients had decreased because of a synergistic combination of targeted pathways of channels and nanoparticle-mediated heat conduction. The pressure analysis helped establish that the pumping penalty still exceeded the acceptable levels, and this established the hydraulic viability of the given design toward compact electronic cooling systems.

In addition to the impact of nanofluids in unlocking the capability of adding-manufacturing to design application-specific and performance-optimized heat exchangers, the case highlights the potential level of performance when nanofluids are deployed in the context of additive manufacturing. With the

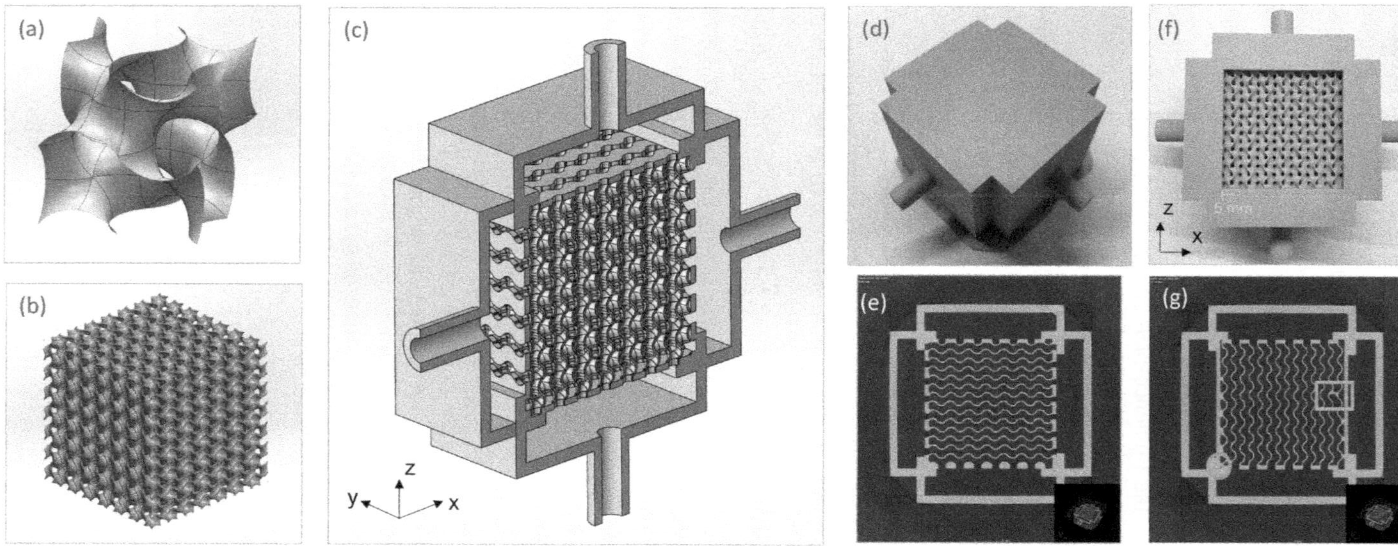

FIGURE 14.1 3D or gyroid lattice compact heat exchanger design: (a) Gyroid Unit Cell Surface Model; (b) Assembled Gyroid Lattice Core with Defined Porosity; (c) Sectional CAD View Showing Core, Plates, and Headers; (d) 3D-Printed Heat Exchanger Assembly; (e) Micro-CT Image of Mid-Plane Cross-Section; (f) Exposed Gyroid Core without Cover Plate; (g) Magnified View of Minor Surface Imperfections [1].

interconnection of 3D printing and nanofluid engineering, it is a new transformational route to fabricating lightweight, miniaturized, and thermally sustainable energy systems that are an important part of present electronics, renewable energy, and EV systems.

14.4.2 Nano-Coatings and Thermal Interface Materials

Nano-coatings, consisting of *nanoscale layers of metal oxides, carbides, nitrides, or carbon-based materials*, are applied to surfaces to *modify interfacial wettability, thermal emissivity, and surface energy characteristics.* When applied to heat exchanger surfaces or inner walls of flow conduits, such coatings enhance nanofluid interaction by:

- *Promoting uniform nanoparticle dispersion and stable wetting*, thus improving convective heat transfer coefficients.
- *Suppressing fouling, scaling, and corrosion*, particularly when using water-based nanofluids in harsh operational environments.
- Enabling *photothermal or thermochromic behavior*, which can be exploited in solar thermal collectors or smart cooling surfaces.

For instance, *graphene oxide (GO)* or *reduced graphene oxide (rGO)* coatings exhibit ultra-high thermal conductivity and act as *interfacial enhancers* that facilitate rapid lateral heat spreading across the coated substrate. Similarly, *alumina or titania nano-ceramic coatings* impart dielectric properties while supporting high thermal conductance, ideal for electronics cooling substrates.

Recent innovations include *self-assembled monolayers (SAMs)* and *layer-by-layer (LbL) deposition techniques*, which allow for sub-10 nm precision in coating thickness while enabling chemical tunability to suit different nanofluid chemistries. Moreover, *smart nano-coatings* with embedded phase change or magneto-responsive properties allow *dynamic control of thermal flux*, creating responsive thermal interfaces in energy-harvesting and thermal switching systems.

14.4.2.1 Advanced Thermal Interface Materials (TIMs) with Nanoparticles

Thermal interface materials serve as *conduction bridges* between solid components, such as CPUs and heatsinks, or battery modules and cooling plates. Traditional TIMs (greases, pads, gels) are plagued by low conductivity, aging, and poor mechanical compliance. To overcome this, *nanoparticle-enhanced TIMs* have emerged, incorporating *high-aspect-ratio fillers* such as the following:

- Carbon nanotubes (CNTs)
- Graphene nanoplatelets (GNPs)
- Boron nitride nanosheets (BNNS)
- Metallic nanoparticles (Ag, Cu)

These nanoparticles significantly reduce interfacial thermal resistance (ITR) by forming percolated thermal networks, thereby increasing effective thermal conductivity (often exceeding 10 W/m·K). The synergistic design of viscoelastic polymer matrices with surface-functionalized nanoparticles allows these TIMs to maintain conformability while achieving ultra-low thermal impedance. Additionally, nanofluid-impregnated TIMs are an emerging class where nanofluids are encapsulated within porous TIM scaffolds or gels, enabling microscale convective heat transfer within the interface region. These "hybrid TIMs" combine the advantages of liquid-phase convection and solid-phase conduction, particularly useful in transient and pulsating thermal environments.

14.4.2.2 Application Domains and Future Outlook

Nano-coatings and nanofluid-optimized TIMs are increasingly being adopted in the following domains:

- Power electronics and IGBTs
- LED thermal management
- Electric vehicle (EV) battery thermal control
- High-frequency radar and RF systems
- Spacecraft thermal tiles and avionics modules

Future trends focus on AI-assisted design of nanostructured interfaces, self-healing TIMs, and 3D conformal nano-coatings fabricated via atomic layer deposition (ALD) or molecular layer deposition (MLD), enabling atomic-level control over interfacial properties. As devices shrink in size and grow in power density, the integration of nanofluids with engineered nano-interfaces will be key to managing heat effectively while maintaining energy efficiency and operational reliability.

14.5 CHALLENGES AND RESEARCH NEEDS

Figure 14.2 presents a structured overview of the key challenges associated with the application of nanofluids in data center thermal management, along with corresponding future research directions. It categorizes the primary obstacles—namely stability, erosion and abrasion, and electrical insulation—and links them to specific investigative approaches. For each challenge, intermediate steps such as nanoparticle characterization, wear and corrosion studies, and dielectric testing are outlined, ultimately converging toward advanced research pathways including hybrid nanofluid development, rheological modeling, and optimization of insulating properties. This visual framework effectively highlights the multiscale and interdisciplinary nature of ongoing research required to enable nanofluid deployment in high-performance, mission-critical cooling systems like those found in modern data centers [2–4].

14.5.1 Stability and Long-Term Performance under Real-World Conditions

One of the most pressing challenges in the deployment of nanofluids in energy systems is ensuring colloidal stability and long-term functional reliability under

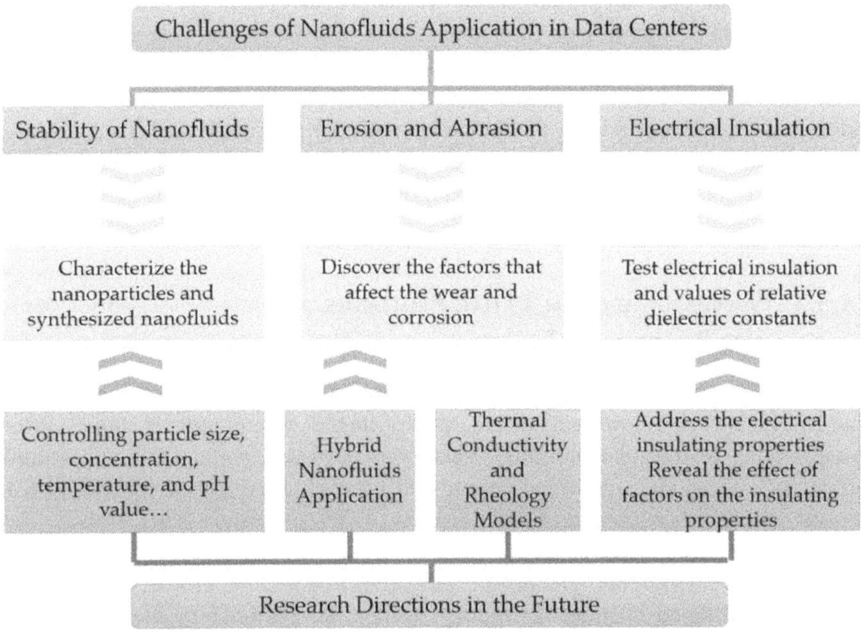

FIGURE 14.2 Next-generation nanofluids: enhancement strategies and scientific outlook [5].

real-world operational conditions. In practical applications—such as solar thermal systems, nuclear reactors, HVAC units, or geothermal loops—nanofluids are exposed to continuous thermal cycling, fluctuating pH, pressure variations, and shear-induced turbulence. These dynamic conditions can lead to particle agglomeration, sedimentation, or even chemical degradation of surfactants or base fluids, thereby drastically reducing thermal performance and potentially clogging the flow channels. Advanced techniques such as surface functionalization, pH tuning, or the use of smart dispersants have shown promise in improving thermophysical stability; however, long-duration field trials and predictive modeling of degradation kinetics remain limited. There is an urgent need for long-term accelerated aging studies, in situ monitoring technologies, and standard stability metrics to validate the sustained thermal conductivity and viscosity behavior over extended operational lifetimes.

14.5.2 Cost and Scalability of Nanomaterial Production

Another critical barrier hindering the widespread adoption of nanofluid technologies is the economic and scalable production of high-quality nanomaterials. The synthesis of engineered nanoparticles—such as doped metal oxides, hybrid graphene-metal composites, or surface-functionalized magnetic particles—often requires energy-intensive, multi-step processes involving controlled atmospheres, expensive precursors, and specialized post-processing (e.g., calcination, size separation, or purification). While lab-scale methods (e.g., sol-gel, hydrothermal synthesis, chemical vapor deposition) offer high control over particle morphology and dispersion compatibility, their translation to kilogram or ton-scale remains commercially unviable without significant capital investment. Furthermore, ensuring batch-to-batch consistency, contamination control, and environmental safety during synthesis adds to the complexity. To address this, future research must prioritize green and continuous-flow synthesis techniques, biotemplated or plant-based nanoparticle production, and on-site nanofluid formulation units that reduce logistics and storage costs [6–8].

14.5.3 Standardization in Testing Protocols and Performance Evaluation

The field currently suffers from a lack of universally accepted standards for thermophysical property measurements, performance benchmarking, and system compatibility testing of nanofluids. Variations in experimental protocols, sample preparation methods, sonication times, and even measuring instruments lead to inconsistent and non-comparable data across different studies. This severely hinders industrial uptake, as designers and engineers lack confidence in extrapolating lab results to real systems. The need of the hour is to develop globally harmonized protocols, validated by inter-laboratory studies, for parameters such as thermal conductivity, viscosity, specific heat, zeta potential, fouling behavior, and corrosion risk. Moreover, standard testbeds simulating real operating conditions (e.g., plate heat exchangers, microchannel loops, solar collectors) should be used to report figures of merit like heat transfer coefficient per unit pumping power or thermal enhancement versus pressure drop under dynamic flow regimes. These efforts will enhance the credibility and comparability of nanofluid performance data for regulators and industries alike [9, 10].

14.5.4 Environmental, Health, and Regulatory Concerns

With increasing interest in deploying nanofluids in energy applications, concerns have grown around their environmental toxicity, occupational exposure risks, and lack of regulatory oversight. Many commonly used nanoparticles, including TiO_2, ZnO, CeO_2, and CNTs, are known to exhibit cytotoxic effects on aquatic and human cells at certain concentrations, raising questions about their environmental persistence and disposal after usage. Additionally, inhalation or dermal exposure during handling or accidental spillage poses health risks to workers, especially in poorly ventilated industrial settings. Currently, most national and international safety frameworks are not equipped with nanomaterial-specific risk assessments or disposal regulations, creating legal ambiguities and hindering mass adoption. There is a need for a multidisciplinary approach combining toxicology, environmental science, and nanotechnology to develop eco-design principles, life cycle assessment (LCA) frameworks, and safe-by-design guidelines. Moreover, regulatory bodies must collaborate with scientific communities to develop nanofluid-specific safety codes, workplace handling standards, and disposal/recycling protocols that address the unique properties and risks of engineered nanomaterials.

14.6 INTEGRATION WITH SUSTAINABLE DEVELOPMENT GOALS (SDGS)

Adoption of nanofluid technologies in energy conservation and thermal management is not only a scientific breakthrough but also a strategic play in attainment of the United Nations Sustainable Development Goals (SDGs). Nanofluids present a multidimensional complement to major SDGs, given their energy efficiency, enabling the design of compact systems, and decarbonization, especially in energy, infrastructure, climate mitigation, environmental resilience, and other areas.

14.6.1 Alignment with SDG 7—Affordable and Clean Energy

Nanofluids are essential in enhancing the productivity of the clean energy systems in total compliance with SDG 7, which supports access to clean, modern, reliable, sustainable, and affordable energy. Indeed, they increase the energy conversion efficiency, minimize energy loss, and improve the renewable energy infrastructure reliability when incorporated into solar thermal collectors, photovoltaic/thermal (PV/T) systems, geothermal loops, and phase-change based thermal energy storage systems. Nanofluids have the potential to enable cost-effective thermal management since nanofluids may enable lower operating temperatures at the same thermal output or higher thermal outputs using fixed energy input, particularly in decoupled, off-grid energy systems in the developing world.

14.6.2 Supporting SDG 9—Industry, Innovation, and Infrastructure

SDG 9 focuses on the areas of developing resilient infrastructure, sustainable industrialization with innovation. Nanofluids usage in compact heat exchangers, smart HVAC, electronic thermal interface, and advanced manufacturing (e.g., 3D-printed nanofluidic systems) results in intensification of processes, miniaturization of systems, and energy productivity. In addition, the current work in hybrid nanofluids, self-healing formulations, and machine-learning-based formulation optimization is representative of technological innovation, in the sense of having industrial applicability. Nanofluids can also support the building of future cooling systems of high-performance computing, data centers, and electric mobility infrastructure, the cornerstones of a green industrial future.

14.6.3 Contribution to SDG 13—Climate Action

Nanofluids contribute significantly to SDG 13 because they can enhance the thermal performance of the energy system and end up mitigating the direct and indirect emissions of greenhouse gases. In other industries such as fossil-based power plants, effective cooling by a nanofluid leads to less amount of fuel used, leading to the minimization of carbon emissions per unit energy generated. They allow more use of low-carbon technologies when used in renewable energy systems or systems used to recover energy. Furthermore, it has a direct application to climate mitigation strategies since they are used in thermal energy storage, waste heat recovery, and cooling in the built environments without the need to use energy, which increases the degree of environmental sustainability of built environments and industrial activities.

14.6.4 Cross-Sectoral Benefits: Healthcare, Water, Environment, and Transportation

Beyond energy systems, nanofluid applications have significant spillover benefits across multiple sectors. In healthcare, nanofluids are being used in localized thermal therapies, drug delivery, and sterilization systems requiring precise thermal regulation. In water treatment, nanofluid-based emulsion liquid membranes enhance selective extraction and pollutant separation, contributing to clean water access and pollution reduction. In transportation, particularly in electric vehicles and aerospace, nanofluids provide lightweight, high-performance thermal management, improving energy density and operational safety. Environmental monitoring systems also benefit from nanofluid-based sensors and responsive heat exchangers, enabling sustainable environmental control and climate resilience.

14.7 ECONOMIC AND POLICY CONSIDERATIONS

The large-scale deployment of nanofluids in energy systems hinges not only on scientific and engineering feasibility but also on robust techno-economic justification and supportive policy frameworks. As nanofluid-based technologies mature from laboratory-scale innovations to commercially viable solutions, a clear understanding of their cost–benefit profile, market readiness, and regulatory landscape becomes essential to accelerate industrial adoption and societal impact.

14.7.1 Techno-Economic Feasibility and Life Cycle Assessment (LCA)

Techno-economic analysis (TEA) of the economic viability of nanofluid integration would entail a full-scale analysis of capital expenses (CAPEX), direct expenses (OPEX), nanoparticle creation and dealing costs, maintenance aspects, and integration intricacy. Because the thermal efficiency and size of a nanofluid-cooled system are more efficient, operation savings through decreased pumping power, energy loss, and equipment space can more than pay back the initial capital investment. But the price of producing the nanoparticles, stabilization using a surfactant, and reliability in the long run need to be looked into.

Life cycle assessment (LCA) is a requisite to measure their wider environmental impact. It reflects effects on the whole supply chain, including the mining of raw materials and production of nanoparticles and fluids, as well as operation, disposal, and the release of nanoparticles to nature. According to a recent LCA research, green synthesized nanofluids and recyclable or biodegradable nanoparticles would exhibit a far lesser impact on the environment as compared to standard formulations. The description of the first hybrid LCA—TEA framework enables decision-makers to have a holistic assessment of trade-offs between environmental achievements and financial gains.

14.7.2 Return on Investment and Commercialization Potential

From an investment perspective, nanofluid-enhanced technologies exhibit high potential in application-specific markets such as high-performance computing, electric mobility, solar thermal systems, and advanced HVAC, where conventional fluids limit system efficiency or thermal density. In these sectors, the return on investment (ROI) is amplified by energy savings, prolonged equipment life, and improved thermal control. For instance, in data centers using nanofluid-cooled systems, studies indicate up to 30–40% reduction in cooling energy, translating into substantial annual operational cost savings. However, for mass adoption, scalable production methods, standardized formulations, and validated reliability data are essential. Startups and SMEs entering this space often face high R&D costs, market uncertainty, and regulatory hurdles, which slow down commercialization. The development of IP portfolios, pilot-scale demonstrations, and public–private partnerships can help bridge the "valley of death" between lab-scale research and full market integration.

14.7.3 Policy Frameworks and Incentives for Nanofluid-Based Innovations

The policy intervention is the key to catalytic the role in mainstreaming nanofluid technologies. Presently, there is no clarity and standardization procedures regarding nanomaterials, which leads to confusion regarding safety, environmental compliance, and industry acceptance. Governments and International Agencies should come up with a system on how nanofluids can be used, handled, and disposed of, along with exposure limits and ecotoxicological monitoring. Along this line, fiscal incentives, including tax breaks, encouraging energy retrofit, and providing limited grant funding to nanotechnology-based R&D, can also decrease barriers sufficiently to allow entry to manufacturers and those at the end use.

Moreover, if nanofluid-based systems are included in the green building certification, the renewable energy credits, or the climate adaptation funds, they would become more appealing in both the developed markets and emerging markets. Innovation cycles can also be accelerated by creating test beds and regulatory sandboxes where pre-commercial testing can take place, so that firms can prove performance, have confidence in their customers, and qualify to export.

14.8 ROADMAP FOR FUTURE RESEARCH AND COLLABORATION

The future of nanofluid technologies in energy conservation is heavily reliant on *multidisciplinary research ecosystems*, *collaborative innovation networks*, and *readiness across technical*, *commercial*, and *policy dimensions*. Given the complexity of nanofluid systems and their cross-sectoral impact, a robust roadmap must address not only technical optimization but also innovation translation, commercialization strategies, and long-term ecosystem development.

14.8.1 Multidisciplinary Collaboration: Materials Science, Energy Engineering, and Data Science

Nanofluid development inherently demands integration across disciplines. Materials scientists are essential to designing nanoparticles with tunable chemical surface properties, crystalline shapes, and stabilities, and the energy engineers perform the optimization to integrate them into real systems (like a solar collector, heat exchanger, or cooling loop). The increase in data scientists and AI specialists comes with the inclusion of predictive thermophysical models, system-level optimization, and real-time monitoring utilizing digital twin architectures. New studies have been undertaken to apply machine learning methods to study nanofluids in dynamic conditions and speed up formulation discovery by screening high numbers. The three fields are likely to produce intelligent and autonomic nanofluids and digitized thermal management systems.

14.8.2 Public–Private Partnerships and International Consortia

The successful commercialization of nanofluid-based solutions will require strong alliances between academia, industry, and government bodies. Public–private partnerships (PPPs) are vital for piloting innovations in operational environments such as data centers, electric vehicles, and renewable energy systems. Moreover, international consortia and mission-oriented programs—such as those supported by the European Union's Horizon Europe or India's DST Nanomission—provide frameworks for harmonizing R&D, accelerating standards development, and validating cross-border applications. These consortia also play an instrumental role in knowledge transfer, policy alignment, and capacity building through training and co-funded infrastructure. The valorization of nanofluid innovations, as outlined in Figure 14.3, demands well-coordinated steps from IP protection and proof-of-concept validation to testing, regulation, and scaling through business model refinement and stakeholder engagement.

14.8.3 Strategic Research Priorities for 2030 and Beyond

To ensure the long-term success of nanofluids, strategic priorities should target both technological advancement and innovation ecosystem readiness:

- Smart nanofluids that are magnetically tunable, self-healing, or thermally switchable, improving system responsiveness under variable operating conditions.
- Green and biodegradable nanofluids, reducing ecological footprint and aligning with circular economy principles.
- Microfabricated heat management systems, such as 3D-printed channels with integrated nanofluid pathways, enabling compact and efficient cooling in constrained environments.
- Life cycle-based performance metrics, integrating techno-economic analysis and environmental impact for informed deployment.
- Global testing protocols and safety standards, harmonized across sectors and geographies, to address regulatory uncertainty and industrial adoption hesitancy.

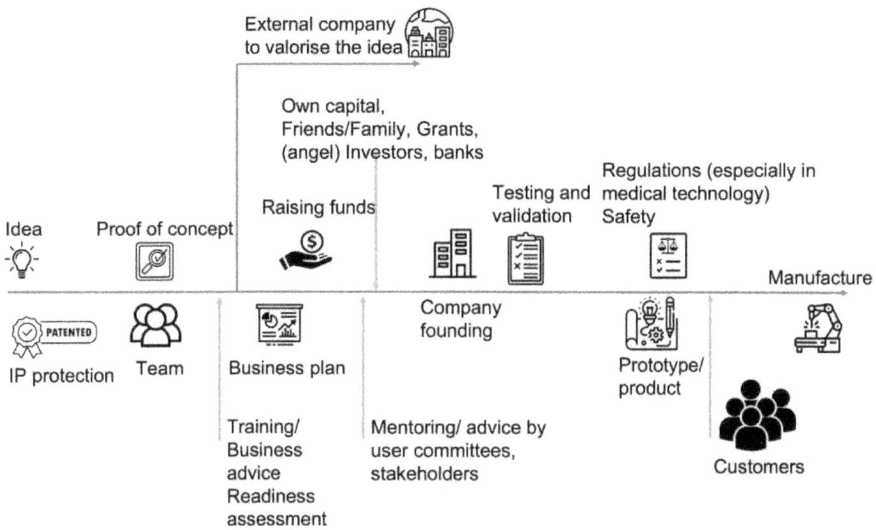

FIGURE 14.3 Roadmap from idea to commercialization of nanofluid-based technologies [11].

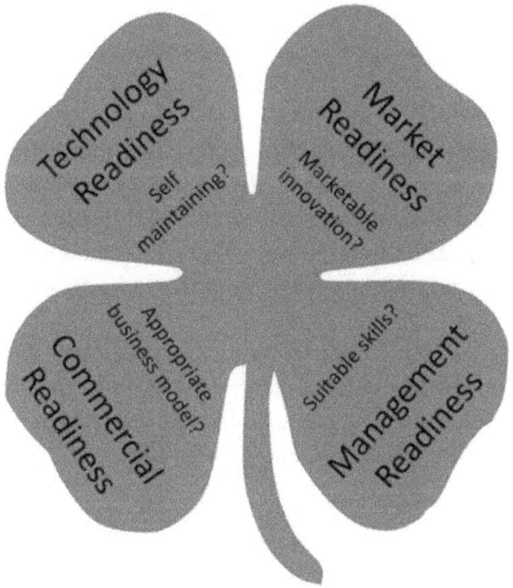

FIGURE 14.4 Readiness dimensions for successful deployment of nanofluid innovations across technical, market, commercial, and organizational domains [11].

However, for nanofluid innovations to move beyond proof-of-concept, they must pass a multi-dimensional readiness assessment. As illustrated in Figure 14.4, this includes the following:

- Technology readiness (e.g., scalability, robustness under real-world conditions)

- Market readiness (e.g., customer demand, competitive advantage)
- Commercial readiness (e.g., business model fit, distribution strategy)
- Management readiness (e.g., talent, leadership, regulatory navigation)

14.9 SUMMARY AND CONCLUDING REMARKS

The application of nanofluids represents a paradigm shift in the field of energy conservation and thermal management. This chapter has elaborated on their multifunctional potential across diverse sectors—from solar thermal systems, electronics cooling, and geothermal loops to next-generation energy storage and smart grid infrastructures. The use of engineered nanomaterials significantly enhances heat transfer, system miniaturization, and process efficiency, making nanofluids a key component in the broader strategy for decarbonizing thermal energy systems [12].

Looking ahead, nanofluids are poised to play a critical role in the global clean energy transition. Their compatibility with renewable technologies, their ability to improve thermal performance at micro- and macro-scales, and their utility in circular energy economies make them ideal candidates for integration into sustainable infrastructures. Moreover, their alignment with Sustainable Development Goals (SDGs)—notably SDG 7, 9, and 13—highlights their relevance not just from a technical perspective but also from social, environmental, and policy standpoints.

Despite their promise, the responsible development and deployment of nanofluids demand continuous innovation, robust life cycle assessments, and well-defined regulatory frameworks. Issues such as long-term stability, environmental safety, cost-effectiveness, and standardization remain non-trivial. Future progress will depend on multidisciplinary collaboration, public–private partnerships, and global research consortia focused on bridging lab-scale breakthroughs to real-world implementations [13].

In conclusion, the future of nanofluids lies in intelligent design, sustainability-focused innovation, and systems-level integration. As we advance toward a net-zero energy future, nanofluids are not just enablers of thermal performance—they are strategic assets in building a resilient, energy-efficient, and environmentally responsible world.

REFERENCES

1. Dixit, T., Al-Hajri, E., Paul, M. C., Nithiarasu, P., & Kumar, S. (2022). High performance, microarchitected, compact heat exchanger enabled by 3D printing. *Applied Thermal Engineering*, *210*, 118339.
2. Sonawane, S. S., & Malika, M. (2025). *Thermo-acoustics of nanofluids and transfer processes*. CRC Press.
3. Sonawane, S. S., & Malika, M. (2025). *Hybrid nanofluids: Heat and mass transfer processes*. CRC Press.
4. Thakur, P. P., Malika, M., & Sonawane, S. S. (2025). Challenges of nanofluid's applications in the chemical engineering industry. In *Hybrid nanofluids for application in the chemical and petroleum industry* (pp. 223–232). Elsevier.

5. Sun, L., Geng, J., Dong, K., & Sun, Q. (2024). The applications and challenges of nanofluids as coolants in data centers: A review. *Energies*, *17*(13), 3151. https://doi.org/10.3390/en17133151.
6. Thakur, P. P., Malika, M., Sonawane, S. S., & Kumar, M. A. (2025). Current overview of applications of the hybrid nanofluids. *Hybrid Nanofluids for Application in the Chemical and Petroleum Industry*, 1–16.
7. Malika, M., Pargaonkar, A., & Sonawane, S. S. (2023). Application of emulsion nanofluid membrane for the removal of methylene blue dye: Stability study. *Chemical Papers*, *77*(7), 3967–3977.
8. Malika, M., & Sonawane, S. S. (2023). Key markets of nanofluids. In *Nanofluid applications for advanced thermal solutions* (pp. 307–325). Elsevier.
9. Krishnan, S., Jai, S., Malika, M., Sharifpur, M., Sonawane, S. S., Mahian, O., & Meyer, J. P. (2023). Progress and challenges in nanofluids research. *Nanofluid Applications for Advanced Thermal Solutions*, 327–348.
10. Malika, M., & Sonawane, S. S. (2023). Few more significant applications of nanofluids. In *Nanofluid applications for advanced thermal solutions* (pp. 267–286). Elsevier.
11. Rambaran, T., & Schirhagl, R. (2022). Nanotechnology from lab to industry—a look at current trends. *Nanoscale Advances*, *4*(18), 3664–3675.
12. Thakur, P., Sonawane, S. S., Sonawane, S. H., & Bhanvase, B. A. (2020). Nanofluids-based delivery system, encapsulation of nanoparticles for stability to make stable nanofluids. In *Encapsulation of active molecules and their delivery system* (pp. 141–152). Elsevier.
13. Bethi, B., Sonawane, S. H., Bhanvase, B. A., & Sonawane, S. S. (2021). Textile industry wastewater treatment by cavitation combined with Fenton and ceramic nanofiltration membrane. *Chemical Engineering and Processing-Process Intensification*, *168*, 108540.

Index

A

B

C

D

E

For Product Safety Concerns and Information please contact our EU representative GPSR@taylorandfrancis.com
Taylor & Francis Verlag GmbH, Kaufingerstraße 24, 80331 München, Germany

www.ingramcontent.com/pod-product-compliance
Ingram Content Group UK Ltd.
Pitfield, Milton Keynes, MK11 3LW, UK
UKHW022213190926
475702UK00003B/332
* 9 7 8 1 0 4 1 0 4 3 6 8 3 *